Learning with Multiple Representations

ADVANCES IN LEARNING AND INSTRUCTION SERIES

Series Editors:
Neville Bennett, Erik DeCorte, Stella Vosniadou and Heinz Mandl

Forthcoming titles in the series

DILLENBOURG
Collaborative Learning: Cognitive and Computational Approaches

KAYSER
Modelling Changes in Understanding: Case Studies in Physical Reasoning

ROUET
Using Complex Information Systems

SALJO, BLISS & LIGHT
Learning Sites: Social and Technological Contexts for Learning

Other titles of interest

REIMANN AND SPADA
Learning in Humans and Machines:
Towards an Interdisciplinary Learning Science

Computer Assisted Learning: Proceedings of the CAL Series of biennial Symposia 1989, 1991, 1993, 1995 and 1997 (five volumes)

Related journals - sample copies available on request

Learning and Instruction
International Journal of Educational Research
Computers and Education
Computers and Human Behavior

Learning with Multiple Representations

edited by

Maarten W. van Someren,
Peter Reimann,
Henny P. A. Boshuizen
and Ton de Jong

Pergamon
An imprint of Elsevier Science
Amsterdam – Lausanne – New York – Oxford – Shannon – Singapore – Tokyo

ELSEVIER SCIENCE Ltd
The Boulevard, Langford Lane
Kidlington, Oxford OX5 1GB, UK

First edition 1998 by Pergamon (an imprint of Elsevier Science Ltd) in association with the European Association for Research on Learning and Instruction (EARLI)

Library of Congress Cataloging in Publication Data
Learning with multiple representations / edited by Maarten van Someren
... [et al.].
p. cm. -- (Advances in learning and instruction series)
ISBN 0-08-043343-X (hardcover)
1. Learning, Psychology of. I. Someren, Maarten W. van.
II. Series.
BF318.L413 1998
153.1'5--dc21 98-8794
CIP

British Library Cataloguing in Publication Data
A catalogue record from the British Library has been applied for.

ISBN: 0 08 043343 X

The paper used in this publication meets the requirements of ANSI/NISO Z39.48-1992 (Permanence of Paper).
Printed in The Netherlands.

Table of Contents

PART 2: Problem Solving and Learning with Multiple Representations

PART 3: General Issues and Implications for Education

Acknowledgement

Producing this book would not have been possible without the support of the European Science Foundation, Strasbourg. The European Science Foundation is an association of 62 major national funding agencies devoted to basic scientific research in 21 countries. The ESF assists its Member Organisations in two main ways: by bringing scientists together in its Scientific Programmes, Networks and European Research Conferences, to work on topics of common interest; and through the joint study of issues of strategic importance in European science policy. The scientific work sponsored by the ESF includes basic research in the natural and technical sciences, the medical and biosciences, the humanities and social sciences. The ESF maintains close relations with other scientific institutions within and outside Europe. Through its activities, the ESF adds value by cooperation and coordination across national frontiers and endeavours, offers expert scientific advice on strategic issues, and provides the European forum for fundamental science. This book is one of the outcomes of the ESF Scientific Programme on "Learning in Humans and Machines".

Contributors

Shaaron Ainsworth
ESRC Centre for Research in Development, Instruction and Training
Dept. of Psychology
University of Nottingham
University Park
Nottingham
NG7 2RD
UK

Laurence Alpay
Institute of Educational Technology
The Open University
Walton Hall
Milton Keynes
MK7 6AA
UK

Peter A. Bibby
ESRC Centre for Research in Development, Instruction and Training
Dept. of Psychology
University of Nottingham
University Park
Nottingham
NG7 2RD
UK

Henny P.A. Boshuizen
Dept. of Educational Research and Development
University of Maastricht
PO Box 616
6200 MD Maastricht
The Netherlands

Rainer Bromme
Universitaet Muenster
Psychologisches Institut III
Fliednerstr. 21
D 48149 Muenster
Germany

Rose Dieng
ACACIA Project
INRIA
2004 Routes des Lucioles
06902 Sophia Antipolis Cedex
Sophia Antipolis
France

Mike Dobson
Program for Advanced Learning Systems
Graduate Division of Educational Research
Education Tower
University of Calgary
2500 University Drive N.W.
Calgary, Alberta T2N 1N4
Canada

Monica G.M. Ferguson-Hessler
Faculty of Applied Physics
Eindhoven University of Technology
PO Box 513
5600 MB Eindhoven
The Netherlands

Alain Giboin
ACACIA Project
INRIA
2004 Routes des Lucioles
06902 Sophia Antipolis Cedex
Sophia Antipolis
France

Anja van der Hulst
TNO Fysisch en Electronisch Lab
Postbus 96864
2509 JG Den Haag
The Netherlands

Ton de Jong
Faculty of Educational Science and Technology
University of Twente
PO Box 217
7500 AE Enschede
The Netherlands

Alan Lesgold
Professor of Psychology and
Intelligent Systems
Learning Research and Development Center
University of Pittsburgh
3939 O'Hara Street
Pittsburgh, PA 15260
USA

Jarmo Levonen
University of Joensuu
PO Box 111
FIN 80101 Joensuu
Finland

Matthias Nückles
Universitaet Muenster
Psychologisches Institut III
Fliednerstr. 21
D 48149 Muenster
Germany

Peter Reimann
Professor for Instructional Psychology
Dept. of Psychology
University of Heidelberg
Hauptstr. 47-51
D-69117 Heidelberg
Germany

Michael Rohr
Dept. of Psychology
University of Heidelberg
Hauptstr. 47-51
D-69117 Heidelberg
Germany

Elwin R. Savelsbergh
Max Planck Institute for
Human Development
Centre for Educational Research
Lentzeallee 94
14195, Berlin
Germany

Eileen Scanlon
Institute of Educational Technology
The Open University
Walton Hall
Milton Keynes
MK7 6AA
UK

Hermina J.M. (Tabachneck-)Schijf
TNO Human Factors Research Institute
Kampweg 5
Postbus 23
3769 ZG Soesterberg
The Netherlands

Julie-Ann Sime
Center for Studies in Advanced
Learning Technology
Department of Educational Research
University of Lancaster
Lancaster
LA1 4YW
UK

Herbert A. Simon
Dept. of Psychology
Carnegie-Mellon University
Pittsburgh, PA 15213
USA

Maarten W. van Someren
Department of Social Science Informatics
University of Amsterdam
Roetersstraat 15
1018 WB Amsterdam
The Netherlands

Hans Spada
Albert-Ludwigs Universitaet
Psychologisches Institut
Abt. Allgemeine Psychologie
D-79085 Freiburg
Germany

Keith Stenning
Human Communication Research
Centre
Edinburgh University
2 Buccleuch Place
Edinburgh
EH8 9LW
UK

Janine Swaak
Faculty of Educational Science and
Technology,
Department of Instructional
Technology
University of Twente
PO Box 217
7500 AE Enschede
The Netherlands

Jerzy Surma
Dept. of Computer Science
Technical University of Wroclaw
ul. Wybrzeze Wyspianskiego 27
50-370 Wroclaw
Poland

Huib Tabbers
Dept. of Social Science Informatics
University of Amsterdam
Roetersstraat 15
1018 WB Amsterdam
The Netherlands

Pietro Torasso
Dipartimento di Informatica
Universita degli Studii di Torino
Corso Svizzera 185
10149 Torino
Italy

Margaretha W.J. van de Wiel
Skillslab
University of Maastricht
PO Box 616
6200 MD Maastricht
The Netherlands

David J. Wood
ESRC Centre for Research in Development, Instruction and Training
Dept. of Psychology
University of Nottingham
University Park
Nottingham
NG7 2RD
UK

Preface

The mission of the "Learning in Humans and Machines" research programme was to advance our knowledge about learning from an interdisciplinary perspective. Researchers from cognitive, computer and educational science worked together. The guiding theme was the analysis, comparison and integration of research on human learning and of computational approaches to learning. The human learning perspective comprised in particular psychological research, but also contributions of other behavioural sciences; machine learning research was the focus of the computational learning perspective, but work from other areas of artificial intelligence was also of interest.

During the years 1994–1997 nearly 100 scientists from 17 European countries and from the USA were involved in the programme. Half the participants were junior scientists. The programme activities were centred around a series of workshops and conferences and involved study visits especially by the junior scientists. Research in the programme was conducted in five task forces on the following themes:

- Representation changes in learning.
- Learning with multiple representations and multiple goals.
- Learning strategies to cope with sequencing effects.
- Situated learning and transfer.
- Collaborative learning.

More extensive information on the programme can be found at:

http://www.psychologie.uni-freiburg.de/esf-lhm.

This book consists of contributions by members of the task force "Learning with multiple representations and multiple goals", and of invited papers by some authors from outside the task force. Contributions were reviewed by the editors, and most of the contributions were discussed during the workshops of the programme. The general theme of the book is a fascinating one. When knowledge in a domain serves different tasks, such as diagnostic problem solving in medicine and explaining/communicating this knowledge to others, knowledge representation and handling pose difficult problems. How are they overcome by humans, how can they be solved by artificial systems? How do different representations of domain complement each other, how do they interact? Which types of external representations facilitate the construction of different forms of mental representations?

This book provides answers to questions of multiple representations in human reasoning and learning. Computational approaches to learning with multiple representations are introduced and discussed. Finally, the role of multiple representations in teaching is addressed. The book should be of interest to a wide audience of psychologists, computer scientists and educational scientists.

Lorenza Saitta and Hans Spada

1

Introduction

Maarten W. van Someren, Henny P.A. Boshuizen, Ton de Jong and Peter Reimann

Almost all forms of learning involve information that is represented in different forms. Human teachers use a range of different representation techniques to present information to students. Besides texts, teachers use diagrams, practical demonstrations, abstract mathematical models and semi-abstract simulations; within these categories of representations they use a variety of more specific representational forms. The environment in which we live is another source of information, that also appears in a wide variety of representational forms. Although some tasks can be performed using a single representation, many tasks require the combined use of knowledge in different forms, or can be performed much more efficiently by the integrated use of multiple representations than by a single representation. What is taught at school should then be combined with what we learn from experience to become really effective knowledge. Our experience in interacting with the world in which we live, including our interaction with other people, shows an immense variety of representational forms. How do people deal with this variety? How are they able to combine information from many sources and in many representations into one integrated knowledge structure? How do they subsequently use their knowledge and select for themselves the right representation form to perform tasks? How can they improve their understanding of and performance in this variety of representational forms? These are the questions that formed the starting point of this book.

1 Multiple Representations in Human Reasoning

These questions have a long history in psychology and education, and also in philosophy, anthropology, logic and computational sciences, in particular artificial intelligence. An important question in psychology, and a topic of much debate, is whether the human cognitive architecture consists of different components that use different representations. These components may be associated with different senses. For example, there are strong arguments that visual information is perceived, stored and processed differently from auditory information and possibly also from verbal information, which does not derive its meaning from direct perception. This raises the question how information from these components becomes connected and integrated to make it possible to perform tasks that involve information of more than one internal representation.

Differences between representations are not only associated with different senses. Different domains and disciplines all have their own terminology and their own concepts. When knowledge from different disciplines must be combined to solve a particular problem, connections must be made: the concepts from one representation must be mapped onto those of another to combine the knowledge from both in performing a task. The first main theme addressed in this book is that we usually acquire "islands of knowledge" which are separated from each other by differences in the "object" of that knowledge (different disciplines and domains), differences in the representation format in which knowledge is expressed (e.g. pictorial, verbal or as formulae), differences in underlying concepts, etc. Several chapters in this book demonstrate that it is difficult to build "bridges" between these "islands" even for people who have adequate knowledge within a particular representation. Schijf and Simon (chapter 11), Dobson (chapter 5) and Ainsworth, Bibby and Wood (chapter 7) show in very different contexts that learners have difficulty relating numerical data, formulae and text to each other even when these all represent the same "objects". Rohr and Reimann (chapter 3) show that students are hardly able to relate knowledge at different levels of aggregation even when visual representations are used to illustrate verbal and mathematical descriptions. In physics, students are taught about thermodynamics both at a high level of aggregation, in terms of concepts like temperature, volume of gas, liquid or solid, etc., and also at the molecular level in terms of speed of molecules. Other chapters demonstrate the difficulty of relating representations that overlap in their "object" but that use very different concepts and vocabulary. Boshuizen and Van de Wiel (chapter 12) demonstrate that when inexperienced physicians reasoned about a diagnostic problem they did not use all the knowledge they had learned, because they failed to switch between several domains that were taught separately. The result is that it took them longer than more experienced diagnosticians to reach the essential conclusions about a patient. A similar result was found by Savelsbergh, De Jong and Ferguson-Hessler (chapter 13) in a study of physics problem solving. More expert problem solvers differed from relative novices in the flexibility with which they can switch between representations. Alpay, Giboin and Dieng (chapter 9) show the problem in controlling collaborative problem solving that involves experts from different disciplines. Bromme and Nückles (chapter 10) study the even more complicated situation in which representations are used which overlap in "object" and in "form" without necessarily having the same meaning in the sense of the same relation between expressions and the object they are about.

A few studies (Van Someren and Tabbers (chapter 6) and Schijf and Simon (chapter 11)) begin to identify the knowledge that "mediates" between different representations and that makes it possible to switch between representations in problem solving.

2 Multiple Representations in Human Learning

As a result of practice and experience in a domain our knowledge changes. It becomes more effective, more accurate and more general. These changes can be explained by changes in the content of knowledge but representation forms also play a role. In the context of multiple representations there are two forms of learning that distinguish it from learning within a single representation form. The first is that the learner constructs the mappings between elements of different representations. For example, a learner may find out which part or which aspect of a simulation of moving objects corresponds to the symbol V for velocity and which corresponds to A for acceleration. This new knowledge makes it possible to use knowledge in different representations for problem solving. The second form of learning is that knowledge is "moved" (translated) into a different form that allows faster problem solving. For example, a diagram can make it easier to find answers to questions that can also be derived by reasoning with mathematical formulae or with verbal descriptions. Reasoning with a diagram requires that, for example, algebraic knowledge is translated into operations on diagrams.

Although this is an interesting form of learning, it introduces the problem of choosing the optimal representation for knowledge. An important criterion for this choice is efficiency. The performance characteristics, such as solution time or memory usage, may vary widely between representations but the choice may depend on characteristics of the task, such as the frequency distribution of problems. Dobson (chapter 5) discusses an attempt to use the "specificity" (the lack of redundancy) of a representation to predict its success in problem solving. The negative results point out that it may be difficult for people to exploit "specificity". Boshuizen and Van de Wiel (chapter 12) use the concept "encapsulation" to describe a transformation on representations to improve smooth problem solving, even if this involves representations from different domains. Van Someren, Torasso and Surma (chapter 14) present a more general discussion of the problem of optimisation in an architecture with multiple representations.

3 Computational Approaches to Learning with Multiple Representations

In machine learning the role of different representations is not well developed. In machine learning, and in artificial intelligence in general, a distinction is made between the concepts representation language, knowledge (that is expressed in a representation language), the meaning of particular knowledge and the meaning of the language in general. A single fact can be represented in different representation languages, denoting the same information. A single expression can have different meanings in different representation languages. In machine learning the main issues connected with representation are the expressiveness of the representation language, the technical difficulties of finding adequate learning methods for complex, expressive languages, and the effect that the language has on the result of learning ("bias"). More expressive languages require more complicated learning methods and tend to be computationally more expensive than less expressive languages. Applying machine learning methods involves selecting or defining an appropriate language. The internal representations of the learning system can be chosen by the designer. In psychology and education, the term "representation" refers to what machine learners call "knowledge" and normally no distinction is made between the language or form and the content of a representation. It is possible that this is due to the fact that machine learning systems are not very interactive. Because these systems do not have an intensive interaction with their environment, they do not have the problem of having to accommodate information in a variety of forms. We may expect that when machine learning is applied to interactive systems, or agents that communicate with users in different representational forms and that involve more realistic forms of perception than current systems, the representation issues will shift in the direction of the social notion of representation.

As yet, machine learning does not have much to say about how multiple representations can be related to perform tasks that involve reasoning with different representations. Van Someren and Tabbers (chapter 6) use the notion of learning as a search to analyse human inductive discovery learning and the knowledge that is needed to combine different representations, in this case qualitative and quantitative knowledge (about physics). A different subfield of machine learning, relevant for learning with multiple representations, is "speedup learning": the construction of new representations of knowledge that enable more efficient problem solving (without changing the solutions that the system finds). Van Someren, Torasso and Surma (chapter 14) apply concepts from this area to model the optimisation of problem solving that uses both complex causal knowledge and stored problems.

4 Multiple Representations in Teaching

These questions are currently even more urgent than they used to be because new technology, in particular multimedia technology, makes it feasible to construct material of any form and present it to learners. New communication technology gives learners access to an immense amount of information. The possibilities made available for teaching by new media and information technology introduce new choices for the designers of educational material. Abstract concepts can be presented in a visual representation. As some of the chapters in this book will show, this is not necessarily a good thing in itself. One representation may well be just better than another and combining multiple respresentations creates new problems for the learner. Understanding how the properties of representations determine their utility and how multiple representations can be related is a prerequisite to the effective use of new technology.

Although most authors do not make strong claims in this direction, the implication of this work for educational practice and for instructional design seems to be that multiple representations are not a good thing *per se*. When information is presented in different ways it is important to teach the relations between these representations because these can be hard for the learners to construct themselves. A close analysis of these relations both in terms of their semantic relations and their performance characteristics is needed to enable the profitable use of multiple representations in learning and problem solving.

5 Learning with Multiple Representations: The Structure of this Book

With this book the editors and authors want to achieve three goals: (a) to present the reader with an overview of the current research on learning with multiple representations; (b) to demonstrate how different research methods such as experiments, analysis of (verbal) behaviour and the use of computational methods can collaborate; and (c) to articulate research questions for the future. Part 1 of this book addresses the role of multiple representations in learning about science domains and mathematics, with an emphasis on learning conceptual, declarative knowledge. The first chapter by De Jong et al. gives an overview of the dimensions of different representations and their role in science teaching. The other chapters describe specific empirical studies. The first chapter in Part 2 gives an overview of the main issues in connection with the role of multiple representations in problem solving and the acquisition of problem-solving skills. The other chapters describe the empirical study of this and a theoretical model for optimising the efficiency in an architecture with multiple representations. Part 3 consists of two chapters that address two issues that go beyond the research covered in Parts 1 and 2: the principled design of instruction that involves multiple representations and the idea of teaching about representations rather than just by representations.

PART 1

Multiple Representations in Learning Concepts from Physics and Mathematics

2

Acquiring Knowledge in Science and Mathematics: The Use of Multiple Representations in Technology-Based Learning Environments

Ton de Jong, Shaaron Ainsworth, Mike Dobson, Anja van der Hulst, Jarmo Levonen, Peter Reimann, Julie-Ann Sime, Maarten W. van Someren, Hans Spada and Janine Swaak

1 Introduction

Trends that nowadays dominate the field of learning and instruction are *constructivism*, *situationism* and *collaborative learning*. More specifically, we can say that in modern education learners are encouraged to *construct their own knowledge* (instead of copying it from an authority, be it a book or a teacher), *in realistic situations* (instead of merely decontextualized, formal situations such as the classroom), *together with others* (instead of on their own). Technology plays a major role in implementing these new trends in education. Constructivism is supported by computer environments such as hypertexts, concept mapping, simulation and modelling tools (see De Jong & Van Joolingen, 1996), realistic situations can be brought into the classroom by means of video, for example in the Jasper series (Cognition and Technology Group at Vanderbilt, 1992), and collaborative learning is supported in environments such as CSILE (Scardamalia & Bereiter, 1990) and Belvedere (Suthers et al., 1995). One of the effects of these trends is that learners are confronted with information from multiple sources (computer programs, books, the teacher, reality, the classroom, peers, etc.) and in many different representations that they have to evaluate, make a selection from, and that they have to integrate into their personal knowledge construction process.

The main goal of this chapter is to explore the different representations that information (learning material) can take, to appraise the virtues of these representations, and to evaluate the use of *multiple* representations in one learning environment. The general context in which we discuss these formats is within computer-based learning environments for science domains and mathematics. This context was chosen since it is in these types of environment that we find the latest developments in the use of alternative and multiple representations. Furthermore, science and mathematics are domains where using and coordinating multiple representations are often considered to be prerequisites for understanding. In the following sections, these "dimensions" for describing representations are used: *perspective*, *precision*, *modality*, *specificity* and *complexity*. *Perspective* refers to the particular theoretical viewpoint taken in presenting material, *precision* refers to the level of accuracy in the description (mainly qualitative vs. quantitative), *modality* is the representation format, e.g. propositional or figural, *specificity* concerns the informational economy of a representation and, finally, *complexity* pertains to the amount of information present in a representation. Although we will speak of "dimensions", the categories we use are not all dimensions in the sense that we can classify representations somewhere in between two extremes. It should also be noted that although the dimensions can be used independently to describe representations, in practice specific values on dimensions will cluster together. In their study on representing properties of gases Rohr and Reimann (chapter 3) indicate that specific modalities are particularly suited to display specific perspectives (ontologies) of the domain. In Dobson (chapter 5) and Ainsworth, Bibby and Wood (chapter 6) we see that modality is related to specificity. It may further be helpful to relate our dimensions to a distinction made by Newell (1982) between the *knowledge level* and the *symbol level*. The knowledge level is a level where descriptions are given in terms of what somebody knows and believes. At this level no attempt is made to specify the symbols or data structures in which knowledge is represented. Representations, i.e. symbolic structures which are subject to manipulation by cognitive operators, come into play at the symbol level. It is here that a physical or mental action is explained in terms of operations over structures such as schemata, propositional networks, mental models or rules. Of the five dimensions introduced above, modality is the one which is clearly linked to the symbol level and so is complexity. Of the remaining three, perspective is closely related to the knowledge level. Precision and specificity are dimensions which can be used for distinctions both at the knowledge level and the symbol level.

2 Dimensions of Representations

In the following sections a number of dimensions of representations are described. Each dimension is illustrated by at least one extensive example from a technology based learning environment.

2.1 Perspective

Domains can often be seen from different perspectives. To take a straightforward and simple example, the different parts of an engine can be considered from a functional perspective that describes the functions of specific parts (e.g. the ignition), or a topological perspective that describes the location of the different parts in the engine. An early illustration of offering multiple perspectives in a learning environment can be found in STEAMER (Woolf et al., 1987). In STEAMER, operators learn how to solve problems in a "recovery boiler" which as such is a common part in pulp and paper mills. STEAMER offers operators a simulation of the recovery boiler and allows them to view the boiler from different angles. A more elaborate early approach is displayed in the work by White and Frederiksen (1989, 1990) on QUEST. In QUEST, based on a computer simulation of simple electrical circuits, three perspectives were offered to learners: a functional perspective, in which a model and its subsystems are described according to their purpose; a behavioural perspective, which is a view on the system at the level of circuit components; and a physical perspective, which refers to the behaviour of the circuit at the level of physics. White and Frederiksen use the perspective idea (together with the dimensions "precision, and complexity", see below) for sequencing instruction in QUEST (see also section 3). In line with the work on QUEST, Van der Hulst (1996), discusses three different (behavioural) perspectives on the physics domain of harmonic oscillations. She analysed a variety of (high school) science teaching material to gain insight into the dimensions which the authors of such material use to organise it. In the teaching material, Van der Hulst found that authors consistently present oscillation by means of three separate perspectives, a *motion*, *force* and *energy* perspective. The motion perspective introduces basic concepts such as amplitude, period and frequency. Such an introduction is generally followed by sections providing diagrams and formulae representing displacement, velocity and in some cases acceleration as a function of time. The motion perspective is followed by a description and illustration of forces acting upon the oscillating mass, showing how these forces affect displacement. Apparently, this is perceived to be the right moment to show by derivation that the period of the oscillation is independent of its amplitude. The final parts of the oscillation material reveal an energy perspective that explains how potential energy is transformed into kinetic energy during oscillation. In some of the texts that were analysed the effect of damping on the total energy is discussed. Later in this chapter (section 3) we will give an extensive description of the work by Van der Hulst, and also of that by Sime (1994, 1998) on the process rig (an industrial heat exchanger). Using a computer simulation in teaching about the process rig, Sime uses six different representations. Sime (1994, 1998) uses *approximation* to describe a particular conceptual view of a model which enables the focus to be on one aspect of the system, say the thermal properties or the spatial relationships. These are not just subsets of the complete model, but alternative analyses of the system that form a conceptually coherent theme. The models of the process rig include representations of the thermal processes and the flow processes within the physical system, as well as a combined representation of both thermal and flow processes. This separation is useful in teaching about the process rig because of the interaction effects between these two views; this emphasises the fact that flow changes and thermal changes cannot be considered in complete exclusion.

A term closely related to perspective is *ontology*. Sime (1994) uses ontology to describe the domain theory or basic unit of representation within a model. For example, the model may use constraints to represent the relationships between variables (Kuipers, 1984), or may represent individual components within the system (De Kleer & Brown, 1984) or may represent the processes (Forbus, 1984). Rohr and Reimann (chapter 3) focus on the ontology of a science domain and discuss the problem of ontological confusion in learning, i.e. where a naive view (representation) of physics interferes with the learning of the taught representation. Ontological aspects of knowledge representation refer to decisions about "how and what to see in the world" (Davis, Shrobe & Szolovits, 1993). Thus, ontology refers to the *content*, to the objects and relations one uses to represent a domain, not so much to the symbols by which objects and relations are denoted. Ontological commitments are quite important in science learning, because it is often observed that pre-scientific conceptualisations are ontologically incompatible with their scientific counterpart (Chi, Slotta & De Leeuw, 1994). For instance, most children see temperature as an attribute of an object or a substance, whereas science has come to see it as a process. Rohr and Reimann (chapter 3) study how pupils struggle with the concept of particles when this concept is used to explain gas phenomena such as expansion of volume as gas temperature increases. Children view gases, and matter in general, as continuous entities and for many pupils to accept that matter is made out of the smallest interacting particles amounts to a radical change of their "theories" about the nature of matter. When confronted in school with the particulate model of gases, most pupils will not give up their continuous model but try to integrate it with the new information. This results in a number of predictable problems when pupils attempt to explain the outcomes of experiments with gases. Rohr and Reimann are particularly interested in mistakes that occur because pupils transfer, by means of analogy, elements and relations from the macroscopic world to the realm of the submicroscopic world with its abstract notion of particles. Based on process models of analogical transfer, they argue that some of the problems pupils experience are probably due to ontological mismatches.

2.2 *Precision*

The *precision* with which information is presented expresses the level of accuracy or exactness of the information. The main distinction we can make here is the one between precise, quantitative, information and less precise, qualitative, information. Sime (1994, 1998) uses the term *abstraction* to describe the difference between a numerical description of a relationship (such as the change in a variable over time, e.g. the flow rate is 0.2 l/s) and a qualitative description of the relationship (e.g. the liquid temperature in the tank is increasing). This determines whether the model is based upon numeric values (i.e. a quantitative model) or abstract symbolic values (i.e. a qualitative model). Within Sime's PRODS learning environment (Sime, 1996, 1998) students learn to make predictions about the behaviour of the physical system using both qualitative and quantitative models. When multiple representations are taught it is important to also teach the meta-knowledge required to select the most appropriate representation for each task or sub-task to be carried out. The selection of an appropriate model for making a prediction is dependent on the initial change in behaviour and whether a qualitative or quantitative prediction is required. A related characteristic distinguished by Sime (1994, 1998) is *commitment*. Commitment refers to the certainty or belief attributed to a relationship or variable used in the model. This captures some of the inherent vagueness of modelling assumptions. For example, in Sime's domain (the process rig, see also section 3) there are differences in accuracy depending on the type of flow meter used. So the output from the simulation used in reasoning about the system is not completely accurate and an expert problem solver may take this into account. In this book the interplay between knowledge at several levels of precision is discussed in a study on inductive learning by Van Someren and Tabbers (chapter 6).

As another example of the role of precision of information in learning we take an example from the domain of classical mechanics by Plötzner et al. (1990). In their work qualitative and quantitative physics knowledge about various concepts in dynamics (e.g. mass and force) and kinematics (e.g. velocity and acceleration) are addressed. For instance, with respect to the resultant force F on an object and the object's acceleration a, four levels of precision can be distinguished:

- Qualitative Level 1: The magnitude of F and the magnitude of a are related;
- Qualitative Level 2: If the magnitude of F increases (decreases), then the magnitude of a increases (decreases), too;
- Semi-quantitative Level: If the magnitude of F increases (decreases) by some factor, then the magnitude of a increases (decreases) by the same factor;
- Quantitative Level: The quotient of the magnitude of F and the magnitude of a is constant: $F/a = m$.

Thus, in the simplest case, the knowledge merely includes the information that two concepts are related along some dimension such as magnitude. Knowledge at the next, hierarchically higher, level of precision takes into account monotonic properties of the relationship. Knowledge that encompasses more precise information defines the next level. This knowledge may take into account, for example, that the respective relationship is proportional. Finally, knowledge about functional relationships in terms of fully developed algebraic and vector-algebraic equations forms the most precise level. In the given example this corresponds to Newton's second law stating that the magnitude of F equals the object's mass m multiplied by the magnitude of a: $F = m \cdot a$.

Research on qualitative reasoning based on discrete representations of continuous variables is an established subfield of artificial intelligence (Weld & De Kleer, 1990). The original motivation for this research was to develop computational models of how engineers and scientists reason about physical systems in qualitative terms rather than in numeric quantitative terms. While one direction of research aims at the development of representational formalisms and computational mechanisms, the other direction focuses on tasks which can be solved on the basis of qualitative reasoning techniques including their applications to simulation-based learning environments (e.g. Van Joolingen, 1995). A relevant finding of artificial intelligence (AI) research is that qualitative reasoning inherently tends to be ambiguous. Although qualitative descriptions of physical systems often suffice to simulate possible outcomes, they tend not to capture exactly the behaviour that will take place. As a consequence qualitative reasoning has to be complemented with a quantitative analysis of the problem under study, to capture all relevant aspects.

We illustrate the didactic approach to find the best possible use of different levels of precision of information by two computerised systems, which complement each other: an interactive discovery learning environment and a diagnosis system. The interactive discovery learning environment DiBi (*Di*sk *Bi*lliard; Spada, Stumpf &

Opwis, 1989; Stumpf, 1990) includes qualitative and quantitative knowledge about functional relationships between physics concepts. It simulates elastic impacts by means of a computerised disk billiard. DiBi enables the design of simulated experiments which allow the investigation of phenomena such as central and oblique impacts on a disk, disks rebounding off the cushion and various collisions of two disks.

The interaction between a student and the learning environment is structured by SPFP sequences (*S*election, *P*rediction, *F*eedback, *P*rocessing of information; Spada, Reimann & Häusler, 1983). Within such a sequence, a student selects a phenomenon to be investigated, designs an experiment and predicts the outcome of the experiment. After students have made their prediction, they receive feedback from the learning environment about the actual outcome of the experiment in a static and/or dynamic form. Based on a comparison between the predicted and the actual outcome, the student may construct new hypotheses about the phenomena or modify existing ones. As an adaptive learning environment, DiBi enables the interaction with a student to take place on that level of precision which corresponds to the students' actual understanding. Very often, however, such passive forms of adaptability do not reach far enough and more active forms are required. For instance, if students lack knowledge about relevant functional relationships on a certain level of precision, then this should be pointed out to them before they move onto the next level. Even more helpful would be the generation of an experiment by the learning environment which exactly addresses the missing knowledge. Thus, one possibility to provide active adaptability is to use a model of the student's actual domain-specific knowledge.

MULEDS (*Mu*lti-*Le*vel *D*iagnosis *S*ystem; Plötzner et al., 1990) has been designed for assessing the domain-specific knowledge of students who interact with the learning environment DiBi. It is applicable with respect to the phenomena of central and oblique elastic impacts on a disk. MULEDS aims at assessing correct, fragmentary and incorrect knowledge about the functional relationships relevant to these phenomena. It takes into account that this knowledge might be expressed at different levels of precision. MULEDS has been implemented as a rule-based system by means of the production system interpreter PRISM (Ohlsson & Langley, 1984). It relies on the bug library approach (Brown & Burton, 1978). The domain-specific knowledge makes up a hypotheses library which contains correct, fragmentary and incorrect hypotheses about the relevant functional relationship on the different hierarchically related levels of precision. The rules out of the hypotheses library make up a space of potential models of a student's domain-specific knowledge. Assessment with MULEDS aims at identifying that subset of these rules which allows the best possible reconstruction of a student's predictions. MULEDS comprises a mechanism which allows for tailored testing. This mechanism compares syntactically pairs of competing student models in order to determinate all those physics concepts which are handled differently within these models. Obviously, an experiment in which such a physics concept is varied would allow one to discriminate between the competing models. Therefore, the set of physics concepts identified by this mechanism is provided to the learning environment DiBi in order to guide the arrangement of further experiments. An assessed student model can also be used to guide the arrangement of experiments which contradict an incorrect hypothesis assumed by the student. In MULEDS, the generation of counter-examples relies on the same mechanism as that employed for tailored testing. In this case, however, the assessed student model is compared with an ideal student model which is made up of exclusively correct rules out of the hypotheses library. The comparison yields a set of physics concepts which is either neglected or handled incorrectly in the assessed student model. This information can subsequently be utilised by the learning environment DiBi to arrange experiments which address exactly the missing or incorrect knowledge.

An empirical study has been conducted in order to test the validity of this diagnostic approach (Opwis et al., 1994). Students ($N = 132$) of secondary school grades 5, 7, 9 and 11 worked on 30 problems addressing the functional relationships involved in central and oblique elastic impacts on a disk at different levels of precision. Many of the students' correct as well as incorrect answers to the problems were reconstructible by the model. With respect to the 12 problems addressing central elastic impacts on a disk, the students' knowledge was not assessable in only 5% of the cases at the qualitative level and in 9% at the semi-quantitative level, but in 28% of the cases at the quantitative level. A further result was that the overall consistency of the assessed knowledge across the different levels of precision and considered phenomena was not as high as expected. An interesting and very frequent misconception was observed at the qualitative level. Fifty-one per cent of the subjects reasoned in the following way: If both the impacting force and the mass of a disk are increased (or decreased), the effect of these two variables on the dependent variable, acceleration (respectively velocity of the disk), compensates. The ambiguity of qualitative reasoning is apparently resolved, but the result is not correct.

Ideally, qualitative and quantitative reasoning in problem solving and learning are processes which complement each other. While qualitative reasoning helps students to understand conceptually the domain under scrutiny and, on the basis of this understanding, to select the appropriate quantitative equations for application, quantitative reasoning allows one to disambiguate, extend and verify the outcomes of qualitative reasoning.

2.3 Modality

Here, we use the term modality in a specific sense, to denote the particular *form of expression* that is used for displaying information. Throughout the chapters in this book we see the use of several modalities including: *text* and *animations* (Rohr and Reimann, chapter 3), *diagrams* (Dobson, chapter 5), *graphs*, *algebraic notations* and *real life observations* (presented in video) (Scanlon, chapter 4), and *formula* and *tables* (Ainsworth et al., chapter 7). As can be seen in these studies, in the ones described later in this section, and in many others in the literature, modality is clearly related to learning. An example can be found in two studies by Rieber and colleagues. In Rieber, Boyce and Assad (1990) undergraduate students received a computer-based introductory course on Newtonian mechanics. The content of the course was supplemented by *static graphics*, or *animated graphics* or *no graphics* at all. Students' resulting knowledge was tested by a 32-item multiple choice test intended to measure the students' ability to apply the rules of the domain. Half of the test items were all verbal, half also contained an accompanying visualisation, and half of the items with visualisation contained an animation. Both correctness scores and latencies were collected. With respect to the correctness scores no major effect was observed with visual representations. With regard to the response latencies a major effect was seen for visual representations, the students who were enrolled in the course with animated graphics took less time to answer the post-test items than the students who had static graphics or no graphics in their course. No main differences between static graphics and no graphics could be detected. In Rieber (1990) an identical procedure was followed, only with younger (10–11-year-old) students, and a similar post-test was used; therefore, the results can be compared. It was hypothesised that animated representations would be more effective in presenting the subject matter than static graphical representations or no graphics. In this study the correctness scores of the post-test did show significant differences between the experimental groups. Students who had worked with animated graphical representations outperformed the students who had worked with either the static graphical representations or with no graphical representations at all. The latter two groups did not differ. No major effects of specific representations were detected with the latency scores in this study.

Rieber (1990) suggests that dual coding (Paivio, 1986) is responsible for the effects of animations and graphics. Dual coding is the principle which proposes that texts are processed and encoded in the verbal systems whereas pictures or graphics are processed both in the image and verbal systems. Dual coding was seen as a way to explain why memory for pictures may be better than that for texts. Graphic representations, however, can be computationally more effective in communicating material, as evidenced in such advantages as locality (Larkin & Simon, 1987), emergence (Kulpa, 1995) and inexpression (Stenning & Oberlander, 1995). These effects most likely outweigh those of simple memory extension suggested by dual coding, and therefore constitute the more likely underlying variant rather than the form of the representation. In this chapter, computational effectiveness is captured under the term "specificity" and discussed further in chapter 5. To further illustrate the role of modality in learning we now outline two examples, one from learning logic, more particular syllogisms, the other from understanding statistical graphs.

The example taken from logic comes from Dobson (1996). The following sequence of diagrams is used for demonstrating the syllogism: all Bs are As, some Bs are Cs, which has the conclusion some As are Cs. The diagrams are available as a direct manipulation interactive microworld much like physical simulation microworlds. There is a fixed semantics and syntax which is designed so that all possible configurations are legal and all forms of the syllogism can be represented. The syntax describes the dynamic process of working with the representational conventions and could be described in four steps.

1. Form representative diagrams for each of the premises (there are four patterns which can be in different orientation depending on the mood of the premise).
2. Register diagrams by adding tokens from each premise diagram into one single diagram.
3. Resolve regions where there is ambiguity and confirmed individuals; for example, where there is a shaded counter separating two regions and one of those regions has an empty counter in it the shaded counter must be moved to the non-empty region.
4. The conclusion is read from the diagram by determining any statement that fits the patterns of the premises between the A and the C terms.

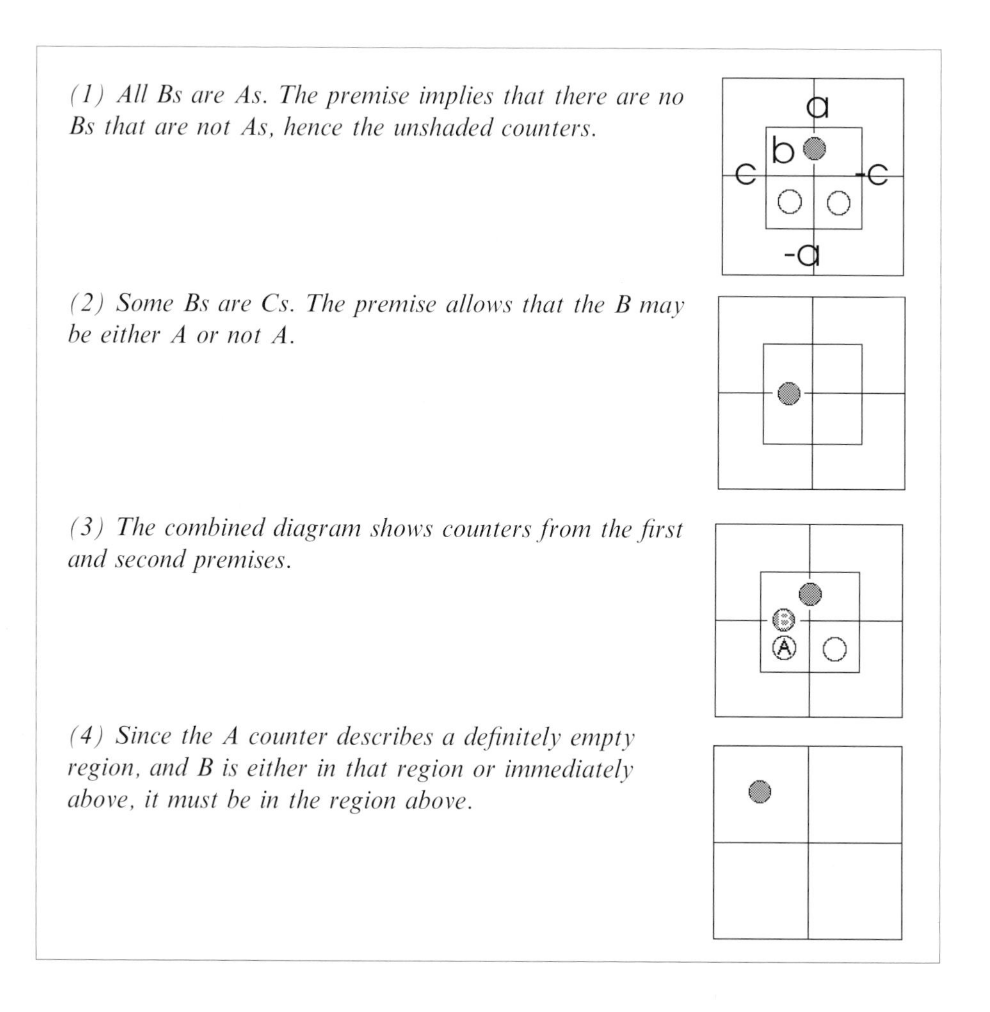

At the end of the four steps the syllogism has been solved. The diagrams help the learner to see the relationships and constraints between premises, but in addition serve a different function from when in the physical domains. The diagrams are actually good representations of the domain without extra linguistic notation. By comparison, in graphical models of structural distortion (Tessler, Iwasaki & Law, 1995) qualitative tendencies in the distortion of pillars and beams under stress are shown. Such representations only show a range of the possible outcomes of applying mass to structures. Inferences made from them may guide the problem solver to the right set of equations but alone cannot provide a precise solution. Improvements might include making the thickness of lines showing the beams relate exactly to the tensile strength of a beam and the length of an arrow proportional to the mass of a source of stress, effectively making the semantics of the diagram more rigorous. These advantages, however, would not add to the functionality of our logic diagrams where the conventions actually prohibit impossible situations, say an individual being both Dutch and not Dutch at the same time. The diagram is therefore a visual language of structural behaviour, either concrete for a physical device or abstract where the diagram represents a domain such as early logic. It serves as a place holder or short-term memory device, allowing the problem solver, designer and learner to sketch out causal effects and conceptual relations from diagrammatic constraints. The diagram allows step retracing to "see" further effect, and can "guide" the learner or problem solver in choosing a next step in their hypothesis space and provide a constrained environment for practice in experiment space. Empirical studies (Dobson, 1996) have shown that this kind of manipulation of interactive graphical systems goes on both when supported by appropriate software and while using pencil and paper. Students do not always find the skill easy, however. The syntax of two circle-based systems for the syllogism used in the studies, VENN and EULER require shading regions adjacent to boundary arcs to indicate either the known existence or known non-existence of individuals. Failure to resolve the continuity of boundaries by poor learners and those who do not practice diectic sketching during problem solving may have to do with human abilities pointed out by Gestalt psychology. Video recordings of students using these systems show successful learners engaging in high levels of gestural communication sketching with pointing devices on screen and on paper during communication with a tutor. Poor learners suffer difficulties in retaining the continuity of diagram element boundaries, which may not be surprising; however, the successful students' strategy of outlining boundaries proved useful and may be transferable.

The second example on the modalities of statistical graphics comes from Levonen (unpublished) who studied the interpretation of statistical graphs (diagrams, charts, histograms, etc.). In general, graphs are multi-layered displays of information, which are commonly used in society at large, in the media and in schools to display information from multiple sources. The availability of graphic displays, however, does not necessarily lead to a better understanding of the phenomenon displayed in the graph. Understanding graphs is a complex and subtle cognitive activity (Pinker, 1990). For example, detecting hidden fallacies in graphs has been shown to be quite a difficult process (Huff, 1954; Wainer, 1984). In graphs we find four distinctive properties of interpretation processes.

First, the modalities of information shift in the process of creating graphs so that the visual construction of a graph represents numerical data which is extracted from particular phenomena (see Dibble & Shaklee, 1992; Janvier, 1987a, b; Tukey, 1977). A graph is often constructed by first sampling a certain data set from a real-life phenomenon, then the sampled set of data is arranged and organised according to particular criteria. The constructed graph consists of, in part, visual representations (a line or a bar) which may reflect the actual real-life phenomenon, e.g. displaying an increase in an amount within a certain period using respectively increasing bars.

Secondly, multiple visual signs within a graph (e.g. background, axes and lines) and textual information (e.g. labels and numbers) must be mapped relative to each other so that variables, scales and quantities form a coherent representation of the displayed information (see Kosslyn, 1989; Pinker, 1990). The use of multiple representations in a single graph often means that the reader needs to interpret multiple representational modalities. For example, in the interpretation process the reader needs to interpret textual information (labels and the title of a graph) and graphical information (lines, curves and bars) and to map the locational arrangements of displayed objects in order to create a mental representation of a graph.

Thirdly, information in a graph is presented simultaneously rather than linearly. The reader is not constrained by the information display into having only one possible way of parsing the display, and yet the locations and relations of visual symbols in a graph need to be perceived and processed before the graph can be understood (see Larkin & Simon, 1987). The reader is required to clarify the relationships between different representational modalities present at a graphic display as they proceed in making sense of the phenomenon represented. For example, Bertin (1983) identified features of the visual variables (size, value, texture, colour, orientation and shape), the "implementations" of a graphic representation (points, lines and areas) and the relationships between different components (e.g. selective, ordered or quantitative variables), which it is essential to understand when processing a graph. In addition, these elements of the graphic sign system show that parsing information from a graph is a non-linear process that could not, for example, be talked about linearly over a phone line.

Fourthly, a single graph can be processed and, thus, understood in multiple ways depending on the reader's level of expertise in the domain the graph represents and on the questions the reader has targeted to the retrieval of information from a graph (see Bertin, 1981, 1983). Experience and knowledge of certain representational modalities (e.g. numeric vs. iconic vs. symbolic modalities) may influence how a graph is perceived and understood. For example, Bertin (1983) in his distinguished work on *Semiology of Graphics* suggested that reading information from a graph may take place using different levels of questions, i.e. different levels of reading a graph. An elementary level of reading addresses a single component which has a single correspondence (e.g. on a given date what is the amount of X?). An intermediate level of reading addresses a group of elements and results in a group of correspondences (e.g. over the first 3 days what was the movement of X?). Finally, an overall level of reading addresses questions related to a whole component (e.g. during the whole period what was the development or trend of X?).

In summary, understanding statistical graphics requires the reader's active construction using the multiple modalities of information presented on a graphic display.

Levonen (unpublished) studied the interpretation and construction skills of interpreting graphs using a Finnish version of the TOGS test (McKenzie & Padilla, 1986; Padilla, McKenzie & Shaw, 1986). Using a cross-national test provided the opportunity to compare the scores of the TOGS-FIN with the results of US students presented in the literature (McKenzie & Padilla, 1986; Padilla, McKenzie & Shaw, 1986). The level of student success according to the sub-skills is presented in grades 7–9. The easiest task areas among the Finnish students were locating a point in the coordinates (81% correct) and identifying a trend by interpolating or extrapolating (69%). The most difficult tasks were selecting a best fit line (44%) and selecting appropriately scaled axes (27%). Similar findings on the location of difficulty in sub-skills were reported by Padilla, McKenzie and Shaw (1986). They found that the easiest sub-skills were plotting points (84%) and determining the X and Y coordinates for a point (84%), and the most difficult sub-skills were using a best fit line (26%) and scaling axes (32%). The results of the study suggested that the most difficult tasks were items which required an understanding of Bertin's (1983) overall level of reading and mental manipulation of whole components (axes) and/or variables.

2.4 Specificity

One of the dimensions that we use to characterise representations is *specificity*, which captures the idea that there are computational properties to representations. A more precise usage of the term "specificity" can be found in Stenning and Oberlander (1995), and in Dobson (chapter 5). As we explained in the introductory part of this chapter, specificity is a dimension that is related to the dimension of modality as discussed in the previous section, and even more specifically, is used quite often in studies emphasising the benefits of graphical representations. There are several psychological approaches to explaining the general phenomena of modality preferences at a cognitive level and two are described here. The first tries to explain the benefit of diagrams in terms of their tendency to be designed for a "special purpose" and to fit habituated human reasoning patterns. The second approach concentrates on computational properties of the information presentation scheme and claims that graphics tend to constrain the information into more vivid and tractable schema.

In the first approach, the advantages of graphical systems include: (a) locality, where information that needs to be used together, or is connected logically, is typically physically close together in diagrams, reducing search and the need for symbolic labels (Larkin & Simon, 1987); (b) psychological emergent properties, due to well-practised perceptual inferences, that make working with diagrams easier than working with sentential systems (Koedinger & Anderson, 1994); (c) computational emergent properties where geometric relations reduce the search space syntactically possible in the sentential systems; and (d) structural constraints, where for example, whole–part relations in diagrams guide efficient knowledge organisation. The nature of the mapping between graphical systems and what they represent, makes them good for searching. However, departure from the formality often requires linguistic annotations, multiple level diagramming or some other added convention. These additional methods detract from the processing efficiency of graphical systems (Larkin & Simon, 1987).

The second approach, properly called "specificity" (Stenning & Oberlander, 1995), is concerned with the properties of representation systems, and refers to how far the conventions used in a graphical system demand specification of classes of information. The approach begins by describing the expressive abilities of a system as its capacity for showing information. This "expressivity" is the ability of a system to characterise relationships and models and to express the information intended. In general, graphical systems are thought to be more specific than sentential systems, which is supposed to explain their pedagogical benefit. The specificity approach describes a way in which information can be enforced within the most fitting representational system. The most fitting representation (most specific) is based on the material (models, relationships, etc.) in the domain and the system's ability to show the material (models, relationships, etc.). Another way to describe the idea of specificity would be to say that the representation which has least superfluous expressive capacity will be most useful in communication, other things being equal.

These two principles appear quite distinct, but are not inconsistent. In the first approach, examples of general features which often occur in diagrams are presented. The idea of "whole–part" relationships being easier to comprehend from a diagram is of course dependent on the choice of representational conventions used. The most obvious choices, possibly some kind of tree diagram, would certainly help to show the relationship referred to. The specificity idea, however, suggests that it is not the diagram which tends to support computational effectiveness, but the relationship between the diagram and the material. In this case the relationship between the tree diagram and the "part–whole" semantics was a good one and so the diagram works well.

Although the two approaches are not inconsistent, they do lead to different emphasis in their application to educational computing. The first approach suggests no particular way of generalising the principle to include new graphical systems, nor to expect that any graphical system would be a good one. By comparison, the specificity approach appears to offer the idea that graphical systems inherently compel information which tends to support learning. Several studies have been run that investigate the factors involved in modality selection (Dobson, in press). The empirical conclusion is that there are very many features in a learning interaction involving media which influence the learning outcomes. These include: the prior knowledge and expectations of the learner, their styles and disciplines of interest, as well as the appropriateness of the representation. Good graphical systems for learning are usually more specific than poor sentential systems. However, neither the relationship between expression and domain nor the aspects of single models and example conventions, can provide enough information to decide between alternative choices of representation. The *tendency* for good representations to be more specific than poor ones is not inconsistent with either model. In Dobson (chapter 5) an empirical study is reported which contributes to the description of specificity and provides empirical results which relate to its implications for educational computing.

An example of using the "specificity" in the design of learning environments can be found in Ainsworth et al. (chapter 7). Ainsworth et al. describe two empirical studies of technology-based learning with multiple representations. The domain considered is number sense; an increasingly important aspect of mathematics education. It can be defined as a "sound understanding of the meaning of a number and of relationships between numbers, a good understanding of the relative magnitudes of numbers and awareness of numbers used in everyday life" (National Council of Teachers of Mathematics, 1989). Two aspects of mathematical understanding linked to number sense are the role of approximate solutions (computational estimation) and considering multiple ways to solve problems. These two concepts underlie the design of the systems. The first environment, COPPERS, teaches 7–10-year-old children that mathematics problems can have many different correct solutions. The system sets users money problems such as "'what is 2 × 20p + 3 × 10p?" and requires children to give multiple decompositions of this total, (e.g. 5p–5p–10p–50p or 5p–5p–20p–20p–20p). COPPERS is used to tell students whether their answers are correct and shows their answers broken down into partial products. This is performed in two representations and, by highlighting, the system encourages students to map between the different representations. The first representation is the standard place value representation common in the primary classroom. This is a propositional representation in which the operations of multiplication and addition needed to produce the total are made very explicit. The second representation is a summary table. Tabular representation is commonly referred to as semi-graphical. This representation is less familiar to the children and the arithmetical operations are implicit. To understand and make use of the information, children must decide what processes are involved and perform them for themselves, hence practising their multiplication and addition skills. The table also displays previous answers to the question. This allows students to compare their answers with those already given and, it is hypothesised, prompts pattern seeking and reflection. Thus, each row in the table is informationally equivalent to the place value representation. However, computationally the representations differ as for example patterns and missing values are more salient in the tabular representation and mathematical operators are more salient in the place value representation. Therefore, these two representations are proposed to be complementary. The information needed to interpret the complex tabular representation is present in the familiar place value representation, and this complex mapping is supported by the use of highlighting.

2.5 *Complexity*

Information can be presented at different levels of *complexity*, complexity referring to the amount of information present. In this book the concept of complexity can be found in the work by Ainsworth et al. (chapter 7) on CENTS, a mathematics world for learning about computational estimation. The complexity of the information is based on the two different dimensions of estimation accuracy — direction and magnitude. Hence, the representations used in CENTS can be chosen either to display direction or magnitude separately or can display both dimensions simultaneously. This is particularly interesting when considering multirepresentation systems because it allows for different levels of (informational) redundancy across representations. In the case of CENTS, three levels of redundancy are possible — no redundancy, partial redundancy and full redundancy. In no redundancy situations, each representation expresses a different dimension of accuracy. Thus, one representation is used to display direction (either higher or lower than the exact answer) and one to express magnitude (either continuously or categorically). When multiple representations are fully redundant, then the same information is derivable from both of them. For example, both representations could express direction and (continuous) magnitude. Finally, multiple representations could be partially redundant. In this case, there is some overlap between the information derivable in the representations. For example, one representation could express magnitude only, while the other both magnitude and direction. Learning environments (or more particularly simulation environments) can be based on models that are of a simple nature, but they can also hide models of an overwhelming complexity, such as in the learning environment HUMAN (on the human body) which contains more than 200 variables (Coleman & Randall, 1986). In presenting information to the learner not all the complexity of the model needs to be shown at once. At first, information can be left out and then gradually introduced as the learner passes through the material. Complex models can be presented to the learner as such, but the complexity of a model can also be presented in discrete steps. In doing this, the natural sequence is to unfold gradually to the learner the complex model by introducing a new variable or variables at each step. There are two modes of model progression in which the model increases in complexity for the learner. The first one is "real" model progression which means that the underlying model actually develops. In the second type the underlying model stays the same, but the access of the learner to the model increases step-wise. QUEST (White & Frederiksen, 1990) is the classical example of where the model actually develops. The second type of model progression can be found, for example, in a study by White (1984). Here, learners were confronted with a simulation on Newtonian mechanics in which they could apply impulse forces to moving objects. Learners were initially restricted in the impulses applied by only allowing four directions. From the learners' point of view there is no real difference between the two approaches. Several studies have looked at the effects of presenting learners with a sequence of increasingly complex models. Quinn and Alessi (1994) performed a study in which students had access to a simulation (on the spread of a disease within a population) with four input variables. One group started off with access to all four input variables, one group worked with three variables before proceeding to the full simulation, and the last group started with having access to two variables, proceeded to three and ended with four. In all cases students had to minimise the value of one of the output variables. Their data revealed that model progression overall had no positive effect on performance. Model progression, however, proved to be less efficient than providing the students directly with full complexity. It should be noted that the domain that was used by Quinn and Alessi was quite simple: the variables in the model did not interact. In another study on a simulation of a multimeter, Alessi (1995) found that gradually increasing the level of complexity of the interface was beneficial for initial learning and for transfer. Also, Rieber and Parmley (1995) found, in the area of Newtonian motion, that subjects learning with a simulation that presented increasing control over variables, scored significantly higher in a test measuring application of rules, than subjects who could exercise control in its full complexity from the start.

In the work by Sime (1994, 1998, and section 3) we find three characteristics that are related to complexity: granularity, generality and scope. *Granularity* refers to the grain size of the model, i.e. the amount of detail within the representation. The models of the process rig used in Sime's work include a representation of the heater and the stirrer as one item. A finer grain size model would include separate representations for these two components. Similarly the cooler and fan are modelled as one so that when the cooler is on, the fan is also assumed to be on. If the fan was represented separately the model would be of a finer grain size. The *generality* of a representation refers to the width of applicability of the knowledge for the purposes of problem solving. A representation such as an operational procedure is highly specific to the domain and to achieving a particular task, e.g. emergency shut-down of the system, it is less general in applicability than a model. Where a procedure is an appropriate representation when carrying out a common task, a model is more appropriate under unusual circumstances such as fault diagnosis. A procedure represents a compilation of a frequently used set of control changes and the resulting behaviour, there is no need to work out the required actions each time it is done. A more general representation might be a set of heuristics for controlling the system under normal operating conditions. At a more general level, fundamental knowledge about facts, concepts and principles have a wider range of applicability and are used in unusual or infrequently encountered situations. The *scope* of a model determines the breadth of the representation and is a term used to define the extent of the model and the limits of its representation. For example, the models of the process rig are defined in terms of the representation of thermal and flow processes but do not include pressure differences or spatial relationships. The scope is the extent of the components of the process rig with no representation of the environment in which it is set, e.g. the temperature of the room it is in is assumed to be a constant 18°C. This is not a true representation of the situation as a change in the temperature of the room may affect the cooling of the liquid.

As examples of using model complexity in instruction we summarise three learning environments from the SMISLE family (now continued in the SIMQUEST system, see De Jong & Van Joolingen, 1995; De Jong et al. 1998): SETCOM, Collision and TeEl. *SETCOM* (see Swaak, Van Joolingen & De Jong, 1998) covers the topic of harmonic oscillations, illustrated by the situation of a mass hanging from a spring. The system includes three levels of complexity. The most simple level is free oscillation, the next level of complexity refers to damped oscillation, and at the third level of complexity an external force is added to the system. The fourth level included in SETCOM is an application of the theory introduced in the first three levels. Therefore, this is not really part of the simple to complex dimension despite the fact that this level does add some complexity to the preceding model progression level. The more complex the model, the more variables can be controlled by the learner. When, for example, after the free oscillation, damped oscillation is introduced, the learner may change the damping constant as well. Figure 1 shows the interfaces of the four model progression levels in SETCOM. *Collision* (see De Jong et al., 1995) treats the topic of simple two-dimensional elastic collisions. In this environment a progression is made from single-particle kinematics to two-particle dynamics. During this progression, variables are added gradually, for example, momentum and the centre of mass. On the final level two alternative perspectives (see section 2.1) on the model are offered, a momentum view and an energy view. *TeEl* (De Jong et al., 1996) contains the topic of transmission lines. In TeEl a sequence of four models is offered. In this sequence different perspectives on the model are offered, and even a system switch is made: the first two levels handle pressure distribution in a pipe, whereas the latter levels handle voltage and current distribution over a wire. The underlying model of these systems is the same. Learners see the same output representation for all model progressions: two arrays of sliders representing the pressure and velocity, or voltage and current, at various positions. With proceeding model progression, learners get more control over input variables, which are fixed at the first level and under learner control at the final level. In the three applications both ways of introducing complexity mentioned above

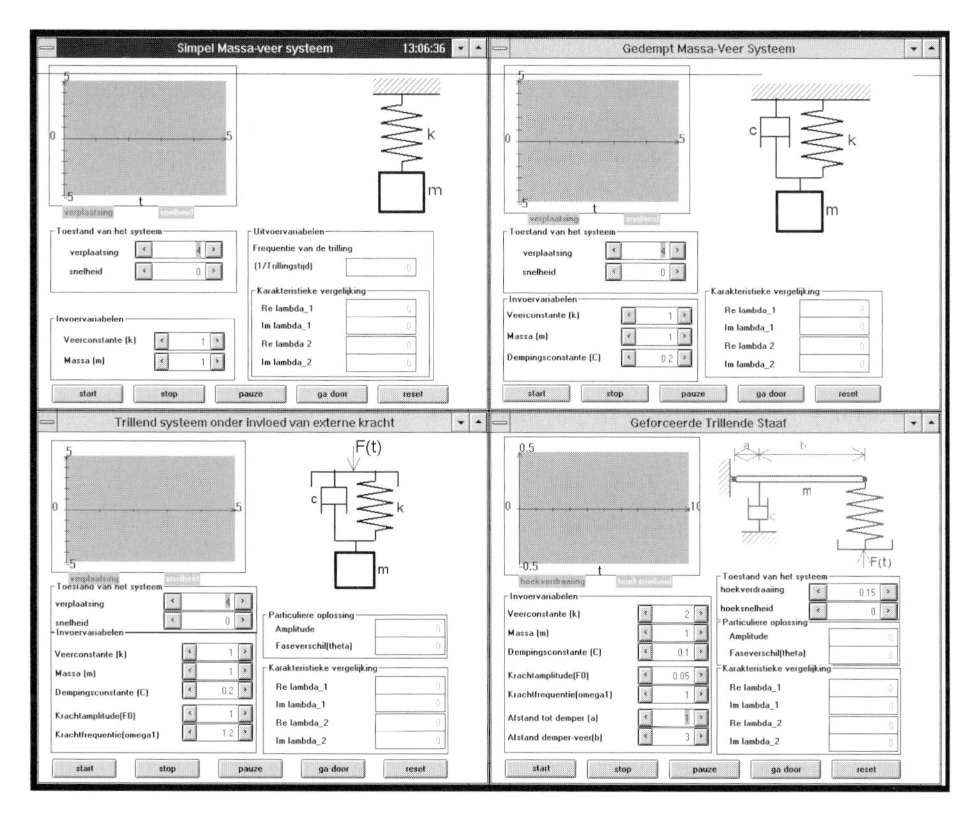

Figure 1: Learner interfaces of the four model progression levels in SETCOM

are used. For instance, each model progression level in SETCOM has its own simulation model. Each simulation model is an extension of the simulation model of the previous level. *TeEl*, however, only has one simulation model, and the model progression levels are created by hiding part of the variables. *Collision* uses both techniques, the five model progression levels use three different models.

Inherent in presenting information in multiple levels of complexity is a sequence from simple to complex. This is not necessarily so, but this is the most natural way of presenting the information. However, there can still be several ways to decide when to proceed from one level to the next one.

- A *free method* from the beginning, all model progression levels are available for the learner.
- A *learner controlled* method, where learners can only go from one level to the level next in complexity.
- A *timed* method, after a certain amount of time, a new model progression level becomes available.
- A *controlled* method where the learner had to perform a specific action specified by the author of the learning environment (e.g. to complete at least one assignment) before being able to go to the next model progression level.

The three learning environments SETCOM, TeEl and Collision were evaluated by having learners work with different versions of the environments and measure their knowledge in a number of different ways prior to and after learning. Next to model progression, SMISLE-based environments carry a number of instructional features, such as the availability of explanations and assignments (De Jong & Van Joolingen, 1995). For assessing the effect of model progression we can use data from the Collision (De Jong et al., 1995) and SETCOM studies (Swaak et al., 1998). In the SETCOM study a learning environment was compared with another environment that differed only with respect to the presence and absence of model progression. Here it was found that in the condition with model progression the students' intuitive knowledge was improved as compared with the environment without model progression. In the Collision study, De Jong et al. (1995) could not find any effects of providing learners with model progression on top of giving them assignments

3 Using Dimensions of Representations for Structuring Learning Material

In the preceding section we discussed several "dimensions" for classifying representations for instruction. In learning material, and also in the examples presented in the previous sections, we often find more than one of these dimensions present. The QUEST system is the classical example of presenting domain information in a sequence following several dimensions, and in this respect it has been the inspiration for the examples by Van der Hulst (1996) and Sime (1994, 1998) that follow in this section. QUEST (White & Frederiksen, 1986, 1990) uses multiple models, which vary along more than one dimension, to teach about the behaviour of electrical circuits. Learners are provided with a problem-solving environment in which they can run qualitative simulations, obtain qualitative explanations and experiment with the design and testing of circuits. Through a series of problems based on models of increasing complexity, the learners gradually transform their own mental model. The learner progresses through a series of upwardly compatible models that correspond to the desired evolution of the learner's mental model. The nature of these models, and the transitions between them, are therefore crucial to the success of the learning environment. Models may vary along one of three dimensions: perspective, order and degree of elaboration. Perspective describes whether the model is a behavioural, functional or reductionist physical model. Order, in the context of behavioural models, can be sub-divided into zero order, first order and second order. A zero-order model reasons about binary states of devices, e.g. the presence or absence of voltage. A first-order model reasons with changes in variables, e.g. when resistance is increasing, decreasing or remains the same. A second-order model deals with rate of change of variables, e.g. rate of change of current, or how much of an increase has occurred. The last dimension is the degree of elaboration and it relates to the number of rules used to simulate the behaviour of the circuit. The simpler models focus on principles that relate resistance to voltage and subsequent models add additional rules, or constraints in quantitative circuit theory, to allow reasoning about current flow within the circuit. However, this addition of rules could result in refinement of existing principles or the introduction of new variables into the model.

In Section 2 we mentioned a study by Van der Hulst (1996) who analysed high school teaching material on the physics domain of "oscillations". This study also made clear that different dimensions might play a role in organising learning material at different levels of granularity. In the oscillations domain, for instance, four types of progressions of the material along different dimensions could be observed: two types of complexity, a perspective and a precision dimension. Each of these dimensions provided an organising structure at a different level of granularity, which explains how all these dimensions could be used in an integrated manner. The following provides an illustration of how different dimensions might be integrated. In the material on oscillation, progression along a complexity dimension was observed to result in a coarse-grained decomposition into three models. A first model representing simple harmonic oscillation behaviour, a second that is damped oscillation and a third where damped oscillation was combined with an external harmonic force. Within each of these models different perspectives on behaviour were presented. That is, various accounts of behaviour are given, starting with graphs showing displacement or velocity of an oscillating object, subsequently force(s) acting upon an oscillating object, and finally an energy perspective showing a shift between potential and kinetic energy in oscillating devices (see Figure 2). Within the separate perspectives a finer grained level of organisation could be observed. First, progressions along a second complexity dimension were found. For instance, within the motion perspective, "displacement", "velocity" and "acceleration" were described in exactly the

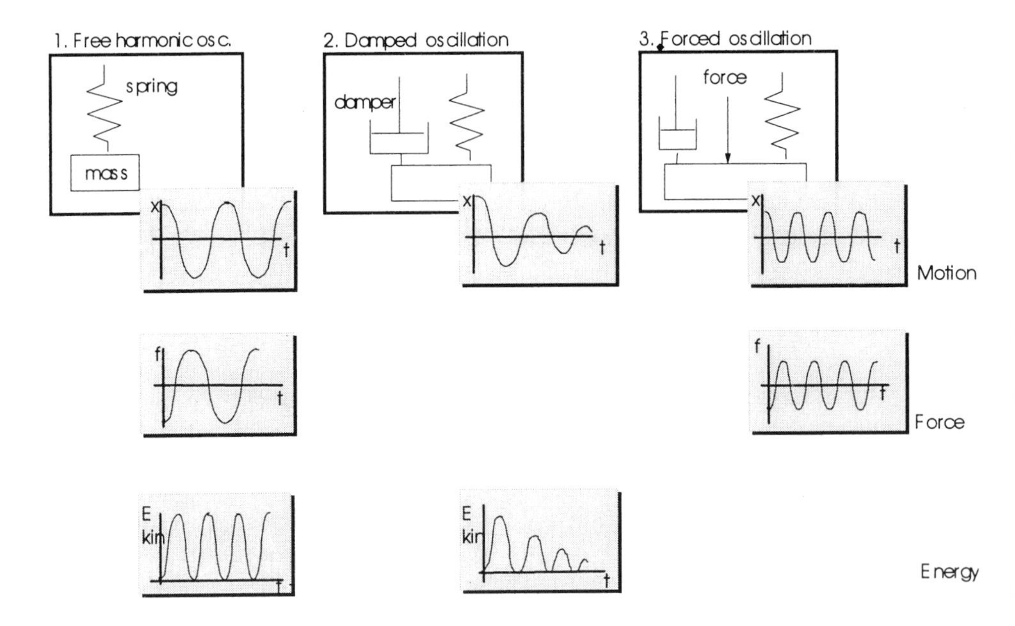

Figure 2: An integration of a complexity and a perspective dimension

given order. As velocity is a derivative of displacement and acceleration is again derived from velocity, the "degree of derivation" apparently formed a basis for an instructional progression. Secondly, at the same fine-grained level, both precise and less precise accounts of behaviour were found throughout the material. That is, formulae were frequently accompanied by imprecise statements such as "The period of an oscillating mass is independent of the amplitude and is entirely determined by the mass and force constant".

An elaborate example of using multiple representations can be found in the work by Sime (1994, 1998). She proposes a framework for the design of learning environments based on multiple representations of a physical system that vary along seven modelling dimensions which include notions of perspective, precision and complexity (see section 2). Its strength lies in the theoretical basis of the instruction and its tight integration with the design of the multiple simulations. This provides coherence to the instructional materials, the simulations and the instruction. This framework is illustrated in the "Model Switching — PROcess Rig Demonstration System" (MS-PRODS), and more recently, in "Cognitive Learner support in the PROcess Rig Demonstration System" (C-PRODS), two simulation-based learning environments designed to foster the acquisition of knowledge for the prediction of behaviour of an industrial heat exchanger, the process rig (Sime 1995, 1998). An empirical study of learners using the MS-PRODS learning environment has been carried out (see Sime, 1996; see also page 33).

There are six different representations of the same physical system that produce six simulations of its behaviour. Each model can be described in terms of seven modelling characteristics: approximation, ontology (see Perspective, section 2.1), abstraction, commitment (see Precision, section 2.2), granularity, generality and scope (see Complexity, section 2.5). These dimensions represent some of the typical analyses within the domain of reasoning about the behaviour of physical systems. The models of the process rig vary along abstraction (qualitative and quantitative models) and approximation (combined, thermal and flow models) but they all have the same level of representation for commitment, ontology (component-based), scope (the physical system), granularity and generality (models). Each model can be described, relative to the others, by the value of these seven characteristics. These characteristics were chosen as factors important in teaching about the physical system, but may not be applicable to teaching in other domains where the important dimensions differ. For a fuller discussion of these modelling characteristics, see Sime and Leitch (1992a, b) and Sime (1994). There are two types of knowledge represented within the learning environment: knowledge related to the selection of a representation, and knowledge about the behaviour of the physical system. Gentner and Gentner (1983) in studying human reasoning, concluded that the choice of model is crucial to the inferences that can be made and hence the whole outcome of the reasoning process. This highlights the importance of the selection of a representational formalism and the importance of this meta-knowledge.

The process rig is typical of an industrial heat exchanger (see the schematic in Figure 3). Liquid is pumped through tanks and pipes being heated and cooled. There are a number of sensors and indicators that provide information (on flow rates, temperatures and liquid levels) and controls that enable interaction (the pump setting, the heater setting, the valve setting and the cooler setting).

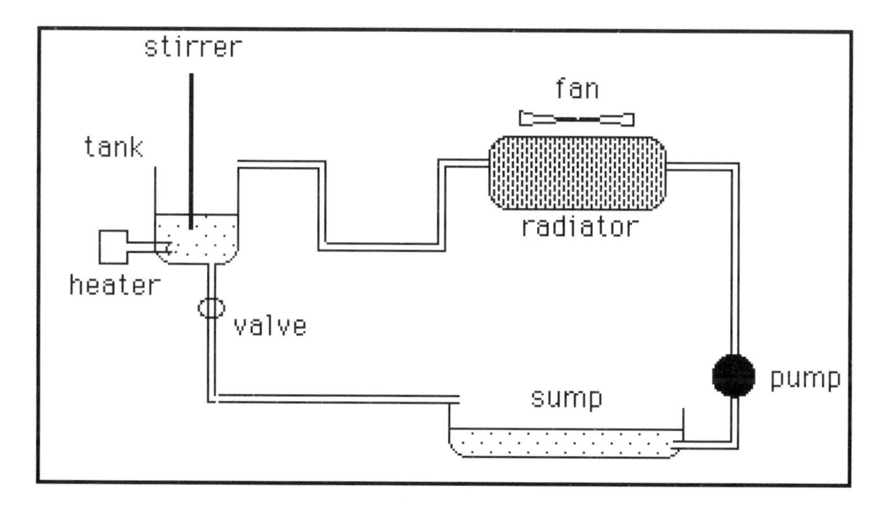

Figure 3: A schematic of the process rig

The learning goal is for the learner to be able to make predictions about the dynamic behaviour of the physical system and to describe the changes in qualitative or quantitative terms using the most efficient model for the task. The learner should be able to predict the behaviour given some change to one or two controls. When a control is adjusted, the system moves from an initial equilibrium state through a series of intermediate states before reaching a new equilibrium state, e.g. a change to the pump setting will result in an increasing flow through the pump, increasing flow through the valve and an increasing water level in the tank, the system will eventually reach a new equilibrium state where the flow rate is constant at a higher rate (increased) and the water level in the tank is constant at a higher level than before. The behaviour may be described in numerical terms, e.g. the flow increases 1.5 l/s, or in qualitative terms, e.g. there is a large increase in flow. The qualitative descriptions of behaviour expected of learners are the same as those produced by the qualitative simulations of behaviour.

The models of the process rig were chosen in collaboration with an expert on the basis of their utility in teaching about the behaviour of the system. The six models are related to each other in two ways: along the qualitative–quantitative dimension, there are three qualitative and three quantitative models; and they represent alternative conceptual views of the physical system, the thermal model, the flow model and the combined thermal and flow model (which includes representation of the interaction effects between the thermal and flow processes). The simulations of behaviour are generated from six models of the process rig: three qualitative and three quantitative. The qualitative models use linguistic, rather than numerical, descriptors of behaviour. For example the cooler is either on or off, temperature is described as three-value: increasing, decreasing or steady; and liquid flow is described using one of five values: increasing greatly, increasing slightly, unchanged, decreasing slightly or decreasing greatly. In White and Frederiksen's terms (1986, 1990), the order of the models is not uniform, e.g. in the qualitative thermal model some variables are binary (zero-order) and others are three-valued (first order). The quantitative thermal model also contains rates of changes in temperature (second order). Distinctions between models are seen as those that have some utility in teaching rather than distinctions based upon modelling characteristics, such as whether the variables are first- or second-order. These two requirements may match or may result in a model containing first- and second-order representations.

The alternative conceptual views of the physical system enable the learner to focus on one perspective of the system, say the thermal properties. These are not just subsets of the complete model, but alternative analyses of the system that direct attention in a conceptually coherent theme. The models of the process rig include representations of the thermal processes and the flow processes within the physical system. This separation is useful in teaching about the process rig because of the interaction effects between these two views, this emphasises the fact that flow changes and thermal changes cannot be considered in complete exclusion.

The design of the multiple models is closely tied into the design of the learning environment and the provision of instructional guidance that is offered to learners. The design utilises guidelines derived from cognitive flexibility theory (Spiro & Jehng,

1990) which are re-interpreted to take account of the multiple representations. Cognitive flexibility theory is a psychological theory of knowledge acquisition and use based on empirical studies of learners and expert problem solvers. This theory suggests that in order to construct flexible knowledge structures, that can be adaptively assembled when problem solving under novel circumstances, one must learn through multi-dimensional explorations of the domain.

Learners using the theory are able to interactively explore the six simulations, browse hypertext material describing the models and modelling dimensions, and solve problems where learners make predictions about the behaviour of the physical system. An automated tutor provides optional guidance to learners in the use of the learning material and recommends a sequence of problems. The relationship between the models is central to the learning material and provides a mental framework for remembering the models. Thus the modelling dimensions serve as links between the models and foster the acquisition of small units of inter-related knowledge which later can be used flexibly. They also provide the information necessary for selecting between the models. Guided instruction within the learning environment suggests a model progression along the modelling dimensions so that the learner moves from problems based on one model to problems based on a related model. In this way, the structure of the knowledge guides the instruction by progression along the abstraction (Precision, see section 2.2) and approximation (Perspective, see section 2.1) dimensions.

MS-PRODS can be compared with QUEST (White & Frederiksen, 1990) in that both are learning environments that use qualitative simulations and are based on an explicit theory of learning. However, both the theory of learning and the models are different. QUEST is a simulation-based learning environment where a progression of qualitative models of increasing complexity are presented to learners. In comparison the MS-PRODS automated instructor produces a non-linear exploration of qualitative and numerical models. This sequence, based on cognitive flexibility theory, means that the models may, or may not, increase in complexity and the models may be re-examined. This feature of re-examining models means that it is not a straightforward model progression but a path which may cross and recross the learning material.

4 Strengths and Weaknesses of Using Multiple Representations

In an earlier section we identified a number of dimensions for classifying representations and presented several examples of learning environments in which multiple representations were used to display learning material. We gave examples of learning environments in which dimensions of representations were used for ordering the learning material. In these studies three reasons are (explicitly or implicitly) used for introducing more than one type of representation in one learning environment. The first has to do with the tuning of the domain information and the representation, the second with the idea that using multiple representations in learning material will result in more flexible knowledge, and the third with the assumption that a specific order of representations will facilitate the learning process.

The first reason for using multiple representations is that *specific information can best be conveyed in a specific representation*, and that for a complete set of learning material, containing a variety of information, a combination of several representations is therefore necessary. The main issue here is that of *adequacy*, which concerns the expressional possibilities of a representation. If a certain perspective needs to be conveyed, this obviously has to be represented. Rohr and Reimann (chapter 3) present a design that follows this principle and also uses specific modalities (such as animations) for displaying specific perspectives. Also, in the modality dimension we see that a specific modality is better suited to express certain qualities of the domain than are others. It is evident that spatial relations are better displayed in diagrams than in texts and we all know the saying that "one picture tells more than a thousand words". A second aspect that can be involved here is *efficiency*, which concerns the expressional power of a representation. Within one level of adequacy, e.g. graphical representations, some may still be more efficient than others.

The second reason for using more than one representation is that *expertise* is quite often seen as *the possession and coordinated use of multiple representations of the same domain*. As White and Frederiksen (1990) write: ". . . a set of mental models that embody alternative, but coordinated conceptualisations of system operation" (p. 100). Expertise is also seen as the ability to readily switch between representations (De Jong & Ferguson-Hessler, 1991; Savelsbergh, De Jong & Ferguson-Hessler, chapter 13), and, when performing a task such as problem solving, to select the representation that is most suited for the (sub)task at hand (Gentner & Gentner, 1983). The most important theoretical background for this stance can be found in cognitive flexibility theory (Spiro & Jehng, 1990). In this theory expertise is viewed as being able to understand the domain knowledge from multiple perspectives. Problem solving is assumed to be based on the selection and use of multiple models and switching between models during problem solving is seen as a means of coping with complexity. One of the learning goals is to learn the strategic knowledge about the selection of the most appropriate model for use in problem solving. In this respect the theory is concerned with learning that goes beyond the introductory level, this is knowledge refinement and tuning, and not simply with-knowledge acquisition. The emphasis is on the development of cognitive flexibility in the learner, i.e. the ability to creatively restructure one's knowledge in response to new problem situations. This means that the learner is encouraged to develop a number of small knowledge structures that can be reused in unforeseen situations. So, following cognitive flexibility theory, learning should foster the ability to adaptively assemble diverse knowledge structures to fit the needs of a particular problem situation, rather than the application of precompiled knowledge structures such as procedures. The work on the process rig by Sime (1995, 1998), discussed previously, was explicitly based on cognitive flexibility theory. In teaching about the prediction of behaviour of a physical system, as is the case with the process rig, it is important to be flexible, and to consider problem solving under unusual circumstances such as when there is a faulty component, sensor or control within the system. The models of the process rig represent different perspectives on the domain knowledge and they are related along the modelling dimensions so that learners can explore the differences between the models, e.g. by exploring the differences between a qualitative and a numerical model or by exploring the relationship between the flow and the thermal processes by comparing the behaviour of the individual models with the combined model.

The third reason for using more than one representation is based on the assumption that a *specified sequence of learning material is beneficial for the learning process*. Three of our dimensions are often used here. When the *complexity* dimension is used this mostly results in a sequence of models of increasing complexity. The work by White and Frederiksen on QUEST (1986, 1990) and on the SMISLE environments reported in section 2.5 are characteristic thereof. The second dimension is the *precision* dimension used in the work on MULEDS reported on p. 14 and also used in the work on QUEST. Here a sequence from qualitative to quantitative models is seen as an optimal sequence for learning. However, the study by Van Someren and Tabbers (chapter 6) shows that this is not always true. They observed students who were learning a physics law that can be formulated as $(A * B) / (A - B) = C$, or alternatively as, $1/A + 1/B = 1/C$. In this case, there are no simple relations between these variables in qualitative terms. For example, neither A nor B has a monotonic relation with C. Van Someren and Tabbers give a qualitative model for this complex equation and then show with an experiment that teaching this qualitative model before the numerical law does not help in this case. Learners who were taught the qualitative knowledge first did not perform better on the induction task than those who received no teaching on this. Unless the learner is able to use more sophisticated mathematical concepts than "monotonic related" to recognise classes of possible equations, qualitative knowledge is not of much use here. For this particular equation, a better sequence is probably to first teach the (simpler) quantitative equation and then its qualitative implications. The third dimension, *perspective*, is used in the PRODS learning environments along with *precision* to produce a number of model progressions that have been compared in an empirical study (Sime, 1996). Three alternative model progressions were compared to assess which had the greatest benefit for learners. The first introduces quantitative models before qualitative models, and presents simple perspectives before complex perspectives. Hence producing a sequence of assignments based on the quantitative flow model, quantitative thermal model and quantitative combined model, qualitative flow model, qualitative thermal model and qualitative combined model. The second sequence maintained the same sequence in the perspective dimension but presented the qualitative models before the quantitative models. Finally the third sequence, influenced by cognitive flexibility theory, introduced the models in a mixed manner although limited by a change along one dimension only between assignments. The greatest increases in learning were found in the second and third progressions. This is interesting in that the first progression is the one most commonly used in the teaching of this material on university engineering courses, where students are particularly poor at answering qualitative exam questions. While the sample size is too small to make strong claims, a new look at the teaching of this material may lead to improvements in the teaching of both a qualitative and quantitative understanding of the learning material. Introducing multiple representations in a learning environment may also jeopardise the learning process. In chapter 7 Ainsworth et al. mention a number of cognitive tasks (which they adequately label as "costs") that learners have to perform to cope successfully with multiple representations. The main "costs" are that learners have to use different forms of syntax in one learning environment and that they have to coordinate the information from different representations. Coordination is always an issue when multiple representations are involved taking each of the dimensions mentioned in this chapter, but empirical studies concentrate on the "modality" dimension. In a learning environment on chemical systems, Kozma et al. (1996), for example, use the following representational systems: "chemical notation, video of reactions, molecular-level animations, dynamic graphs, displays of absorption spectra, and tabular data" (p. 47). Kozma et al. hypothesise that connections between different representations will emerge from the mere copresence. They enhance this copresence by having more than one dynamic representation run synchronously so that changes in one representation have concurrent displays in another representation. In the SETCOM environment discussed previously (see Figure 1) the same principle is used when an animation of a mass hanging from a spring moves synchronously in time with a dynamic graph producing the sinus function characteristic of oscillating movements. Quite a few studies, however, indicate that learners have problems integrating representations. In chapter 4 of this volume, for example, Scanlon concludes that students have great difficulty connecting tables and graphs. Tabachneck, Leonardo and Simon (1994) found that novices learning economics did not attempt to integrate information between line graphs and written information. Yerushamly (1991) found that few children in his study used two representations and those that did were just as error prone as those who employed a single representation. Mayer and Sims (1994) made a detailed study of the cognitive task involved in connecting two representations that are presented together and they refer to this as the "contiguity effect". In their study, in which learners were presented with figural and verbal information on the principles of a bicycle tyre pump, they found that this contiguity effect was high for high spatial ability subjects, but not for low spatial ability subjects. The conjecture is that low spatial ability subjects have to spend so many cognitive resources in spatially linking the two representations that they do not find the cognitive room for making referential links between the representations. This same phenomenon of cognitive load effects when using multiple representations is also mentioned in other studies. For example, Kozma et al. (1996) collected protocol data and on the basis of this they conclude that: "One problem with the multiple, linked representation approach is the excessive demands on the limited capacities of visual perception and attention" (p. 56). Gruber et al. (1995) studied the learning behaviour and gains of learners working in a game-like simulation in the area of economics (a "jeans factory"). They found that adding multiple perspectives to a simulation on an economics system was detrimental for students' performance. Gruber et al. (1995) also attributed this effect to an increasing cognitive load.

5 Effects of Using Multiple Representations

Now having seen the potential advantages and disadvantages of using multiple representations in a learning environment it is interesting to have a look at empirical studies evaluating the use of multiple representations. An extensive study in this area was carried out by White (1993). She had sixth graders work with a simulation on Newtonian motion called "Thinker Tools". In Thinker Tools a variety of representations for motion are used. The representations in Thinker Tools can be classified along (at least) two lines: there are dynamic representations and static representations, and concrete, "intermediate" (p. 16) and abstract representations. The intermediate representations share features of both the real world and higher order abstractions. Furthermore, the alternative representations for motion are linked together in the microworld. A major instructional goal of Thinker Tools entails the understanding of alternative representations of motion and, also, the ability to translate between these alternative representations (p. 32). In order to assess students' abilities White used several tests. In one test consisting of 16 open questions, the items showed students one representation of motion and the students were asked to indicate how this motion would look in an alternative representation. This test directly addresses students' ability to translate between the representations used in the microworld. A second test was developed to measure knowledge of the effects of impulses and continuous forces on motion as displayed in the microworld. This test contained 13 multiple choice questions, using only text, in which students were asked to make predictions. The last test also consisted of 13 (actually 18 as 5 questions consisted of two parts) multiple choice questions. Here students were expected to apply knowledge of the laws of force and motion in real world contexts (instead of in the microworld context). In this transfer test both text and accompanying pictures were used to pose the problems. Moreover, the test is composed of "notorious" questions that had been used by researchers studying misconceptions among physics students (p. 33). The first two tests were applied to the 42 students who had worked with Thinker Tools. On the whole the students scored higher on the test assessing ability to translate between representations than on the test measuring understanding of the laws of force and motion within the Thinker Tool context. White calls this result "not unexpected", as "understanding the representations is prerequisite to acquiring the laws" (p. 33). The transfer test was not only completed by the Thinker Tool group but also by several control groups. The Thinker Tool group outperformed all control groups, even a control group consisting of high school physics students, who were on average 6 years older and had been taught physics in the traditional way. According to White, one main difference, between Thinker Tools and traditional methods of teaching physics, lies in the representations used. In traditional instruction students are not given practice at mapping between abstract representations of motion and concrete representations and therefore have difficulty in applying abstract principles to everyday phenomena. In contrast, in the Thinker Tools environment, the intermediate representations provided help students to map abstract principles to concrete phenomena.

Several studies, as presented in this volume, have made comparisons between different representations and between using multiple and single representations.

Dobson (chapter 5) compared three different representations for syllogisms. Representations that differed with respect to specificity (this is the expressional power of the representation, see section 2.4) were compared. Although it was predicted that the representation with the highest specificity would result in most learning, this did not turn out to be true for a group of mathematics students. For a group of students from the humanities, in all three representations, there was even a decline in performance from pre- to post-test. Dobson tries to explain these surprising results in term of deficiencies in the metric used, improvements in the graphs still to be made (e.g. adding animations), and characteristics of the prior knowledge of learners. Rohr and Reimann (chapter 3) studied the effects of self-explanations on students' understanding of the particulate nature of matter. They compared three experimental conditions. One group only used text, one had text and static pictures, and one had text and animations of the particle behaviour. Students were asked to self-explain the textual and/or visual materials. Rohr and Reimann applied various pre- and post-tests. One paper and pencil test contained items with text and pictures that required students to explain the phenomenon displayed in the picture. In addition, in the test phases, students observed a particular movement of particles under a microscope, followed by two animations, and were asked which animation explained the movement of particles under the microscope better. By means of these measurements Rohr and Reimann were able to assess any naïve conceptions students held on the subject matter. On the basis of results from two students, Rohr and Reimann cannot conclude that there is a simple overall effect of representation used and suggest that the effect interacts with properties, in this case prior knowledge, of the learner.

Ainsworth et al. (chapter 7) looked at 10–11-year-old children using CENTS, a technology-based program supporting children performing computational estimation. CENTS contained either mathematical representations, pictorial representations or a combination of both. Ainsworth et al. measured the children's ability to perform estimation problems and the insight the children had into the accuracy of these estimates. The two-part test was given to the children both before they worked with CENTS and after they had worked with the program. They found that the children of all three groups improved their estimation skills. The insight into the accuracy of the estimates improved in the mathematical group and the pictorial group, but not in the mixed representation group. Ainsworth et al. believed that the children that used both the mathematical and pictorial representations failed to translate properly between the two representations. In another environment on simple arithmetic (COPPERS) Ainsworth et al. compared two conditions, in one condition a single, traditional representation was used, in the other both the traditional (place value representation) and a tabular representation were presented alongside. It appeared that children in both conditions learned a lot, but the gain in the multiple representation condition was higher than in the single representation condition. The multiple representations in COPPERS were designed to be easy to translate and support for this translation was provided by highlighting common elements.

Scanlon (chapter 4) evaluated a multimedia program which intended to support exploratory learning about motion. The Multimedia Motion program asks post-16-year-old students to examine videos of several types of motion, such as people jogging, trains crashing, etc. and then to run experiments, to collect data in tables and to plot graphs. As a consequence both tabular and graphical representations are used. Scanlon used videotape records of teacher and students using the program, teacher's and students' self-reports, and questionnaires of students' use of the program. Scanlon found that students have difficulties in connecting tables (generated by running the experiments) to graphs. In a second study Scanlon (chapter 4) looked at 12–13-year-old students using CCIS (Conceptual Change In Science). CCIS covers the subject of Newtonian mechanics and contains a simulation and graphical representations. By means of the predict–observe–explain methodology students are expected to gain understanding of the laws of force and motion. In order to assess student performance several measures were applied, open questions as pre- and post-test, interviews, videotape recordings, and students' completed worksheets. A main finding was that the students performed well in their understanding of the graphical representations, and that they could map the simulated behaviour with the (active) graphs.

Van Someren and Tabbers (chapter 6) compared two situations, one in which learners were offered qualitative information before they started to work in a quantitative discovery environment, and one in which learners started to work in the quantitative environment directly. Van Someren and Tabbers did not find a facilitating effect of the priming with qualitative information on both a quantitative and qualitative post-test. From these studies no clear picture of using multiple representations arises. One reason may be that the effects of using multiple representations are not straightforward and have an intricate relation with other factors such as the characteristics of the learner and the domain used. Another reason may be to do with the qualities of the performance test used. It is too simple to say that representations used in the learning phase should also be used in the assessment phase (which can take place both after or during the learning phase). Moreover, it does not usually make too much sense to assess whether, for example, text, diagrams and graphs used in the learning phase are available and/or can be reproduced during assessment. Generally, the goal of learning with specific or multiple representations does not entail learning those representations by heart, but how to use those representations, as tools, to acquire knowledge of a certain quality. Swaak and De Jong (1996) argue that using multiple representations may lead to a certain quality of knowledge that could be measured. De Jong and colleagues (this chapter), and Swaak and De Jong (1996) study learning and the resulting knowledge of students interacting with simulation-based environments. They argue that both the multiple representations in simulation-based learning environments (the information is usually displayed in more than one representation: a dynamic, graphic representation of the output is generally present next to animations and numerical outputs) and the fact that knowledge has to be extracted, inferred — as no "direct view" on the variables of the simulated domain is given — lead to complex inference and selection processes which in turn result in knowledge with an intuitive quality. In the authors' conception of intuitive knowledge "quick perceptions of meaningful situations" play a central role. Therefore, to assess knowledge with an intuitive quality they tried to elicit from their students quick responses to meaningful situations, in which perceptions are important. The test format they developed especially for this purpose is called the WHAT-IF test format.

Within these types of tests, items basically consist of a description of a situation, a specific state of the simulation, the introduction of a change, and a number of resulting situations the subject must choose between. This choice has to be made as quickly as possible. So far, the WHAT-IF tests have been applied in three studies in which different configurations of two simulation-based discovery environments were tested (see De Jong & Van Joolingen, 1996). The two simulation environments were in physics and concerned the topics "oscillation" (a pilot study and a replication study with more advanced students) and "collision". In these studies parallel versions of WHAT-IF tests were administered as pre- and post-tests. The results of the studies show that the WHAT-IF tests are, as contrasted with definitional knowledge tests, far more sensitive in spotting knowledge acquired (pre-test, post-test differences) during discovery environment interactions. Another important finding was that a trade-off was not found, in any of the studies, between the correctness and the completion time of items. There even seems an indication of the reverse: the quicker an item is answered, the higher the chance it is correct. The suggested direction is clearly in line with our ideas on the intuitive quality of conceptual knowledge as the "quick perception of meaningful situations". Moreover, the WHAT-IF format seems sufficiently sensitive to detect variations in the learning environments that were introduced in the discovery environments. For example, some learning environments contained assignments, whereas others did not. The WHAT-IF tests appeared more able to discern these variations, compared with other knowledge measures we used (e.g. definitional tests, hypotheses lists, concept maps).

Looking at the studies reported here we can say that the additional value of one modality compared with another or using multiple representations is not straightforward, even when close attention is paid to assessment. However, without investing in assessment, the implications of using specific or multiple representations for instruction will remain unclear.

6 Conclusions

In this chapter we have presented an overview of the different dimensions that play a role when designing and using representations for learning. This overview shows that the possibilities when designing a learning environment are large. Owing to the many differences between environments, it is too early to present an integrated view on the effects of using multiple representations. However, as an overall picture we have to conclude that the use of multiple representations does not lead to better performance in all cases. In this respect we share Tergan's (1997) conclusion, based on an analysis of the use of multiple perspective in hypermedia environments, that: "All in all, assumptions holding that studying material with hypermedia in a number of different contexts and from a variety of perspectives would be advantageous for all students have been too enthusiastic" (p. 13). On the basis of the present chapter and the following chapters in this volume we can draw further conclusions. We have seen that the use of multiple representations may serve three goals (see Section 4). First, multiple representations may be used because the information to be learned has multiple characteristics, so that each are most adequately conveyed by a specific representation. Secondly, multiple representations may be used to induce in the learner a specific quality of knowledge that may be labelled as "flexible". Both these approaches lead to a *concurrent presentation* of multiple representations. Thirdly, there is the assumption that a specific sequence of representations is beneficial for learning. This approach leads to a *transitional presentation* of representations. The effects of using multiple representations seem to be mediated by the following factors:

The *type of test* used is partly responsible for the effects. For example, Mayer and Anderson (1992) found only an effect of presenting animations and text together on a problem-solving test and not on a retention test. Work by Swaak and De Jong (1996) reported on p. 37 is also illustrative of this phenomenon.

The *type of domain* in the learning environment may also be of influence. Mayer and Anderson (1992) studied two domains, pumping a bicycle tyre and an automobile braking system. In their discussion they state that the effects found on combining animations and texts may well only be valid for "how it works" type topics (as they used) and not for narrative or descriptive information. The difference in results on combining qualitative and quantitative representation by Spada and colleagues on the one hand (see p. 13 and following), and Van Someren and Tabbers (chapter 6) on the other hand, may also be caused by differences in the domains used.

The *type of learner* using the environment also influences the effectiveness. Again Mayer and Anderson (1992) say that their effects are only valid for the type of learners they used (inexperienced students) since experienced students may also have profited from text and animations not presented together. Kozma et al. (1996) and Kozma and Russell (1997) conjecture that for novices the external representation is closer to the internal one than it is for experts who translate surface characteristics into "deep" characteristics.

Finally, the *type of support* present in the environment also plays a role. Most environments simply assume that the copresence of more than one representation will prompt the learner to integrate the information. For example, Kozma et al. (1996) take this as an explicit starting point. However, after experimentation they remark that the "mappings across representations" is only potential, it is not clear whether students will actually do this and they suggest that learners may need more guidance in this. In contrast other environments provide full linking between representation such that a user acts on one representation and sees the results of these actions on another (e.g. the Blocks Microworld, Thompson, 1992). A further possibility is to provide partial support for a user's mapping across the representations. The COPPERS environment designed by Ainsworth et al. (chapter 7) is an example where this level of guidance was given. The two representations used in the COPPERS environment were connected by the use of "highlighting". The effects of different types of support for translating between multiple representations remains an open research question. Some researchers propose that many benefits follow from dynamic linking as the computer reduces the cognitive load for the user (Kaput, 1992). However, it may also be the case that over-automation does not encourage a user to actively translate across representations.

There is certainly a long way to go before we fully understand the conditions under which multiple representations should be used in learning and can apply these to the design of multiple representation learning environments. Earlier we mentioned some examples of using dimensions of representation in the structuring of the learning material. Section 5 examined the issue of evaluating the use of multiple representations and concluded that there is no clear picture. It is hypothesised that this could be due to the complexity of the problem and to other factors such as the learner and the domain used. The work presented in this chapter and in the rest of this book sets out some paths that need further exploration in determining the best use of multiple representations in technology-based learning environments.

3

Reasoning with Multiple Representations when Acquiring the Particulate Model of Matter

Michael Rohr and Peter Reimann

1 Introduction

We describe the nature and the use of multiple representations in the subject area of the nature of matter. Akin to a distinction made by Newell (1982), we argue that multiple representations occur here on the *knowledge level*, concerning different conceptualisations of the domain, as well as on the *symbol level*: knowledge in this domain is typically represented in propositional form as well as in form of mental models, as analogue, dynamic representations. In this chapter, we focus on the first dimension where multiple representations arise due to different ontological commitments. The acquisition of physical science concepts is particulary interesting for understanding how learners coordinate multiple ontological representations ("theories") because scientific conceptualisations of physical entities are often radically different from the "naive" conceptions and because pupils, and also science experts, often hold multiple theories about the same physical phenomenon, one more closely related to everyday reasoning, the other closer to the scientific opinion.

First we describe the naive models children develop until they are instructed formally on the structure of matter. For the sake of simplicity we restrict the domain to gaseous substances. The chapter continues with a short description of the particulate model of matter as it is normally introduced in school lessons. This is followed by a discussion of possible cognitive mechanisms for acquiring a discrete model of matter. We focus in particular on analogical reasoning mechanisms. After the discussion of these basic mechanisms of cognitive restructuring possible ways for integrating the different representational formats are described. The final part of this chapter is a description of a first study which explores the ways different representations can be used.

In science domains, representations of various kinds are used and combined. De Jong et al. (chapter 2), for instance, distinguish between different representations according to the dimensions of perspective, precision and complexity. We focus on a fourth dimension, close, but not identical to perspectives (as introduced by White & Frederiksen, 1990): *ontology*. Ontological aspects of knowledge representation refer to decisions about "how and what to see in the world" (Davis, Shrobe & Szolovits, 1993). Thus, ontology refers to the *content*, to the objects and relations one uses to represent a domain, not so much to the language by which objects and relations are denoted. Ontological decisions are more fundamental than decisions for a particular language in the sense that they determine the range of objects and relations between the propositions in a language.

In science domains, it is often the case that pre-instructional conceptions and scientific concepts are incompatible ontologically in the sense that they belong to different categories (Chi, Slotta & De Leeuw, 1994). For instance, physics experts are said to perceive problem statements differently from novices; experts represent a problem situation in terms of its "deep" structure, i.e. in terms of theoretically relevant categories, whereas novices represent the problem with much reliance on surface features — the features most salient in the problem description (Chi, Feltovich & Glaser, 1981; for an overview Reimann & Chi, 1989). Furthermore, in the area of science teaching it has been observed that pupils do not find it easy to adopt the scientific view. The most notorious example for this conceptual inertness is studies showing that, even after extensive exposure to Newton's mechanics, pupils still hold on to an impetus theory of movement (Di Sessa, 1982). More precisely, under some circumstances, pupils come up with predictions and/or explanations that are in accordance with one or other theory of motion: they oscillate between these two theories. Eventually, the scientific model will be accepted by some learners at least, which does not by any means indicate that the everyday conceptions have been forgotten completely.

The ontological commitment that is part of every representation can be distinguished from the computational aspects of representations. These concern issues such as how a specific representation language (for instance, logic vs. rules) influences how easy or hard it is to perform certain inferences, come to specific conclusions, etc. It is Larkin and Simon (1987) who deserve praise for stating clearly that although representations can be informationally equivalent, they can, and usually will, vary along the computational dimensions: informational equivalence does not imply computational equivalence and *vice versa*. For instance, although a graphical and a tabular representation of a mathematical relation may contain the same information, the two representations afford different inferences. For instance, looking up a specific value is easier in the table; extrapolating from the data points is easier with the graphical representation (for further examples, see Ainsworth, chapter 7).

Ontologically different representations by definition differ with respect to information: they express a phenomenon by referring to different concepts. Whether the representations also differ with respect to the computational aspect is a matter of circumstances. For instance, when looking at various conceptions of the nature of matter, the domain we focus on later, both everyday conceptions and what one learns about the scientific conceptions could be represented in a propositional format, as a mental theory so to speak. Or both could be represented as mental models, in a dynamic, non-symbolic form. Most probably, such representational formats will be combined by humans and will be used opportunistically for different purposes. For instance, in order to predict the outcome of a chemistry experiment, a mental model may be used; in order to explain the outcome to the teacher or to a peer, a propositional representation may be utilised. The "decision" when to use which representation, given that more than one is available, is probably based on cost/effect considerations: *ceteribus paribus* that representation will be used which imposes the least demands on cognitive resources, in particular on working memory.

In addition to cognitive demands, the selection of representational formats (such as graphical vs. propositional reasoning) will also be influenced, and strongly so, by the format in which the information is presented to learners. Textual input affords primarily a propositional representation, graphical input primarily a mental image (or both propositions and images, see Paivio, 1986), animation/video input induces a (dynamic) mental model (with many limitations, see for instance Mayer & Anderson, 1991).

Animation sequences as instructional input are of particular interest to us because they hold the potential to overcome two problems at once: they may be of great help for students in order to reach the right conceptual understanding of scientific concepts and they can help to overcome the cognitive difficulties involved in using these concepts. That is, they address issues both at the knowledge level (information) and at the representational level (or symbol level, as Newell (1982) phrased it), they help to overcome ontological as well as computational problems.

Animation enhances the chance of improving students' conceptual understanding, particularly if the animations show clearly how events at the macroscopic level (for instance, expansion of volume) "emerge" out of processes at the particle level (such as an increase of particles' kinetic energy). Processes which involve many entities (such as particles) are hard to describe in words and to comprehend that way, but are easily understood when displayed visually. "Seeing" the correlation between events at the particle level and on the macroscopic level should make it considerably easier to understand and accept explanations which stem from the particle level.

It is important to note that we do not believe that mental models, as induced by animations, are less valuable than propositional representations. In particular, we do not want to imply that mental models are pre-scientific and that experts in scientific domains will always use the explicit, propositional format. Among others, Clement's research (e.g. Clement, 1994) shows that experts (in physics) often make use of imaginistic reasoning, in particular when confronted with hard problems. Generally speaking, it seems to be indicative for competent performance in many areas that experts can flexibly use different representations and translate quickly between them (a comparable argument is made by Kaput (1989) for mathematical understanding and by Tabachnek-Schijf, Leonardo & Simon (1997) for economical reasoning; see also Schijf & Simon, chapter 11).

We are particularly interested in the first stages of the acquisition of scientific concepts. In this phase of knowledge acquisition, different mental representations of phenomena and of explanatory models co-exist and influence each other. The differences between the everyday conceptions and the scientific conceptions can be ontological, at least this is the kind of difference we will focus on in this chapter. In addition, the co-existence of multiple representations can result from different presentation formats, for instance, propositional ones (acquired from text input) and mental models (acquired from animations, for example). From a psychological perspective, it is necessary to consider ontological *and* computational aspects at the same time because content and format are not independent. Some forms of external representations will make understanding easier than others; the computational aspects of representation formats will affect the ease with which certain inferences can be drawn. Hence, the format both of the external representations from which one learns and of the mental representation will, presumably, interact closely with the ontological dimension. We consider it a weakness of most of current research on conceptual change that the only kind of analysis usually performed is with respect to differences in concepts' content (their ontology), but not their cognitive–computational demands.

In order to study the interaction of learning problems that result from ontological differences between everyday and scientific conceptions with learning problems that originate from cognitive–computational demands, detailed process models of conceptual change are required, process models which take into account the variety of representational formats which have to be coordinated at this stage. We are attempting a first step in this direction by analysing how pupils react when their everyday conceptions of matter (in particular gases) are confronted with a first approximation of the scientific view of matter. It is known that in this domain everyday conceptions differ significantly from the scientific view; we will provide the reader with a summary of the research looking into these "misconceptions" showing mainly that pupils hold a continuous theory of matter whereas the scientific model is based on discrete entities, the particles.

Our main goal is to come up with a process model of how pupils relate these two ontologically different representations of gas phenomena. We assume that the way everyday concepts are represented will affect the understanding and representation of the scientific model mainly by means of analogical transfer. Hence, our process model will eventually have to provide an account of how this transfer works. In order to come up with a process model, we have to formulate assumptions about the representational formats that learners use to encode their everyday knowledge and to encode information about the scientific model. Owing to the fact that learners in science domains are provided with a variety of symbol systems which are supposed to convey the instructional message (text, graphics and sequences of demonstration experiments and/or animations and/or simulations) multiple representations of domain information have to be taken into account.

Understanding better the cognitive processes that lead to conceptual change is a prerequisite for realising a second goal: to improve on instruction and learning not only for the case of the acquisition of the particulate model of gases, but if possible for the general case where pupils have to relate everyday conceptions and observations from their environment to abstract, theoretical notions. The particulate model of matter is an interesting case in this respect because in most school systems it is one of the first times that pupils are introduced to the abstract idea of the discrete nature of matter. Hence, this is not only an exercise in acquiring a particular scientific model, it is also one of the first contacts pupils have with the notion of a "model" in the natural sciences. Therefore, on top of the ever lurking ontological misinterpretations, and in addition to the cognitive complexity involved, pupils may also have to overcome epistemological difficulties, in particular difficulties in accepting the notion of abstract entities (particles) as a core concept to explain the nature of matter.

2 The Development of Naive Conceptions of Air

Because the conceptual change from a continuous to a discrete model of matter depends to a large extent on pre-instructional models of matter, we begin by describing important and empirically stable characteristics of pre-instructional conceptions about gases, of air in particular, which students have developed when they come into the first formal lessons in chemistry or physics. Compared with other areas of sciences, for example mechanics, only a few studies have been conducted which focus on students' conceptions of matter (e.g. Benson, Wittrock & Baur, 1993; Brook & Driver, 1989; Novick & Nussbaum, 1978, 1982; Sèrè, 1985; Stavy, 1988).

Piaget (1929) made observations on very young children: young children associate air with spirits or dreams that can come from and go everywhere and that can make things happen without being perceived (see also Sèrè, 1985). According to Brook and Driver (1989) children of about 5 years old know that air exists "everywhere". But many identify air only in dynamic situations when it is perceived easily, for example in the form of wind and clouds or in ascending balloons. Some know that air is involved when they breathe. But nearly half of their children said that there is no air in a closed container and only a third appreciate that air takes up space. A majority of children of around 8 years old think that air exists even in a closed container (where it cannot be perceived), and that air is involved in breathing. Between the ages of 5 and 8, an interesting conception develops: while most children at the age of 5 years compare the weight of air in containers in terms of their volume, by 8 years a quarter of the children analysed no longer attribute weight to air; some of them attribute negative weight to air. It seems that these children induce the absence of weight from the observation that air moves upwards or floats (Brook & Driver, 1989; Stavy, 1988). Most 8-year-olds show another characteristic conception in the area of pressure differences: they assume that air in a partially evacuated container (for example in a drinking straw which is used to drink some juice) is "pulling" the liquid (Brook & Driver, 1989).

By the age of 12 years there remain some naive conceptions which are at odds with the scientific model: more than 75% of the subjects in the study of Brook and Driver (1989) still considered air as having no weight or negative weight, and half of them described air in a partially evacuated container as "pulling". Also, only a few children recognise that air is transmitting forces in all directions. At the age of 16 years many incompatible naive conceptions seem to have disappeared; but in the study of Brook and Driver (1989) a considerable number of students remain with beliefs expressing that air does not transmit in all directions, that it has negative or no weight, and that air in a partially evacuated container is pulling actively.

Studies such as this corroborate the conclusion that pupils hold pre-scientific beliefs about the nature of matter even after extended exposure to the scientific model. Why is it so difficult for pupils to accept the scientific model? Before we look into research on conceptual change which has addressed this question in other areas of science, we describe the pre-instructional model of gases with the particulate model taught normally in school.

3 The Particulate Model of Matter for Beginners

The discrete model as described here is not one of the scientific models used in modern physics or chemistry nor is it an exact reconstruction of a historical model such as Dalton's, but is a simplified version which serves as an introduction to scientific models in school, in particular in chemistry lessons. In this short introduction into the scientific model we focus on those points which we consider decisive for the process of acquiring the discrete model of matter.

The first model normally taught in school can be described as follows (see for example Tausch & von Wachtendonk, 1996):

1. All substances are composed of very small particles, which cannot be observed.
2. The shape of the particles of a substance remains unchanged.
3. The particles are of spherical shape.
4. Particles of one substance have the same properties, particles of different substances show different properties.
5. The particles of substances are in constant movement.
6. Particles in solids are vibrating, but stay in fixed configurations. In fluids this fixed order is broken, but the particles move around one another in a bulk. In gases the particles are moving freely in space with large distances between the particles.

Obviously, there are differences between the scientific model of matter and the beliefs pupils typically hold before receiving instruction, as reported in section 1. For instance, the particulate model of matter is an abstract model because its building units and their behaviour cannot be seen, but have to be inferred from the observable behaviour of the gases. Because of its abstract nature, the relationship between what can be observed about gases and how observations are to be explained in terms of the interaction of particles becomes complicated. In particular, the abstract particles do not correspond directly to entities that can be observed. Transferring macroscopic properties of gases to the particulate level will therefore result sometimes in misconceptions. For example, the thermal expansion of gases cannot be explained by thermal expansion of its particles because there would be a contradiction with Postulate 2 above whereas the fact that gas has positive weight can be explained by the positive weight of its particles.

We assume that the discrete model of gases is difficult to acquire and to use for explanation and prediction when properties of gases are involved which cannot be explained by simple analogy between the macroscopic and the submicroscopic level. Therefore, from an educational and psychological perspective two kinds of gas properties need to be distinguished (following De Vos, 1990):

1. *Properties of matter which can be easily mapped from the macroscopically continuous nature of substance to the smallest units.* The most prominent properties of this type are mass and volume which are already mentioned in Dalton's theory as properties of particles as well. Also of this type are the properties electrical charge and kinetic energy as well as time-related behaviour. By way of hypothesis, it is

plausible that these properties, which are common to both levels, are acquired quite easily by students. This is different for the next type of properties.

2. *Properties which cannot be transferred from the macroscopic level to the particles of the submicroscopic level.* There are at least two subclasses of properties in this group: one subclass contains properties that are existing at the macroscopic level but not on the particle level. A prominent example is the macroscopic temperature of substances: in the kinetic theory of gases particles do not possess the feature of temperature. Macroscopic temperature corresponds to the average kinetic energy of particles. This important correspondence is not part of the pre-instructional model held by students before formal instruction. This is a model relation which increases the abstract character of the particulate model of matter and most probably contributes to the difficulty of acquiring the scientific model of matter.

   The second subgroup of macroscopic features is related to submicroscopic properties in an even more complicated manner. Quite a lot of simple macroscopic properties are not related to various submicroscopic features of single particles, but to properties which emerge out of the interaction of whole *systems* of particles. An example which is relevant for beginners' chemistry is the explanation of the macroscopic states of matter in terms of the submicroscopic interactions of particle systems. These interactions are influenced by the relation of potential and kinetic energy of the particle systems. The model relation between the state of a single piece of matter and the interaction of a system of particles again increases the abstractness of the model and one has to expect quite severe learning difficulties for the properties of the scientific model of matter.

In short, students with little exposure to the scientifically accepted model may transfer many macroscopic properties to the submicroscopic level, more than is sanctioned by the scientific theory. For example, students describe smallest particles as being distorted when the substance melts (Ben-Zvi, Eylon & Silberstein, 1986) or, in the case of children of age 9–10, even as disappearing when a solid or liquid evaporates (Stavy, 1990). These kinds of false analogies can account for a substantial amount of initial misinterpretations of the particulate model of matter. Because of the fact that the particulate model is only indirectly related to observations, some of these misconceptions may not change over time: contradictions between observation and pupils' theories are not realised as such by pupils and/or the particulate theory may be too obscure to make theory change admissible (Chinn & Brewer, 1993).

4 Acquiring the Particulate Model of Matter

4.1 Mechanisms for Conceptual Change

In science domains, theories have to be revised or changed often if new data cannot be assimilated into the existing theories. Chinn and Brewer (1993) analysed such situations and proposed several factors that can affect the reaction of people (including scientists) to anomalous data:

- Characteristics of prior knowledge (entrenchment of the prior theory, ontological beliefs, epistemological commitments, background knowledge).
- Characteristics of the new theory (availability, quality).
- Characteristics of the anomalous data (credibility, ambiguity, multiple data).
- Processing strategy (deep processing).

In our case, students are provided with the new theory (particulate model), hence availability is not a problem. Also we do not focus on the the quality of the anomalous data (in our exploratory study we try to design experiments with clear and acceptable observations). We will address the problem of the processing strategy later and concentrate first on the influence of prior knowledge. We discuss two recent theories which focus on the entrenchment of the prior knowledge and on the ontological beliefs.

4.2 Conceptual Change as Ontological Change

Chi and colleagues (Chi, Slotta & De Leeuw, 1994) have developed a theory that can explain the severe difficulties students often show when learning certain scientific concepts. Chi's model is based on the assumption that we perceive all things in basic ontological categories called *ontological trees* (Figure 1 shows trees of Chi's ontological classification system). We use these trees to classify entities and processes. Ontological trees are defined by ontological attributes. These are attributes which can be meaningfully assigned to only one of the ontological trees. For example, the

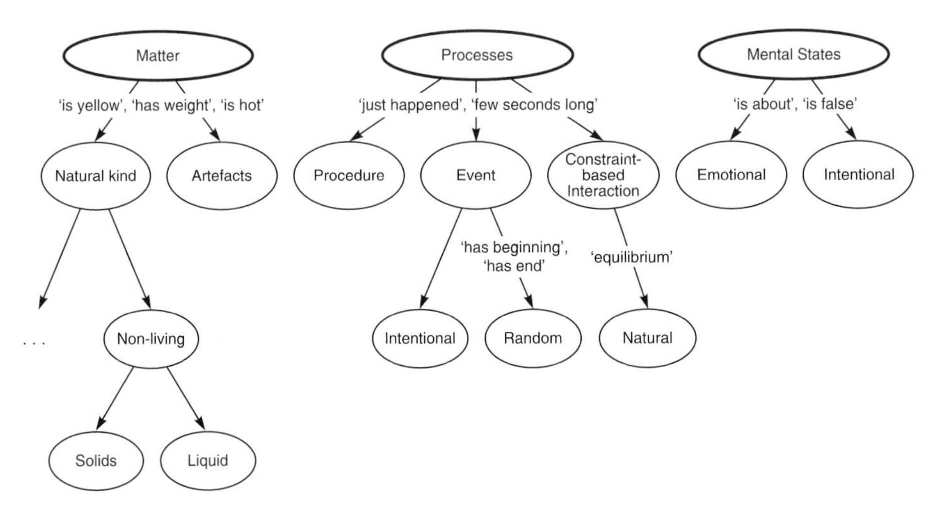

Figure 1: Relevant ontological trees similar to the classification by Chi (Chi, Slotta & De Leeuw, 1994)

attribute "yellow" only makes sense in connection with a concept of the substance tree but does not fit to the tree of processes. It should be remarked further that ontological attributes are not equal to defining (necessary) attributes and that they can be removed from concepts or other ones can be attached.

Chi argues that the most important empirically observed difficulties in learning science concepts arise when so called *ontological shifts* have to be achieved. That means that a concept changes its position from one ontological tree to another. If concepts undergo such an ontological shift they change their meaning radically because some ontological attributes have to be deleted and others have to be added. Let us illustrate what (radical) conceptual change is about by looking at two pre-instructional concepts which undergo radical conceptual change when they are explained in terms of the particle model: heat and state of matter. Let us assume that heat is classified as a concept and state of matter as an attribute of a substance. In a pre-instructional conception heat and state of matter belong to the ontological category of *substances*. But when a student accepts the kinetic theory of gases, heat and state of matter receive an ontologically different meaning. Heat is now an "epiphenomenon" connected to the kinetic energy of the particles of the substance, hence a property which belongs to the *process* category. State of matter is no longer a simple attribute of a substance but an "epiphenomenon", connected to the ontological category of *interaction* between a *system* of particles.

Chi et al.'s (1994) theory explains many findings from misconception studies in an elegant and parsimonious manner. However, it is not comprehensive enough to account for all aspects of the acquisition of a discrete model of matter. Chi assumes that ontological shifts take the form of moving the ontological properties of concepts from one tree to another, i.e. they are limited to "horizontal" replacement. This view does not account completely for conceptual changes. Let us look again at our example of learning the particulate explanation of the state of matter: In order to acquire the kinetic interpretation of the state of matter the naive macroscopic concept, which says that state of matter is an (ontological) *attribute* of a substance, has to be "substituted" by a *process* view. The state of matter has to be explained by a complex interaction of particles in their attracting forces and movement. This is not an ontological *shift* of a concept, but a change in the meaning of an ontological category, because an ontological *attribute* from the ontological tree of substances (i.e. state of matter) changes to a *concept* in the ontological tree of processes. That means that in the process of building a mental particulate model of matter a new ontological categorisation is developed.

But this interpretation of conceptual change is still too simplistic. Looking at our example of learning the concept of heat, it seems questionable that a naive substance-concept such as heat is simply substituted by a process-concept. At least for the case of learning a new abstract model of matter, it is not very useful to mentally "delete" the macroscopic meaning of temperature, assuming for the moment that this is possible psychologically, because this interpretation is still useful in many everyday situations which do not require deep, systematic interpretations, but fast information processing. It seems much more probable that interpretations of properties of matter in terms of macroscopic, continuous concepts *co-exist* along with microscopic, particle-based interpretations and that they are used in reasoning in an opportunistic fashion. If we assume substitution of the macroscopic model by the hypothesised particulate model of matter this would mean that the macroscopic reference knowledge which is necessary to evaluate and develop the submicroscopic model is no longer accessible. If this is a realistic characterisation of the state of affairs then successful learning means enabling students to *relate* macroscopic properties to particulate concepts. In this sense, we could say that students use *multiple representations* when reasoning about matter; the abstract particulate model is called upon only in situations where systematically proofed scientific explanations and predictions are necessary (e.g. when writing a test in school or university). In most other situations, pupils stick to their naive, but for everyday reasoning completely satisfying, models of matter.

Chi's theory is focused rather narrowly on the assumption that the psychological existence of ontological categories is decisive for explaining the principal difficulties observed in (initial) learning of science concepts. But many other factors can influence the acceptance of new theories (see Chinn & Brewer, 1993, for an overview). For example, Chi does not discuss the interaction of other entrenched beliefs with science concepts nor does she mention incremental cognitive processes that lead to correct scientific concepts. Some of these factors are discussed in Vosniadou's (1994) theory of conceptual change.

4.3 Conceptual Change as Theory Change

Vosniadou (1994) understands ontological shifts as only one of several possible difficulties in the process of learning science concepts. She assumes, in a Piagetian tradition, that children establish naive framework theories as the basis for a more specialised individual ontology and epistemology. Difficulties in acquiring science knowledge are expected if new knowledge cannot be assimilated to the existing theories, if new information is partially incompatible with theories currently held by learners. Under such circumstances, the cognitive structure needs to be modified in order to accommodate new information. In this case, Vosniadou expects that children construct new theories. The new theories are not necessarily correct, of course. They often contain aspects which are at odds with the scientifically accepted position but are, as of yet, not in conflict with learners' knowledge and experience.

In the domain of the particle model of matter many empirical observations on students who have received formal instruction can be interpreted in this way. For instance, Ben-Zvi, Eylon & Silberstein (1986) asked students about copper atoms after they had received formal instruction about the particle model of matter and the properties of copper. One observation was that students think that copper atoms are malleable. This belief is contradictory to the particulate view of copper atoms. In order to assimilate the information that copper is built of the smallest particles and the knowledge that copper is malleable, students may have simply mapped the macroscopic copper properties to the copper atoms. Another example is the finding of Pfundt (1981) that students accept the scientific interpretation that matter is built of small particles, but they have difficulties in accepting that these particles exist before any process of subdivision. Presumably, learners try to grasp the notion of particles by relating them to the macroscopic world and to explain their existence in terms of operations on macroscopic entities (segmentation for instance).

One would predict that learners make most of their mistakes along the "interface" of their macroscopic experiences and the microscopic model. Therefore, and somewhat counter-intuitively, students should not have problems with "abstract" scientific concepts in general, but mainly with those that relate directly to their pre-scientific beliefs. There may be abstract concepts which are so abstract that they do not relate to any experience and are therefore integrated into the cognitive structure mainly by accommodation. This seems less problematic when the learner tries to assimilate certain information. A prominent method of assimilating scientific concepts is to reason analogically, in our case to map from the macroscopic, continuous models of matter onto the microscopic, particulate model. The problem with this analogical inference is that learners often do not know enough to decide correctly which properties can be mapped and which cannot.

In the next section, we will elaborate the notion of analogical transfer and argue that even though it is a powerful and essential way of learning about new concepts, it is also not without problems under certain circumstances. We will illustrate these problems with two case studies.

4.4 Mapping from the Observable to the Abstract: Problems of Analogical Reasoning for Learning

In those cases where pupils attempt to understand the nature of atoms by drawing analogies to what they know about the macroscopic world (for a teaching strategy aiming a different kind of understanding see for example Buck, 1990), knowledge acquisition can be seen as being divided into two phases. Initial analogical transfer and, following that, tuning of the analogical mapping. Students go through the first phase when they accept for the first time the assumption that matter is composed of pre-existing small particles. Assuming that pupils have not gained any knowledge about the particular model of matter from their daily experience (which is quite a safe assumption, see for instance Stavy, 1988), they have to add this new assumption to their macroscopic concept of matter. It is plausible that children gain knowledge about the properties of the smallest units by means of transfer from other macroscopic domains, because it seems impossible to derive the properties of a discrete model of matter from observations only (Löffler, 1996). A child may, for instance, assume that the smallest units look and behave like billiard balls when it hears the first time of the discrete structure of matter. The assumption of initial analogy when acquiring the particulate model of matter could explain two empirical findings: first, the frequently reported finding that macroscopic properties are transferred to the submicroscopic level which are incorrect from a scientific view (Brook, Briggs & Driver, 1984; Novick & Nussbaum, 1978, 1982; Stavy, 1988). Secondly, the use of analogies can explain why after the introduction of the particulate model of matter some macroscopic conceptions (for example zero or negative weight of gases: Stavy, 1988; active sucking character of vacuum: Brook & Driver, 1989) become much less frequent: if students use the analogical reasoning in the other direction, from the submicroscopic model as base domain to the macroscopic model as target domain, then incorrect assumptions about properties such as weight and the character of vacuum should become less frequent after acquiring a particulate model of gases.

This transfer process can be captured by different models of analogy. One candidate is *Structure Mapping Theory* (Gentner, 1983) and the SME model by Falkenhainer, Forbus and Gentner (1989). Another model of describing the transfer from macroscopic areas (e.g. billiard balls) to submicroscopic models of matter might be the ACME model of Holyoak and Thagard (1989). Both theories explain how the acquisition of the particle model of matter (and the special case of gases) depends on the extent and correctness of the macroscopic knowledge about gases. If, for example, students hold the belief that air under low pressure in a syringe is actively drawing back the pistol, then the transfer of a billiard ball model to the domain of matter should be more difficult than if they believe that a vacuum does not actively suck because the concept of actively sucking air is not compatible with the view that air is built of small balls behaving in a similar way to billiard balls. The same holds for the macroscopic "misbelief" that air has no positive weight or has even negative weight, which is often observed in children aged between 8 and 16 years. Such a belief is not compatible with the view of air composed by particles behaving like massive balls and should hinder the construction of an initial particle model of gases. If pupils, however, hold correct and detailed knowledge about macroscopic behaviour of air (for example about the behaviour under pressure differences or about the positive weight of air), then the transfer to the particle model should be less problematic.

In the second phase of the knowledge acquisition process (which does not have a clearly defined end point) the outcome of the analogical transfer is modified. Assuming that in the first stage a more or less complete transfer of the source (for instance, from billiard balls) to the target is achieved, some of the mappings will most probably be wrong (Brook & Driver, 1989; Novick & Nussbaum, 1978, 1982; Stavy, 1988). That means the transfer of macroscopic properties concerning model relations that are non-trivial leads to incorrect model properties which have to be modified later. This process of model modification is possible for every part of the model and can be repeated every time new macroscopic facts enforce the modification of the model of matter. This important process of analogy revision (after analogical mapping) receives increasing interest. Gentner et al. (1997) propose the following processes of modifying or enriching analogies:

(a) Processes to combine several analogies; the model of Burstein (1986) and that of Spiro et al. (1989) might be candidates to formalise the combining of several analogies. Burstein's computer model CARL for example tries to connect different base domains via similarities of their causal descriptions to construct a mixed model for examples that cannot be explained by one base domain on its own. To explain the molecular mechanism of the Brownian movement of smoke particles in air, the simple model of static billiard balls as the smallest parts of gases must be combined with models of different types of balls moving and colliding in a container. The combination of both models suggests that collisions of the submicroscopic air particles with the particles of the smoke cause the macroscopic shivering of the smoke particles which can be observed under a microscope.

(b) Incremental analogical reasoning; that means that an existing analogy is changed if new knowledge about the target domain is brought in. The models of Burstein (1986), Keane (1990) and the extension of the SME called I-SME (Forbus, Ferguson & Gentner, 1994) are possible formalisations of capturing this incremental building of analogies. In Burstein's CARL, for example, the program receives a correction of wrong answers on simple questions about programming in BASIC. CARL is able to use another analogy from the base domain to adapt its own target domain model to the teacher's correction. An example of the domain of matter models may be the correction of a naive molecular model of the thermal expansion of gases. One can imagine that a beginner constructs a model where when heated the smallest particles of a gas are expanding themselves like the macroscopic gas. When the teacher corrects this model, the learner has to substitute the assumption of an expansion of the particles; instead, particles are moving faster when heating the gas. The harder collisions of the fast moving particles against movable walls of the container can now explain the expansion of the macroscopic gas.

(c) The system PHINEAS (Falkenhainer, 1990) can describe the process of testing and re-representing of analogies once made.

Finally, research on case-based reasoning addresses problems and methods of "fixing" analogical mappings (Kolodner, 1993). In particular, empirical findings which describe inconsistency of students' behaviour (studies which describe inconsistent students' behaviour in various domains are for example: Engel Clough & Driver, 1986; Pfundt, 1981) can be explained by interpreting students' behaviour as reasoning from specific cases.

5 Two Case Studies

The general goal of this first explorative study is to examine the process of acquiring the basic structure of the particulate nature of matter. In this study, subjects received a short explanatory text on the particulate model and then saw two experiments on which they had to comment. We shall focus on the cognitive processes which lead to the first construction of basic model properties and consider the influence of three different forms of external representation of the particular model on the learning process. In one experimental condition subjects learned only from text, in another from text and pictures and in the last one from text and animations of the particle behaviour.

5.1 Method

The study was carried out with the following design:

Table 1: Design of the explorative study

Condition 1 (2 subjects)	**Condition 2 (2 subjects)**	**Condition 3 (2 subjects)**
Knowledge test	Knowledge test	Knowledge test
Pre-test	Pre-test	Pre-test
Learning phase:	Learning phase:	Learning phase:
Supported by text	Supported by text and pictures	Supported by text and animations
Post-test	Post-test	Post-test

5.1.1 Pre- and Post-Tests

The initial knowledge test contained paper and pencil versions of the test items used by Brook and Driver (1989) to investigate students' knowledge on the macroscopic behaviour of gases and one paper and pencil item of the test used by Nussbaum and Novick (1978) to study students' knowledge on the particular nature of gases. In addition, in the pre- and post-test subjects observed the Brownian movement of Indian ink particles in water with a microscope. Then they looked at two animations presented on a computer. The first animation showed moving particles in continuously swaying material; the second animation showed collisions of the ink particles with small water particles. The students were asked to decide which animation explained the observed Brownian movement better and to give reasons for their decision. The answers to all questions were given verbally to the experimenter.

5.1.2 Learning Conditions

In the learning phase, subjects had to read an introductory text on the particulate nature of gases and after reading two life chemical experiments were shown, accompanied by explanations of the experimental phenomena. The explanations took different forms in the different learning conditions. In condition 1, explanations were given only in the form of text, in condition 2 explanations were given in the form of text accompanied by pictures (snapshots from the animations of the third condition), and in condition 3 the explanatory text was accompanied by animations of the particle movements.

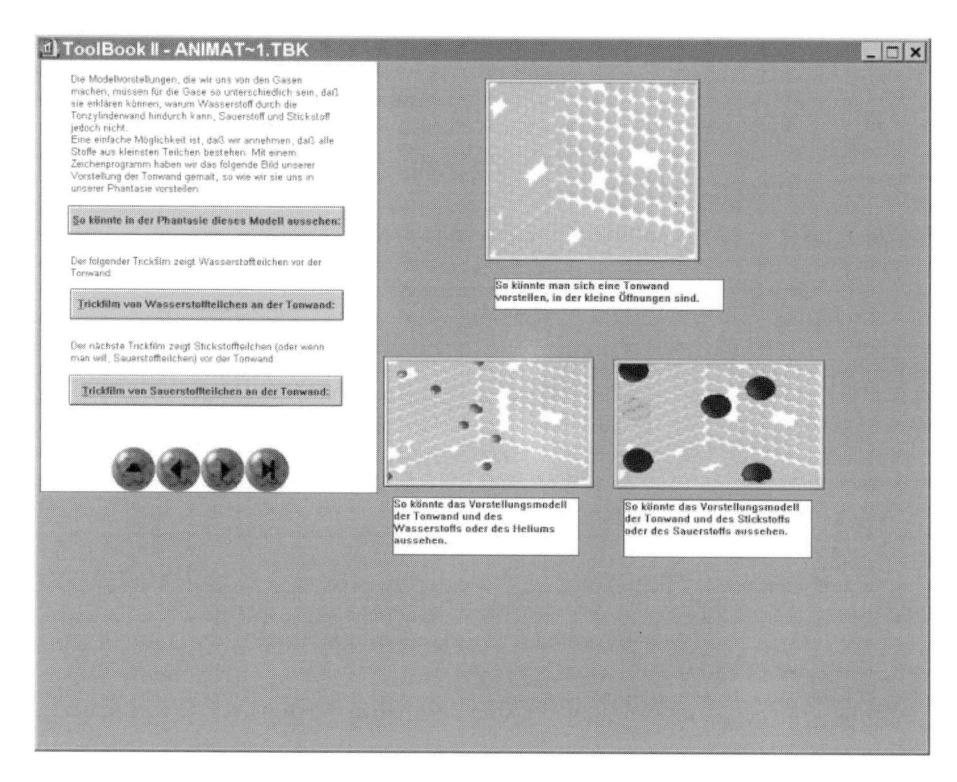

Figure 2: Screenshot of a page of the learning program (learning condition: with animations)

In the first demonstration experiment (Figure 3), hydrogen, oxygen and nitrogen diffused into a cylinder made of porous clay. In the context of the particle model the different behaviour of the three gases can be explained by different sizes of the spherical-shaped gas particles. This particle property can be transferred from similar macroscopic phenomena such as riddling.

The second experiment (Figure 4) showed the diffusion speed of bromine gas under two conditions. In the first condition the bromine diffused slowly into a flask containing air under normal pressure. In the second condition the bromine diffused quickly into the partially evacuated air flask. In the context of the particulate model of matter the difference between the two conditions is explained by the assumptions that the gas particles are moving all the time and that after the evacuation there are fewer air particles in the air flask which could prevent the bromine particles from entering the flask by collisions. The assumption of air particles always being in motion even if the macroscopic gas does not seem to move is not transferable from the macroscopic gas behaviour.

During the learning phase the students had to read the texts and to explain to themselves every new phrase (for the rationale of this method which should lead to

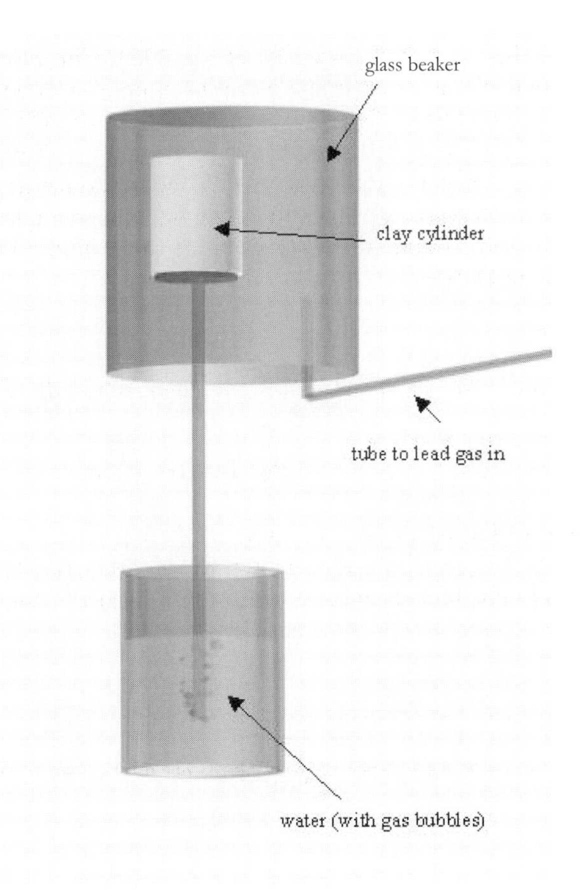

Figure 3: Design of experiment 1

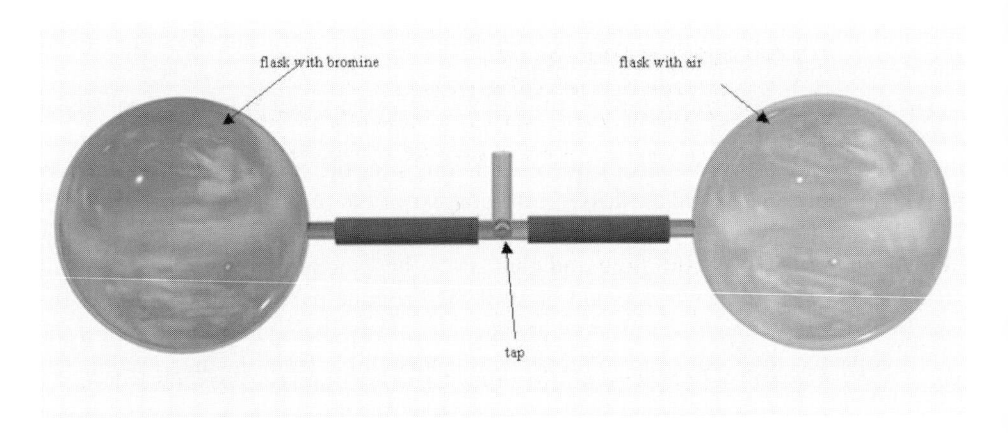

Figure 4: Design of experiment 2

deep processing of the information see Chi et al., 1989; Chi, Slotta & De Leeuw, 1994). In the text plus pictures and in the text plus animation condition the subjects were also asked to self-explain the pictures and animations. The six sessions were filmed on videotape. Two students were investigated under each condition and each student worked for between 1.5 and 2 hours with the experimenter.

We now report two case studies which illustrate our research questions. Student 1, let us call her Diane, learned in the text and pictures condition; student 2, called Carl here, used text and animations.

5.2 Case 1: A Student with Incorrect Knowledge about Some Macroscopic Properties of Gas

5.2.1 Answers to Pre-Test Questions

We chose three out of the 10 questions ("tasks") of the pre-test in order to illustrate naive conceptions of weight of air and of pressure differences. These naive conceptions of macroscopic phenomena influence the acquisition of a particular model of matter. Task 4 tests conceptions about mass of air.

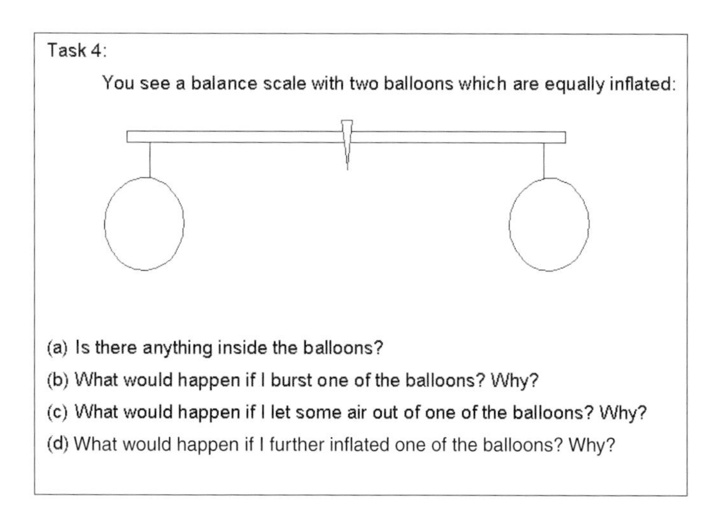

Diane answered the questions as follows:

Question (a): "Yes, air"

Question (b): "I think it [the side with the burst balloon] is going to be a bit lighter."

...

"Perhaps the air which is leaving is a bit more heavy than the other air."

Question (c): "I think nothing would happen. "
Question (d): "Nothing would happen either."

The answers indicate that Diane does not have a clear concept of the weight of air. In the second answer she speculates about two types of air having different weight. In the third and fourth part of the question she assumes that the air in the balloons does not have any effect on the balance; this probably means that she does not attribute weight to air inside the balloon. These conceptions conflict with the particle view of air. The belief that air has no weight should complicate the acquisition of a particle view because every solid air particle has a certain weight.

The following two tasks test beliefs about pressure and vacuum.

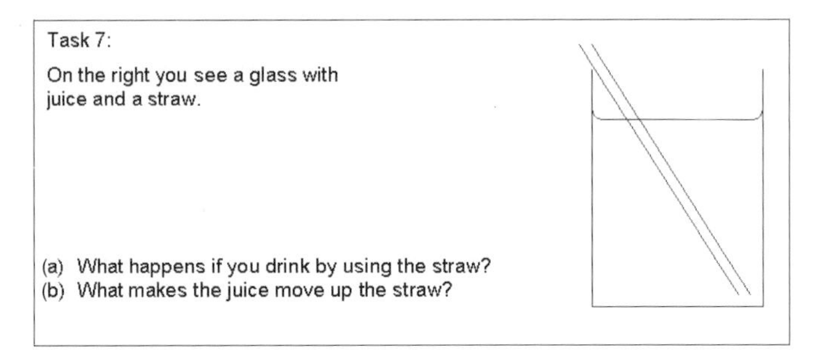

Diane answered as follows:

Question (a): "The juice is moving up."
Question (b): "At first one is pulling the air out of the straw. And if there is no air in the straw then something else must come in. And instead of air it is taking the juice."

...

"Because the fluid is not a solid material"

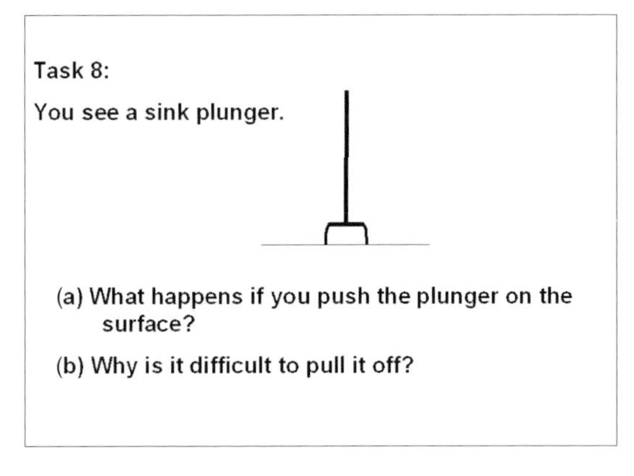

Here the student's answers indicate that she believes that a vacuum is an active agent that tends to disappear whenever possible. Again this belief is not compatible with the mechanistic particle view of air and one should expect difficulties if pressure differences have to be explained with the particle model. Because those questions arise in the second experiment of the learning phase, we expect difficulties for Diane there.

Diane's answers to Task 8:

Question (a): "That here is rubber and it is elastic. And if you press against it the air is going out. And if one stops pushing it sticks to the floor. There is little vacuum inside."

Question (b): "Because it stuck to the floor, because the air in it is no longer there. If the air is no longer there it sticks to the floor."

In the second question about pressure differences the subject again utters that she sees a vacuum as an active agent which actively tries to disappear.

5.2.2 Reactions to Experimental Observations

Experiment 1. We now turn to the reactions of Diane when she observed the demonstration experiments (see Figures 3 and 4) and read the explanations. After the first part of experiment 1 (Figure 3: diffusion of hydrogen, oxygen and nitrogen through a clay cylinder) where hydrogen diffuses through the clay:

> "I do not understand why it goes just through the clay cylinder."

...[student sees the experiment with the other gases and begins to study the text and pictures about the experiment].

> "Yes, perhaps [the reason why the hydrogen is going through and oxygen and nitrogen not] the oxygen [certainly here oxygen is mixed up with hydrogen] is able to decompose into particles which are so small that they can go through the wall and then together again. And the oxygen and the nitrogen are not able to do that."

(...) [she looks at a picture of a particle model of the clay wall]

> "That means, that is where the hydrogen is decomposing"

[she looks at a picture of a particle model of the clay wall and the bigger particles of oxygen or nitrogen]

> "Yes. And these ones cannot decompose as the hydrogen particles do."

Here the subject utters a naive conception which was described by Pfundt (1981): she sees smallest particles not as just existing but as created by a process of subdivision. We interpret this answer as a result of a process of analogy where the divisibility of many macroscopic materials is used to explain the arising of little particles of hydrogen which can diffuse through the clay. Probably she uses the experience that other materials (like steel) are divisible only with difficulty to explain that oxygen and nitrogen are not subdividing and cannot diffuse through the wall. This answer illustrates how difficult it is for pupils to accept the notion of smallest particles because this is an abstract assumption which cannot be transferred by analogy from known materials. Using analogies from familiar substances in such a situation where an abstract property must be invented often results in inferences which are wrong from the scientific point of view.

Let us take a look at the intended effect of the picture: the student is able to integrate the content of the picture (which shows the spherical particles of the clay and the oxygen) with her naive conception. This means that the picture of the particle model of the experimental situation does not lead to the intended conception of gases. It still allows enough interpretation possibilities to conserve to naive assumptions.

Experiment 2. This experiment (Figure 4) deals with diffusion of bromine into a container with air under two conditions: firstly the air is under normal pressure, in the second condition some air is evacuated. Diane has seen the experiment and is looking at the model picture, where the experiment with the partly evacuated air container is shown shortly before the separation between the containers is removed:

> "When the air is being pumped out, then the red particles [the air particles] will be sucked away, somehow. Yes, if the air is gone, then the red particles are gone, too. And then the green particles [bromine particles] are being sucked in, because . . ., if the air is sucked out, then it contracts. And if one removes the separation between the containers, then they cross over. So many [cross over], until it is no longer contracted, until there is something in again."

[Student looks at the picture of the experiment with air under normal pressure]:

> "If the air inside is not moving then the particles are not moving also and therefore they do not mix up."

[After she has read the particle explanation of the experiment]:

> "Yes, because the red particles form something like a wall, then the particles collide and they can not go through. Because, if in the one container is a lot and in the other also, then it cannot mix up".

(...)

> "Yes, and if there is no more air, then they do not form a wall and . . . the gas can cross over without colliding with obstacles."

The student shows in the first part the naive conceptions which we expected after the pre-test question about pressure differences. She transfers her assumption of the sucking character of vacuum to the particle level. When it comes to the explanation of the very slow diffusion in the case of air under normal pressure she accepts the particle explanation given in the text and even adapts it for the former case of the partly evacuated air container. It remains unclear what happened to her assumption of the active character of vacuum. Probably both the explanation with the collisions and the one with the active role of the vacuum exist beside each other.

If we look again at the influence of the picture we see that in the first case the picture is compatible with the naive conception of the sucking vacuum. In the second part of the explanation text and pictures lead to a new explanation scheme on the particle level.

5.3 A Student with Correct Knowledge about Macroscopic Gas Properties Particle Model of Gas

5.3.1 Answers to Pre-Test Questions

Carl's answers to Task 4 are as follows:

Question (a): "Air or some other gas. I would prefer air because there are gases that make balloons move up."

Question (b): "If one burst one balloon then the weight which is in here is away and the balance would go down on the side with the balloon that did not burst. . . .
Because here a counter weight is missing and the intact balloon has a certain weight."

Question (c): "...
If you close the balloon again the balance will go down at that side where the more filled balloon is fixed. Because in that balloon there is more air than in the other."

Question (d): "If one inflated this balloon then there is more air and a bit more weight in the balloon. The balance is going down that side"

Here Carl indicates quite clearly that he attributes weight to air. In contrast to Diane the assumption that air is weighing something is compatible with the particle view of air. So from this aspect no difficulties should be expected in arguing with the particle model of gases.

Carl's answers to Task 7:

Question (a): "One has to suck out with a certain force the air that is in the straw, and then the juice comes behind it."

Subject 2 argues with two effects: first, he probably attributes an active role to the vacuum (like subject 1), but in addition he uses the pressure from the surrounding air. In this case we do not expect many interpretation difficulties with experiment 2 because he probably has only to change his assumption about the active role of vacuum to come to a mechanistic explanation of pressure differences.

Carl's answers to Task 8:

Question (a): "If one has in his hand and there is air in the plunger. And if one pushes the air is pressed out. And because the rubber tries to go back in its former shape a vacuum is created. And the plunger is fixed then."

Question (b): "It is difficult because the vacuum is there and the rubber is close to the floor and there air cannot go in. Therefore you need a certain force to remove it again."

In Task 8 he is only arguing with the vacuum and not considering the pressure from outside. Perhaps this is the case because he only assumes a very small force from the air outside and has made the experience how hard it is to remove a fixed plunger.

Carl seems to be in an intermediate stadium concerning his knowledge about the pressure differences of air. He still holds beliefs about the active role of vacuum but additionally argues with the pressure from the surrounding air. This cognitive situation may be easily interpreted using the theory of Vosniadou (1994).

5.3.2 Reactions to Experimental Observations

Experiment 1. While Carl observes the experiment with nitrogen he utters:

> "Probably the particles — if one could take the gas into pieces — of the hydrogen are smaller and can go through the pores of clay and come out on the other side. And the ones of nitrogen are probably bigger and cannot go through."

The student utters the correct particle interpretation without having read any text or having looked at the animations (later when observing the animations there is no change in his belief). His macroscopic knowledge about gases is probably so compatible with the particle hypothesis that he can produce the particle interpretation after a short theoretical introduction into the particle model and the demonstration of the experiment.

Experiment 2. Before the experiment and when he begins to read the introduction to the experiment Carl says:

> "I assume that all matter is decomposable into smallest particles."

[After the experiment and after he had looked at the animations which show the situation before the removal of the separation]:

> "Oh yes, clear. There have been only very few oxygen particles and on the other side many bromine particles. Now I have got it."

[He observes the animation of the complete experiment under normal pressure]:

> "One can see how the separation is removed. And the bromine particles are going into the one direction and the oxygen into the other one."
> "I am *wondering* about it."

...[Repeats the animation]

> "Now the air is going to be sucked in. The gas is going to be sucked in."

...[He looks at the animation with air under low pressure]

> "But one can see nicely, that the particles are sucked by the low pressure and then get mixed with the rest of the air."

The subject confirms his assumption that all matter consists of the smallest particles. He also states that if there is less air in a container then fewer particles should be there. His difficulties may stem partially from his belief that the partial vacuum is actively sucking. This leads to problems when he tries to integrate the contents of the animation into his explanation.

The confusion the student is going through may be caused by the animation: there is no reason why the partial vacuum is causing the fast disappearance of the bromine particles (they are not accelerated). The only reason is a mechanistic one: fewer collisions with air particles. The student is still not in a position to correct his belief that a vacuum is an active agent.

6 Conclusions

6.1 Cognitive Processes which Lead to the Construction of Basic Properties of the Particulate Model of Matter

With respect to this issue, we obtained the following results from our pilot study (reduced to the two case studies here). The two cases indicate that the first acquisition process contains a transfer from (beliefs about) macroscopic gas properties to the submicroscopic level. One example is the assumption of the first student, Diane, who assumes that hydrogen divides into particles, and that nitrogen does not, when diffusing through the clay. This belief is part of a continuous model of gases: material is divided into parts by subdivision processes. Other examples are the assumptions about the effects of a vacuum: both students clearly utter in the second experiment that on the submicroscopic level the bromine particles are sucked into the container with the air under low pressure. The required conceptual change which would result in seeing the vacuum as passive in nature did not take place in these two subjects.

These findings indicate that conceptual change is only taking place to the extent that is required to account for the observations at hand. For instance, Diane and Carl easily accept the idea of the existence of smallest particles but they do not overtake other assumptions which are not immediately needed to understand the experimental events. These results of the first step of the learning processes were expected from our theoretical considerations.

The correction of scientifically incorrect elements, which were initially transferred from the macroscopic to the particle level (e.g. that particles are not pre-formed but created by subdivision processes) cannot be observed. A possible reason is that in our setting not enough experimental evidence or instructional input on that point was provided to trigger these changes. To change, for example, the assumption that particles are not preformed, one can provide students with evidence that two chemicals can be combined into a compound and afterwards separated again into the same elements as before. After some evidence of this kind we would expect a change in the assumption that particles are not preformed, although there can never be strong proof that particles really exist before such events.

6.2 Influence of the Different External Representations

The case studies do not suggest that there is a simple overall effect of pictures or of animations; instead, the external representations interact with subjects' beliefs in an intricate manner. If one looks for instance at Diane during the first chemistry experiment: when she observes the experiment she has a problem in explaining why hydrogen goes through the clay, despite the previous theoretical introduction of the particle model. But when she looks at the picture she gets the idea that hydrogen is subdividing into particles and is able to build a mental model of the submicroscopic events in the experiment. She does not give up the idea that in "normal" situations hydrogen is of a continuous nature. We assume, partly in correspondence with the theory of Vosniadou (1994), that only those assumptions are changed which are incompatible with the information given in the experiment and with the pictorial illustration of what is happening on particle level. Obviously, the pictorial illustration helps Diane to come up with a locally correct explanation, but she does not give up her larger theory about matter.

Carl is even more resistant. Despite the fact that he observes in the animation sequence of particle behaviour, which is not in accordance with his belief that the vacuum is actively sucking, not only that bromine particles moved into the container with the air particles but also that oxygen particles moved into the container with bromine particles and the bromine particles were not accelerated, he does not change his conception of a vacuum. He hesitates and wonders for a moment when watching the animation at this point, but does not elaborate further.

Despite these observations, we still believe that animations hold a lot of potential for conveying the particulate model of matter. One has to consider that different representations are not isomorphic, but contain different information. Texts normally contain only partial process descriptions written in a strongly condensed fashion. This gives students degrees of freedom to adjust, instead of reject, their naive model of matter. When pictures are presented, more information about the static properties of the particles is conveyed, thus preventing some of the possible misinterpretations based on the naive model of matter. But only animations contain easily accessible information about the dynamic behaviour of the particles, which seems necessary to build a complete mental model. Our subject Diane, for instance, would have found it harder to sustain her argument about the behaviour of particles had she seen the animation.

Because a dynamic model of particles contains the information necessary to explain and predict many macroscopic phenomena it seems admissible to support the construction of such mental models. But one has to consider the enormous cognitive load involved in mentally simulating the dynamic behaviour of particle systems. Probably only some aspects of dynamic particle behaviour can be stored and recalled, but never the complete dynamic behaviour. That would make it necessary to support an appropriate chunking of the dynamic particle behaviour. In other words: external dynamic representations should be designed in a manner to support the student to store strongly compressed mental models which contain a correct particle model and make it possible to argue about matter phenomena without overloading working memory.

6.3 Instructional Consequences

The following aspects make the acquisition of the particulate model of matter difficult. An abstract model has to be understood which has only a few direct connections with the observable world. That means that the central properties of the smallest particles cannot be transferred from the base domain, the macroscopic properties of matter. This also holds for the base domain often used in instruction: billiard balls. If a macroscopic analogue is used, the learner is initially almost forced to build a partially incorrect model and has to correct the mistakenly transferred properties — provided he or she is presented with evidence that does not fit together with the mistakenly held beliefs. From our review of the effects of chemistry lessons, we can conclude that this correction does often not take place or it at least is not completed: pupils hold on to some of their misbeliefs, many of which can be explained in terms of analogical transfer.

If the cognitivist and constructivist position is correct, that knowledge acquisition means knowledge construction and is always performed in terms of what is already known, then analogical inferences cannot be avoided. Pupils will attempt to understand scientific concepts and models introduced to them in terms of what they know, which is to a large extent what they know from everyday experience. Hence, we cannot avoid erroneous analogical transfer — the correct model cannot be implanted directly in learners' minds. What can realistically be done is to try to constrain the initial analogical inferences and/or to provide the learners with opportunities to revise their analogical mistakes.

Besides running more controlled experiments, which should help us to discern the effects of ontological misbeliefs and representational format on the acquisition of the particle model, we are planning to construct a learning environment which constrains analogical transfer. In this learning environment, the student can construct simulations instead of only observing pictures and animations. Instead of building the abstract model by reasoning with multiple analogies or incremental analogical reasoning, students will be able to construct a particle system on the screen particle by particle and property by property. The particle model as constructed by the student can, at any point, be "run" by means of a simulation engine and its dynamic behaviour displayed. Note that we intend to simulate the theoretical model as well as the corresponding macroscopical experiments so that the connection between the two becomes evident for the learner.

The component by component manner of constructing the model is important because it allows us to provide the student with feedback (either evoked by the student as a side-effect of running a model or initiated by a teacher/tutorial component) on model components and their properties; every construction step can be discussed, its consequences can be observed and compared with macroscopic phenomena. This can be done alone or in groups. In such a system analogies have their function when constructing single particles or their properties. But instead of a more or less inadequate base domain model students can use an externalised animation of a particle system. Instead of relying on a concrete analogy students can rely on the visualised particle behaviour on the screen and build appropriate knowledge derived from the model constructing activities on the screen.

4

How Beginning Students Use Graphs of Motion

Eileen Scanlon

1 Introduction

In this chapter some aspects of the use of representations by physics problem solvers are considered, in particular the role played by the interpretation of graphical representations in learning physics and the use of physics knowledge in problem solving. The interpretation of graphical representation has particular relevance for physics learners, as many of the tasks involved in developing physics understanding involve the facility to move easily between algebraic and graphical representations. This chapter reviews the experience of three research projects involving novice physicists in the use of representations in science learning — a protocol research study and two instructional experiments. Representations of abstract ideas in physics are hard to understand and manipulate. The protocol research study is based on novice physics students' behaviour on a number of written physics problem-solving tasks including a graphing task (Scanlon, 1993). The data sources consist of think-aloud protocols of 259 solutions to a variety of kinematics and dynamics problems. These protocols, the work of 35 subjects with varying degrees of familiarity and expertise in problem solving, working individually and in pairs were analysed. The protocols were examined for a number of features which are hypothesised to play a role in a subject's success or failure in solving a particular problem. The graphing task was designed to prompt students to use both an algebraic and a graphical representation of the problem and to prompt the expression of the interrelationship of these in their paper and pencil working, so the protocols raise the interesting issue of how such switches of representation were accomplished. Modelling student behaviour on the problems proved to be a useful analytical tool. Some aspects of the use of representations by these physics problem solvers are considered, in particular the role played by the interpretation of graphical representations in learning physics and using different types of physics knowledge in problem solving.

In recent years it has been assumed that some of the tedium of graph drawing and problems of graphical and algebraic interpretation can be ameliorated with the aid of the computer. The instructional case studies deal with students' experiences with dynamically produced graphs of motion. The first instructional experiment involves observations of 12 16-year-olds working together in pairs on *Multimedia Motion*, a CD-ROM designed to teach them about dynamics (Whitelegg, Scanlon & Hatzipanagos, 1997). The second instructional experiment uses data from a conceptual change in science project which tracked the use of specially designed computer simulations to improve the understanding of mechanics of a mixed group of 29 13-year-olds (Scanlon et al., 1993). These instructional experiments highlight a number of issues including the role of contextual features in graphical understanding, and the interaction between data, and algebraic and graphical accounts of phenomena.

1.1 Problem Solving

For students, particularly in the physical sciences, the development of a problem-solving ability is a core skill. Educators are becoming more explicit about what such skills include. For example, Bolton and Ross (in press) offer the following list of skills which underpin problem solving ability:

- the synthesis of information given explicitly or implicitly,
- the identification and labelling of variables,
- the conversion of a real world situation into a model that is amenable to quantitative analysis,
- the selection of relevant principles or laws,
- basic mathematical and/or computational skills,
- the use of approximations or order of magnitude estimates,
- the exploration of alternative approaches and their use in checking solutions, and
- interpretation skills (e.g. checking solutions against experience).

This list is fairly comprehensive but it underplays a key attribute for physics problem solvers which is the ability to switch between alternative representations of the same physical situation. Educators are also conducting anthropological studies of the use of representations by scientists and science learners. For example, Hayes-Roth (1997) has completed a study of the role played by high-school textbooks in the appropriation of authentic scientific graph-related practices. Chi, Feltovich and Glaser (1981) and other commentators (e.g. De Jong & Ferguson-Hessler, 1986; Zajchowski & Martin, 1993) discuss the key difference between expert and novice problem solvers as being illuminated by the way in which they categorise problems. Various claims have been about the role that the choice of representation (e.g. Larkin & Simon, 1987), and the understanding of the nature of alternate representations used by problem solvers (e.g. Larkin, 1983) play in problem solving.

1.2 Problem Solving with Graphs

The choice of representations for presenting a problem has been shown to influence how well students learn. For example the use of graphical representation can impact on learning and problem solving (Larkin & Simon, 1987). Properties such as specificity can be studied to investigate and suggest improved forms of representation (Stenning & Oberlander, 1995). There is also a body of research on students' understanding and skill in graphing conducted by mathematics educators. For example, Leinhart et al. (1990) in a comprehensive review of such work present a complex picture of students' understanding of graphing. It seems that a student's understanding is strongly influenced by intuitions derived from prior experience and by the heuristics acquired from prior teaching. De Jong and Ferguson-Hessler (1991) suggested that good problem solvers were more successful than poor problem solvers at a reconstruction task which involved a change in modality (e.g. from words to figures). Many areas of physics require students to use a number of different representations.

Consider the challenge that a simple physics problem presents to a student. Taking the example of a simple motion of an object which a student is trying to understand. Educators such as Champagne, Klopfer and Anderson claim

> "mathematics is the medium of analysis and communication in the study of mechanics and that proficiency in mathematics provides the necessary and perhaps sufficient condition for success in learning physics". (Champagne, Klopfer & Anderson, 1980, p. 1076)

However, other commentators such as McClelland (1985) claim that instructors often introduce concepts in motion perfunctorily and then rush to algebra too soon. Some educators use graphs as an intermediate step between students' experiences of motion and their introduction to calculus. Current educational approaches to physics instruction involve students in constructing their own understanding from appropriate contexts using everyday and school-based knowledge. A key task is to understand motion from everyday examples of it. McDermott, Rosenquist and Van Zee (1987) reported problems for students in interpreting graphs, both in their connections to the real world and in their connections with other graphs of the same situation (velocity vs. time and position vs. time graphs for example). Graphical skills are often pivotal in physics problem solving.

2 The Research Study: Protocols of Pairs Problem Solving

The main research study we review in this chapter is a study of pairs of subjects who were recorded working together on a variety of physics problems. The data sources for our analysis consisted of think-aloud protocols collected during the production of 259 solutions to a variety of kinematics and dynamics problems. These protocols, the work of 35 subjects with varying amounts of familiarity and expertise in problem solving working individually and in pairs, were examined for a number of features which are hypothesised to play a role in a subject's success or failure in solving a particular problem. Two of these problems required subjects to work with representations of motion in graphical form.

First, some comments about the subjects' approach to using graphical representations. One strategy which was in common use was that of drawing a graph — often the school children in the study drew graphs when they couldn't think of what else to do. Graphs would have the meaning of their axes shifted in full flight, e.g. a subject would draw a distance–time graph and then treat it as a velocity–time graph and *vice versa* or would draw a distance–time set of axes but fill in details of velocity vs. time. Having drawn the graph, the subject would not know what to do with it. The type of knowledge which was securely known by all the subjects is extremely over-generalised, e.g.

- you can read "something" off from a graph using lines drawn across,
- gradient equals "something" and
- "something" is the area under the graph.

Such over-generalised rules allow enormous scope for continuing to work on a problem while being confused about representations. There was no sign of subjects being aware of any connection between graphs and the equations of motion they represent. Indeed, when one of the more experienced subjects worked out the relationship between the equation $s = ut + \frac{1}{2}at^2$ and his distance–time graph, he was very surprised and pleased with himself. This arose in the problem we describe next which required the average velocity to be read off from a velocity–time graph which the subjects had constructed.

2.1 Problem 1: Graph Drawing and Interpretation

The task presented below was designed to prompt students to use both an algebraic and a graphing representation of the problem and to prompt the expression of the interrelationship of these in their paper and pencil working, so the protocols collected here raise the interesting issue of how such switches of representation were accomplished.

Draw a graph of velocity against time for a body which starts with an initial velocity of 4 m/s and continues to move with an acceleration of 1.5 m/s² for 6 s. Show how you would find from the graph (a) the average velocity and (b) the distance moved in 6 s.

Analysis of 11 pairs of novice protocols revealed several distinct correct ways of tackling how to find average velocity (as well as several error producing ways) and were used to construct models of behaviour which were then evaluated by comparing the model traces to a further 20 individual problem solutions.

The *correct methods* observed were:

- To use the equation average velocity = total distance divided by total time. Total distance travelled in this case is calculated from the area under the graph. The result is calculated average velocity = 51 m divided by 6 s.
- To proceed as in the previous method but total distance is calculated using the equation $s = ut + \frac{1}{2}at^2$, not using graph skills.
- To use the equation average velocity = (initial velocity + final velocity) divided by 2. This leads to the calculation average velocity = (4 + 13) divided by 2, i.e. concentrating on the notion of average rather than velocity.
- To directly read answer from the graph using fact that average velocity is given by the mid-point of the velocity–time graph.

The *incorrect* methods observed were:

- To use the equation average velocity $= (v - u)$ divided by 2 (i.e. $v = 4.5$ m/s).
- To use average velocity $= v/t$ (i.e. $v = 13/6$ m/s).
- To use the fact that average velocity is the gradient of some graph (i.e. v is $(13 - 4)$ divided by $6 = 1.5$). (In fact it is the gradient of the distance–time graph and not the velocity–time graph which subjects were asked to draw here.)
- To use the equation average velocity $= dv$ divided by dt (i.e. treating v as similar to a).
- To use the equation average velocity $= (v + u)$ divided by 2 with an incorrect final velocity of 9 (i.e. $v = 6.5$).

All subjects correctly proceeded to calculate the distance from the area under the graph. This suggests that this one piece of knowledge had firmly fixed itself, perhaps because it is a strange idea to novices that the area under a graph might mean something. One pair of students however revealed a particular elaboration on this as they said you *"can't have it [the graph] going through the origin 'cos then you can't do the area is distance under the graph."*

The attempt of one of the 11 pairs went as follows:

Oh average velocity, yeah (pause) well, average velocity is just
Distance
Top velocity minus bottom velocity, divided by six.
9 over 3, 3 over 2, 1.5
9 yeah 1.5 . . . That's quite clear
Yeah

The pair was attempting to use the fact that average velocity is the gradient of some graph. In fact it was the gradient of the distance–time graph not the velocity–time graph which the subjects had drawn here, so they calculated the velocity as $(13 - 4)$ divided by 6 giving 1.5 using the third of the incorrect models given in the earlier list.

Further details of student performance on the problem which illustrate some more of the correct and incorrect methods described above, based on the first set of data, are given in Scanlon (1990, 1993), and our initial attempt at modelling our pilot data in Scanlon and O'Shea (1998).

2.1.1 Analysis via Production Rule Modelling

Nine distinct prototype production rule models of student behaviour on the problem were built. These models replicate the behaviour described in the list of correct and incorrect methods given at the start of section 2.1. One feature of these models is the distinction made between recalling an equation and using it. This distinction also applies to some of the graph relationships, for example, between marking the mid-point of a graph and knowing its physical interpretation. All the models can be run with or without an incremental graph-plotting rule. When this rule is present the velocity–time graph is plotted second by second. This takes longer than the usual way of plotting a graph with two points but has the advantage that it is not necessary to use an equation to compute the coordinates for the second point. Also, in the course of the incremental plot the mid-point and other features are marked on the graph during the drawing process. The models also include rules for ordering unknowns. These come into play when the rules are read as strictly ordered. In this case every possible way of finding a value for the first unknown is tried before trying to find a value for the second unknown (having started on the second unknown these models will all solve for the first unknown also if a possibility arises). All the models can be run in either ordered unknown or unordered mode depending on the reading of the problem. The choice taken depends on the data being matched.

In many ways the models are very similar. The largest model has 26 rules and the smallest has 18. One model for correct behaviour only involves 6 rule firings and one model for incorrect behaviour results in 18, but all the other pairs or subjects are modelled with 10 or 11 firings. The average number of rules in the models for correct behaviour is 23.6 and in models for incorrect behaviour is 20.9. This is not surprising as the pairs and subjects who fail to solve the problem correctly generally fail to recall or do not know some of the relevant equations or graph relationships. The average number of rule firings before halting is 10.3 in the case of correct termination and 11.4 in the case of incorrect termination. There are two main differences between the models which generate correct and incorrect solutions. First, the models for correct behaviour involve much less switching backwards and forwards between the graph medium and the equation medium. On average, in the behaviour generated by the models for incorrect behaviour there is twice as much switching (3.6 changes per model run) as there is in the models for correct behaviour (1.8 changes). Secondly, all the models for incorrect behaviour involve an initial attempt to find the unknowns in the order suggested in the problem statement. This only occurs in the running of one of the models for correct behaviour (model 3, which replicated the third correct method in the list at the start of section 2.1), and causes the most switching between media for a model for correct behaviour.

Of the models for the correct behaviour three out of four are opportunistic and quickly take advantage of intermediate results. In model one, having computed the distance from the area under the graph, the model then recalls the equation relating average velocity with time and distance. In model two, the graph is drawn but all the values are computed using equations. Model three matches the approach taught in many British schools and generates graph interpretation events and equation manipulation events at the appropriate points. Model four is the most economical and opportunistic and involves effective use of the graph and no equation manipulation at all.

The models for generating the five types of wrong behaviour are all initialised with the ordering information. All the models for incorrect behaviour involve the omission or masking of at least one important correct rule. Three of the models incorporate a single additional overgeneralised rule based on the simpler types of equations often encountered in schoolwork. Another two models incorporate a single additional overgeneralised rule based on graph interpretation experience. One model is just a subset of the set of correct rules but generates solution attempts in the right order before giving up. Another variant of a model computes average velocity with the inappropriate rule but while assigning a value to average velocity still retains it as a sought unknown and then "recovers" when shifting to the graph and reads the value from the mid-point. The models for incorrect behaviour are constructed using the same framework and differ from the model which embodies the standard approach taught in schools by a small number of rules. Only one incorrect model incorporates two overgeneralised rules, the rest incorporate one. So, it is not necessary to appeal to the notion of malrules very often to account for variations in behaviour, although the overgeneralised rules could be considered to be a type of malrule.

In summary, the faulty models of solving this problem have been constructed with the addition of very few new rules. The ruleset of all correct rules consists of 27 rules and the incorrect models require the addition of only a few extra rules. For example, the behaviour modelled in one incorrect model is achieved by modifying one rule and adding three new rules which could be incorporated in a correct model without difficulty. The subset of correct rules for another incorrect model also contains a faulty rule which applies $v = u + at$ in inappropriate circumstances.

2.1.2 Discussion of the Models

The initial intention of constructing this set of graph models was to explore the problem-solving patterns of students using more than one representation. It was assumed that the subjects would find an extra representation helpful. It is surprising to find that to successfully describe the four different examples of correct behaviour we must assume that the students do not use all the available sources of knowledge or necessarily use both the equation and the graph medium when solving the problem. It is also interesting to see in detail the benefits of ignoring the instructions given in the problem statement on the order in which the unknowns should be determined. Furthermore, three out of four of the models of correct behaviour, in completing the average velocity part of the problem, restrict themselves to either equation manipulation skills or graph interpretation skills in the determination of the average velocity. The models for incorrect behaviour involve more changes of medium and in these cases the students are using more than one source of knowledge when they recall equations as well as being subject to the normal problems of overgeneralisation seen in novices. The models match the sequence of actions made by the subjects and give the same answers. The number of rules, firings and switches from medium to medium is summarised and the detailed comparison of protocols and models are given in Scanlon (1990, 1993). It also appears that the consistency of style criteria for the rules put forward by Young and O'Shea (1981) are satisfied in these models. None of the rules is noticeably more complex than the others.

One feature of the incorrect methods used by subjects was that in each case the equation that was recalled for average velocity involves some overgeneralisation from a piece of knowledge about graphs. For example the use of the equation $(v - u)$ divided by 2 will produce the correct answer for the average velocity of a graph of motion where $u = 0$. It is possible that students have used this incorrect equation successfully for graphs which pass through the origin and have overgeneralised this success to other straight-line graphs.

The problem set was taken from a textbook where an instructional model had been adopted making the assumption that the subjects would find being encouraged to use an extra representation helpful. The conclusion of the modelling work is that this encouragement was not helpful. Modelling the successful and unsuccessful problem solving behaviour of pairs of students on this graph problem led therefore to the statement of a cognitive economy hypothesis: *Novices solving a physics problem can more easily achieve success when they restrict themselves to using only one representation of the problem.*

One important component of expertise in problem solving or learning is the ability to control and hence minimise shifts from one type of knowledge representation or problem-solving technique to another, even though experts are able to make such shifts correctly. As novices acquire this type of expertise, their efficiency as problem solvers or learners increases. According to this view, experts try to be as economical as possible in the number of transitions from rule set to rule set. However, novices often behave in a spendthrift way and flip from one rule set to another. For example a novice may go from reading the problem to solving it, from problem solving to using expectations about school work, e.g. *this is the type of problem in which you usually use a graph* from using graphs to using equations and so on. The uneconomical path of least resistance leads the novice to change representations rather than back track when stuck.

2.2 The Graph Interpretation Problem

Also in the data set was another problem which required subjects to use their skills in interpreting graphs. This time the requirement was that subjects describe the motion displayed in the graph and interpret that motion in terms of energy. The problem was as follows:

The lines AB, and BC shown in the following graph (Figure 1) represent graphically the motion of an object which was projected up a rough incline with an initial velocity of 18 m/s and returned to its starting point 9 s later with a velocity of 9 m/s

- *Describe as fully as possible, stating any values you can determine the motion, if any, of the object (a) between A and B (b) at B (c) between B and C.*
- *Determine the distance the object travelled up the incline.*
- *Name and explain the differences between the types of energy the object possessed due to its condition at A, B and C respectively.*
- *Explain how you would conclude that the object has less energy at C than at A. Account for this loss of energy.*

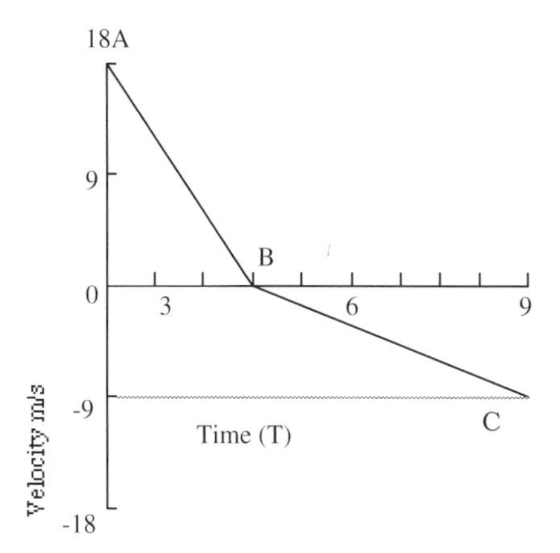

Figure 1: Graph of velocity vs. time for problem 2: the graph interpretation problem

Fifteen pairs of novice physicists subject protocols were completed on this problem, but only five of them solved it completely correctly. The others had difficulty in interpreting the situation presented in graphical form.

For example one pair showed some confusion about the meaning of the negative numbers for velocity on a velocity–time graph and the meaning of a negative slope and argued:

He's going backwards
No, he's not he's going in the opposite direction

Many of the subjects were sure of the positive benefits of dealing with graph problems e.g.

A graph — oh this'll be easier

Students found it very difficult to switch between the graphical representation and their mental picture of the energy transformations involved in motion. Several students had difficulty with the information that as at A and C the body is at rest, there is no kinetic energy. An example of this is shown in the protocol reproduced in Figure 2.

This loss of energy is accounted for by using the notion of friction, taking the hint in the question that this is a rough incline. The "missing" energy is accounted for in the heating up of the object and surface it moves over.

2.3 Instructional Implications

A number of themes emerge from this review of the protocol research study. Subjects exhibited difficulties working with graphical representations. Although positive views were expressed by these beginning physicists in the subject pool about the usefulness of graphs and the relative ease they felt working with these representations compared

It must be less energy at C than at A because at A it's about to move and at C it's just come to a halt completely.

Wait a minute at A is it at rest?

Just put, at A, A is about to move — wait a minute, at A nothing's happened

Yeah, that's the trouble you see, you don't know.

I reckon it should be at A nothing's happening, yeah, as its about to move and at C nothings happening at A it's just going to gain K.E.

You can't conclude that it's got less energy at A than at C. A, A it is.

Unless you've seen what has gone on in between at A it's at rest yeah.

It's at rest at C as well.

Yeah, so there's no energy at A or C.

There is but you can't conclude that as you don't know what's going to happen afterwards.

Yeah then you can say that there's gonna be more energy at A as it moves so that will be lost through friction which you were right going up and down the slope on heat, yeah.

Yeah, so that's wrong K.E. at A.

No, it has got K.E. at A, its been given K.E. at A, B the one AB.

Oh, it's been given K.E. at A otherwise it wouldn't move OA! It's just saying how would you conclude the object had less energy.

Oh, all right.

And you can't conclude unless ..... what was the other bit.

Account for the loss of energy. Energy lost to friction. At least I presume it is.

Through friction and heat.

Yeah.

Everything goes to heat in the end doesn't it.

It must be less energy at C than at A because at A it's about to move and at C it's just come to a halt completely.

Figure 2: Extract from protocol on problem 2: the graph interpretation problem

with working with algebra in equations, the behaviours they exhibited suggest that they overestimated their skill, in both using and interpreting graphs and overgeneralised particular pieces of graph knowledge. This tendency to overgeneralisation was in many cases exacerbated by working collaboratively. We have many examples of pairs jointly constructing an interpretation of the physical situation or a method of using the graph they have constructed. We have very few examples of one of a pair challenging the other's faulty assumptions about how to proceed. There is a similar tendency among instructors not to challenge the assumption that the provision of dynamically generated graphs will help novice students. Studying the protocols of beginning students as they attempt to reason using both graphical and algebraic representations suggest that such an approach may breach our tentatively suggested cognitive economy principle.

The research study suggested that instructional designers should not underestimate the problems novice students have in developing their understanding of graphical representation. This suggests caution for instructional designers making use of multiple representations. Results presented elsewhere in this volume by Tabachneck-Schijf and Simon (chapter 11) with a multiple representation learning task in economics also suggest that the use of multiple representations for learners should be treated cautiously. Kaput (1987) argues that using multiple representations should lead to a deeper understanding but Ainsworth, Wood and Bibby (1996) argue that a key issue in the design of instructional systems is to arrange for support of the users' coordination of multiple representations — this coordination will not appear on its own.

3 Instructional Case Studies

Representations of abstract ideas in physics are hard to understand and manipulate. The following instructional experiments illustrate how difficulties with graphing were approached in two cases, and present data from a sample of students whose performance was monitored extensively.

One way these difficulties with graphing have been addressed is to involve technological devices such as computers, videos or motion sensors in instruction. Simulations of motion have been designed with graphing facilities built in (see e.g. Preece, 1989). Linn, Layman and Nachmias (1987) reported early experiments to use computers to improve graphical skill development. Motion probes have been used with computers to allow real examples of motion to be the object of study. For example, Mokros and Tinker (1987) have used probes which allow students to capture graphs of motion as they occur, referring to these as Microcomputer-based Labs (MBLs). This gives possibilities for designing instruction where students interpret graphs produced by the computer, or students predict and then observe the shapes which appear. Students can even use their own motion to make graphs by moving themselves or other objects in front of the motion detector. Mokros and Tinker (1987) explored misconceptions in graphing and how micro-based computer laboratories can have an impact on understanding. While they identified the common misconceptions of "graph as picture" and the slope/height confusion (subjects confusing maximum rate of change with maximum value of the graph, see also Janvier (1978)) they consider the term misconception applied to these confusions as inappropriate because they found these misconceptions were *not resistant* to instruction. They attribute the power of MBL to four possible reasons:

> "MBL uses multiple modalities; it pairs, in real-time, events with their symbolic representations; it provides genuine scientific experiences and it eliminates the drudgery of graph production." (Mokros & Tinker, 1987: p. 381)

By multiple modalities they mean combining physical and visual experiences, and by genuine scientific experiences they mean the opportunity to gather and analyse real data. The real-time pairing with symbolic representations has been further explored by Brasell (1987) who describes an experiment successfully using MBL during a single class period to clear up many confusions about motion graphing in a middle school class. She used MBL in two treatments, one of which built in a delay before the data was graphed and displayed. She claims that most of the positive effect was attributable to the effect of real-time graphing as opposed to delayed graphing of data and speculates that this is due to poor study skills with students doing nothing during the delay period. The removal of drudgery from graph drawing is a positive feature shared by other instructional treatments.

Other approaches have involved students analysing videos of motion. These videos share many characteristics with computer simulations in that they are replayable, and they are two-dimensional representations of three-dimensional phenomena.

Boyd and Rubin describe these approaches as follows:

> " . . . the common learning goals are for students to understand how the shapes of curves correspond to particular characteristics of motion (e.g. going faster and faster, going at a steady pace then slowing down) . . . how position vs. time curves are related to velocity vs. time curves and acceleration vs. time curves." (1996, p. 58)

Di Sessa et al. (1991) involved students writing Boxer programs simulating real-life motion and followed them as they invented a graphical representation of a journey. Boyd and Rubin (1996) examined the characteristics of interactive digital media as a medium in which motion is presented to students learning graphical representations and conducted a case study of one student constructing graphs of her own design from a video image. Hatzipanagos (1995) has produced a helpful categorisation of the dimensions on which different simulations and interactive video presentations differ, e.g. in terms of fidelity and complexity of the interface.

Both case studies described in the next sections illustrate how using technology to engage students with the construction of graphs can impact their understanding. The first instructional experiment deals with attempts to make students aware of the connections between graphs of motion and reality.

3.1 Multimedia Motion

This instructional experiment features the Multimedia Motion CD-ROM designed by Gil Graham and David Glover which enables users to chart and analyse a range of movements — displacement, velocities, accelerations, momentum, impulse, kinetic energy of a variety of people and vehicles. It is used as part of the teaching of the Supported Learning in Physics Project (SLIPP), an Open University-led curriculum development project to support teachers and develop open and flexible learning materials in physics, self-study material for use by post-16 students in schools and colleges (see Whitelegg, 1996) as part of the teaching material on "Physics on the Move" and "Physics for Sport". (The instructional approach of the overall project is to teach physics content using real-life contexts.)

The Multimedia Motion CD-ROM allows students to select data from moving bodies (such as space rockets, tennis players, etc.) and explore how that data can be displayed in tabular form and then graphically and what the relations are between distance moved, velocity, acceleration, impulse, momentum, etc. It is used in four "explorations" to help teach dynamics and statics and complements the lab-based practical work and text material which together make up the course. The simulations are based on video sequences presented as Quick Time movies from a video source which can be played repeatedly on a small window. Data is collected by clicking on the screen with a mouse while watching the video frame by frame, and leaving a visible trail of overlaid points. The data can then be saved in a spreadsheet which can be used to generate graphs of position, velocity and acceleration against time. By switching between the sequences and the graphs the users can connect significant events with features of the graphs.

An evaluation of the Multimedia Motion CD-ROM-based activities was conducted in conjunction with large-scale developmental testing which was conducted in 16 schools and colleges. The evaluation consisted of observations of teacher and student use of the material in two schools with 12 students engaged in around 60 hours of computer-based activity. The resulting data is in the form of videotape records, observation schedules and student self-reports. Analysis of the data raised a number of issues about how exploratory learning can best be supported by multimedia and this is reported elsewhere (see Whitelegg, Scanlon & Hatzipanagos, 1997).

To obtain a bigger sample of students and probe further their perceptions of working with the CD-ROM, in addition, questionnaires about students' use of it were sent to four further schools. The questionnaire asked the students to describe their use of CD-ROMS, their perceptions of their use of the Multimedia Motion disk and the particular sequences they used for exploration. Teachers reported and the evaluators observed the expected benefits of increased motivation for learners because of access to more realistic applications of the laws of physics illustrated in the disk. For example one student wrote:

> "(the experiments) were useful in that they related actual events, e.g. rockets taking off and cars crashing to the theory we had studied in the classroom."

The strategy of teaching physics using real-life contexts adopted by the course producers is currently very popular (see e.g. Hennessy, 1993). The *Physics on the Move* unit uses the safe transportation of people and goods as its linking context to discuss motion, and the *Physics for Sport* unit uses the contexts of rock climbing, springboard diving and scuba diving to teach forces, vectors, oscillations and pressure. Students can examine videos of car and train crashes, people jogging, playing various sports, etc. and so reinforces the real-life approach introduced in the text materials. Using the CD-Rom, students can also examine examples of lab-based experiments, i.e. the linear air track to teach momentum and kinetic energy. Students ran experiments, collected data and plotted graphs.

The instructional approach here was to work with messy data and real-life contexts to increase student motivation and allow them to perceive a need for simplified graphical representation. The graph drawing facility on the CD-ROM allowed students to be supported in their graph construction. In general they found the CD-ROM supported them in doing interesting practical work. As one student commented

> "the programs are not very time consuming and yet easy to understand. Classroom practicals give hands on experience but do not always show theories work successfully."

However, they had great difficulty in moving from tables of data from experiments to satisfactory interpretation of the graphs they generated. We found some students tried to work with algebraic formulae on their tables of data as they found the graphs they plotted "confusing" particularly where students chose to generate graphs from data points separated by very short time intervals. The facility to go back and quickly collect more data points to include on the graph was useful.

> "Do you want to take more readings
> Let's take more to make it more accurate since it doesn't take long."

Figure 3 shows a screen and Figure 4 shows an example of a student-produced graph generated by clicking on video frames of a woman walking.

In the post-test administered after the use of the CD-ROM we included Problem 1 from the research study. This example was performed correctly by all but one of our subjects, suggesting that the immediate impact of the instruction was positive. It is our intention to conduct a more extensive instructional experiment with the SLIPP materials with a group of students working over a longer time period towards their final examination period, to observe how the combination of mathematical, practical and computer experiences combine to develop students' graphical skills and physics understanding.

3.2 Conceptual Change in Science

The second instructional experiment reviewed here is the Conceptual Change in Science (CCIS) project which aimed to improve the science understanding of a mixed group of 13-year-olds using specially designed computer software. The concepts to be studied were those from Newtonian mechanics. The software consisted of scenarios implemented in DM^3 (the Direct Manipulation of Mechanical Microworlds) computer program written in Smalltalk-80. Our curriculum involved a scheme of work based round four scenarios ("the Rocket Skater", "the Parachutist", "the Speedboat", and the "Supermarket"). These scenarios were brought to life in simulations of horizontal and vertical motion under forces in

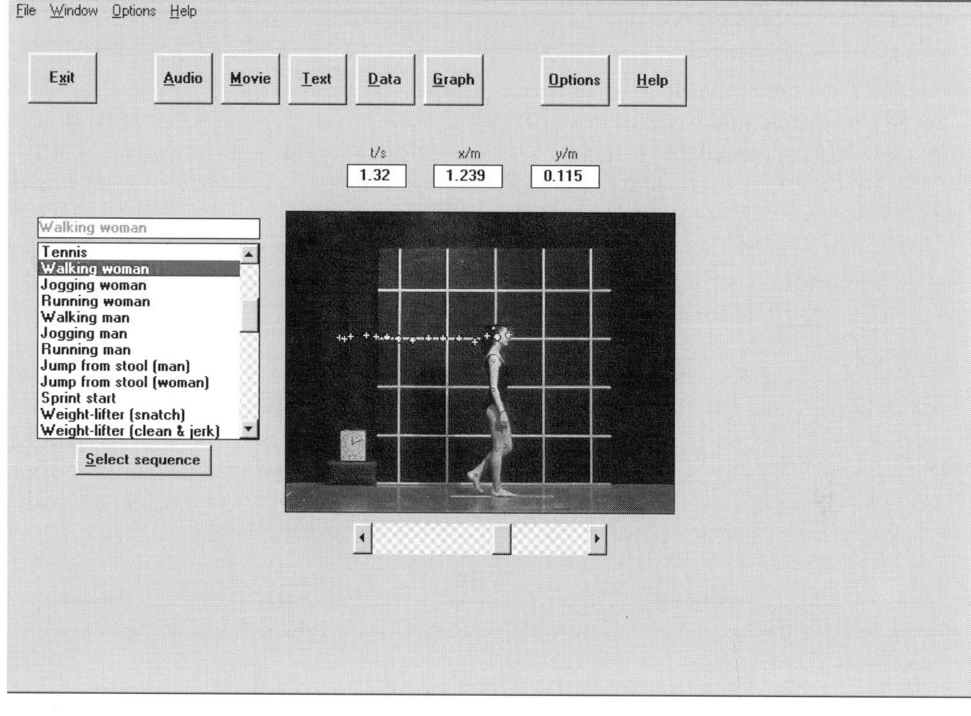

Figure 3: Screen from Multimedia Motion

various contexts: the system contains meters, a dynamic graphing facility (normally used for velocity–time and distance–time graphs) and an on-going screen video recorder to allow replay. Figure 5 shows a sample screen for one of the scenarios. The interfaces are hybrid screens where the representation of the real world is combined with counter and control buttons which start and stop the motion, reset the system when the user wants to alter the values of the variables and output devices such as the graphing facility where the user can observe synchronous graphs of the speed of the displayed motion.

Children could use the scenarios both in structured experiments and in an informal way. The scenarios together with support materials, e.g. worksheets, form an integrated teaching package which also includes "real" practical experiments. All this practical work involves pupils first making predictions about the outcomes, providing their reasons, and then finally checking the outcome against their predictions. The choices made about the contexts to be used in the scenarios involved us in much discussion. The curriculum was based on a constructivist approach (see Twigger et al., 1994). Starting from pupils' prior conceptions, we aimed to help them form new concepts which they could use to interpret their experiences in mechanics. We needed to provide counter-evidence to their predictions from experiences either in the "real" or computer world. On other projects students have been observed to not believe this

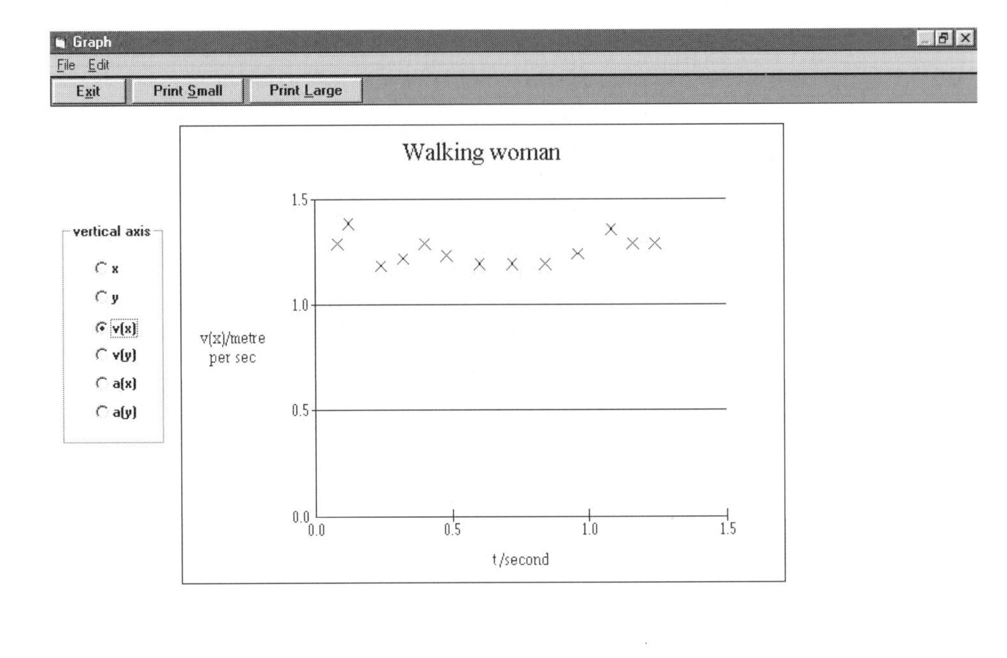

Figure 4: Graph produced to analyse woman walking

sort of evidence or to consider it faulty data or forget it and revert to previous beliefs. We reasoned that children need an alternative way of conceptualising the situation before being prepared to reconsider their current ideas. The scenarios were constructed to provide opportunities for pupils to construct a Newtonian set of rules for force and motion in a supported software environment. The scenarios helped pupils to learn through observation of patterns of events and data on the screen. Learners also needed to assess the validity of their new conceptions by doing and discussing experiments in the real world. This led to the decision to incorporate "real" with computer practicals and the adoption of the Predict/Observe/Explain methodology which meant encouraging students to make a prediction about the outcome of a real or computer experiment, giving an explanation for their prediction, and then trying to explain the outcomes. Our purpose was to see whether we could promote conceptual change in a real school setting. We were fortunate to work with an experienced classroom teacher in a local comprehensive school who was prepared to take on board the curriculum we had prepared and allow us to observe the classroom work in progress. In doing this we hoped to be able to make some inferences about the role that the computer software had in effecting any conceptual change we detected.

The curriculum occupied 29 children aged 12–13 years for 7 weeks. This was a mixed ability class of 13 boys and 16 girls. The classroom time spent was 5 hours per

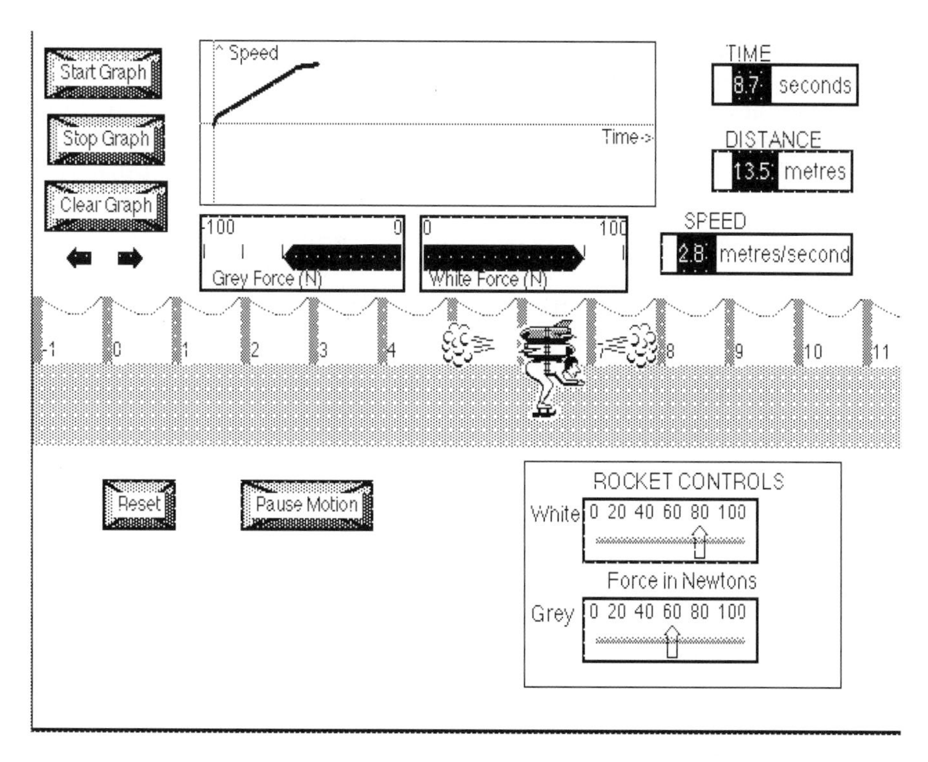

Figure 5: Screen from a conceptual change in science scenario

week with approximately 1 hour homework per week. Work was organised so that pupils worked half-time on practical work and half-time on the computers. All of our subjects had prior experience of using computers and using a mouse. The pupils had no previous secondary level science instruction in mechanics or physics. The data collected included diagnostic tests, a pre-test called "your ideas about motion", which also was given as an immediate post-test and then one month later as a delayed post-test. We collected interviews with individual children, and video- and tape-recorded transcripts of the children at work together with observer's notes, and copies of homework and the children's completed worksheets. Two speed–time graph interpretation tasks were included in the test in order to assess development of skill in this area.

The class was divided into 10 small groups, nine triads and one pair. Our group work consisted of working in fixed triads on a specially designed curriculum with time equally divided between the "real" and computer practicals. Children worked in affinity groups chosen by the class teacher and were operated for a 2-week trial period before the main experiment. The effectiveness of the curriculum was measured by comparing individual student performance on pre-test and post-test and we experienced some significant improvement in pupil performance (see Scanlon et al., 1993).

Reviewing our audiotapes and videotapes of the groups as they worked at their tasks, to try to characterise these differences has allowed us to produce micro genetic case studies of learning in the groups. We found that these younger students performed well in their analysis of graphs of motion in comparison to our expectations based on the protocol research study in particular. In our pilot testing, we found that 12–13-year-olds were limited in their ability to generate speed–time sketch graphs representing the simulated horizontal motion under friction they observed. We found that the interaction with dynamically generated graphs were very beneficial for this. For example, in the following extract from a group of three girls working on one of the scenarios we recorded the following interchange:

Start graph and watch

(pause while they observe the screen)

It just goes on and on

It's gone off the screen

It got faster but then not any slower

No for that you'd have to cut the force in half for that

The graph has gone off too

(Laughs)

So is the answer true or false

(disagreement)

Let's check it

Let's put it back to the beginning

(pause)

Oh now I see

Figure 6: Extract from protocol of a pair of pupils working on a conceptual change in science scenario

This interchange is typical of the way that the provision of the hybrid screens allows the group of students to shift focus between watching the motion of the objects and the graph. The possibility of quickly redoing the experiment when asked to make predictions provides opportunities for continuing the discussion based round the scenario until an agreement is reached. The instructional focus is on using dynamic graphing in a qualitative way. Students were required to connect simulation behaviour with graphs of motion but were not yet being required to connect algebraic and graphical representations. It was not a key aim of this project to explore in detail how children used graphical representations, but the naturalistic setting for the experiment provided some interesting vignettes of students' graphing behaviour to be considered.

4 Conclusions and Discussion

The research study showed a number of misconceptions related to interpreting the meaning of the graph. The misunderstandings were more elaborate than the common "graph as picture" or slope/height confusions reported elsewhere in the literature. Good performance was related to restricting attention to equation manipulation skills or graph interpretation skills, while incorrect performance was related to switching modality or applying features of the school problem-solving context such as over-generalising aspects of graphical knowledge.

The implication from this study was that instructional approaches need to be carefully considered and designed to help students to understand motion and to develop facility with graphical and algebraic representation. The literature suggests a list of desirable features for such instruction. For example, authentic real-life contexts should be provided for students to explore to help them understand motion. This can go as far as gathering and analysing real data, or it can refer to analysing more interesting examples of motion. A number of properties of computers can be helpful in such instruction. Simulations of motion combined with tabular or graphing facilities were the approach adopted in the both instructional experiments we reviewed which met with some success.

With Multimedia Motion the students conducted experiments on video extracts of real interesting examples of motion. They switched between videos of motion, data, tables and graphs. In the Conceptual Change in Science project students were able to explore a number of simulations of carefully chosen real-world contexts while on the same screen graphs were displayed simultaneously. One aspect of instruction which receives strong support in the literature which we have not yet explored is the desirability of using motion probes to give students kinaesthetic feedback related to the visual representation of motion they see in a simultaneously produced graph.

Key aspects of the design of appropriate instruction in motion which remain to be explored are:

- the relative benefits of "real" motion (i.e. practical experiments) as opposed to the analysis of video segments or simulated experiments, and
- how time should be divided between different types of activity. A range of introductory and exploratory activities which help students understand the reason for the adoption of graphical representations by scientists and practising the graphical interpretation skills thus developed could be appropriate.

The protocol study and both instructional experiments highlight the importance of one particular approach to studying learners. This is the value of using collaborative working on problems to produce think-aloud protocols of groups working. These give researchers access to information about the procedural decisions that learners make while solving problems and while working with computer programs (see also Moschkovitz, 1997).

The results reported here are focused firmly on the use of representation in understanding motion. However, there are suggestions from other domains of similar findings. For example, Ainsworth, Wood and Bibby (1996) reported some unhelpful consequences of mixing types of representation in their experiments exploring children's understanding of computational estimation, and concluded that design of effective learning environments using multiple representations needed careful consideration. More instructional experiments on a range of topic areas are required. Instructional designers should not underestimate the problems novice students have in developing their understanding of graphical representation.

Acknowledgements

I would like to thank the Office for Technology Development and the Open University Research Committee for funding support. Thanks also to Elizabeth Whitelegg, Stylianos Hatzipanagos and Tim O'Shea, and the members of the CCIS project for their collaboration on different aspects of the work described here and the editors of this volume for their helpful comments on drafts. The Multimedia Motion screens in Figures 3 and 4 are reproduced with the permission of Cambridge Science Media, 354 Hills Rd, Cambridge CB1 3NN, UK from whom further details of the program are available.

5

Toward Decision Support for Multiple Representations in Teaching Early Logic

Mike Dobson

1 Modality Assignment

Choosing graphical systems to use in teaching has often been more an art than a science. New ways to show material from a subject discipline usually emerge from continued thought about the material by researchers working on the discipline. Educators have in some cases been concerned with teaching difficult subject material and this has forced new graphical communication systems to be built. Introductory logic is one such area where a series of teachers, who were themselves experts in the domain, had the task of teaching an elementary area of logic to a young student. A succession of solutions, from different parts of the world, emerged and each creator was aware of the previous author's solution. These facts would lead most anybody to think that each system was inevitably an improvement upon the last. If any general rules about choosing between alternatives should come from this story, they might be expected to come from looking at the improvements made during this sequence of development. However, a theory of cognition with diagrams suggests that such a sequence of improvement was not the case and that general principles are responsible for the benefit of graphical systems. The principle suggests that underlying the primacy of graphical systems over text in communication is a difference in the abilities of the languages used to communicate. Graphical systems are said to be limited in their ability to express information and are therefore more appropriate when limited material is being taught. Since the principle is based on the communication capacities of the languages involved rather than on the type of language (graphical or linguistic), it should be extendible to comparisons between languages of the same type. It happens that each of the systems used in the series of teaching beginning logic is graphical and so provides an interesting test of this cognitive principle.

The problem of appropriate presentation of information to a user or learner has been treated under the banner of many different research disciplines. In Instructional Design, Dijkstra (1997) has recently advocated emphasis on the development of decision support systems to help authors decide when to use alternative media in instructional systems. This view of media selection has its history in choices between macro-media such as books, radio broadcasts and videotapes. The view claims that there is no best medium but that there may be some media more adequate than others for certain skills. It may be that at this level of analysis, such pessimism is warranted and it is not clear where the information would come from to help make decisions of this nature. The discipline of human computer interaction has generated interesting guidelines. Norman (1993) uses the term cognitive artefact to describe the abstraction of key information into representations. These artefacts are really what make humans intelligent. Without such representations, he says the power of the unaided human mind is highly overrated (p. 43). In common with the instructional design guidelines, this work is difficult to draw hard rules from. The strongest guide for designing or choosing between alternatives, and this is not the main purpose of the work, is that perceptual and spatial representations are more natural and therefore to be preferred, but only if the mapping between the representation and what it stands for is natural and analogous to the real perceptual environment (p. 72). In representing logic, Johnson-Laird and Byrne (1991) have also taken naturalness to describe the merits of their own notation. It is difficult to determine just what is meant by naturalness in either of these guidelines and they are probably both suggesting different phenomena. While Norman's work is an acknowledgement of the complexity of supporting human problem solvers, Johnson-Laird and Byrne's purpose in describing naturalness is to support his view of human deduction. Johnson-Laird and Byrne's mental model notations are sufficient but not necessary to describe his theory of natural deduction.

Assigning information to a modality is not even a well-understood problem but the benefit of a solution is recognised by many disciplines. This work is especially interested in the applied problem of modality assignment for instruction. The information enforcement or specificity principle is a cognitive approach to describing communication with graphical elements and provides the basis for the following description and study.

2 Applying Information Enforcement

The communication of domain knowledge to a student in a learning situation requires a language or combination of languages in which the objects and relations of that domain can be represented. Such tools as used by an instructor, include a range of formalisms normally used in the domain. The effective adoption of the target material will depend on the choice of tools made by an instructor at critical times of intervention during the learner's progress. The successful learner will eventually share his model of the domain with that of the tutor. In many cases, for example in work with Hyperproof (Barwise & Etchemendy, 1995), there has been shown to be benefit from providing the learner with multiple or heterogeneous representations to view the domain material.

This chapter considers choices between alternative formalisms by prescription from quantifiable features of the formalisms. The goal is to determine appropriate choices within a modality of communication based on learner attributes and domain characteristics which would be implemented in a rigorous enough way as to be part of a teaching component in an intelligent tutoring system. The objective is to develop a modality decision support component for the tutoring model in the three part tutoring paradigm described, for example, by Elsom-Cook (1990). This work is an evaluation of the specificity principle and the idea is largely derived from the SIGNAL project (Stenning & Oberlander, 1995; Stenning & Tobin, 1993). Their work outlines a working theory of graphical representations in reasoning and operates on a number of premises including the following, that: (a) graphical representations are one sort of representation which exhibit "specificity", which is to say that they compel specification of classes of information in contrast to systems that allow any abstraction; (b) this is said to mean that such representations are relatively easy to process; and (c) this specificity helps to explain why graphical techniques such as Euler's circles for teaching abstract reasoning are so didactically effective.

There are many ways of demonstrating syllogistic reasoning for a student. We have studied three which can be thought of as finite state machines and since others have infinite numbers of states, these systems with limited states are less expressive. While counting the Venn system the cube is used as an intermediate representation. The cube shares some key features with Venn and provides a very interesting example of using a separate representation to help solve a problem phrased as being about another representation. Euler and Carroll are examined in a similar way.

The intention here is to provide a quantitative interpretation of the specificity principle. The goal is to access the critical factors in systems for the communication of domain facts and relations and to capture them in such a form that they will be beneficial in learning systems design for the future. We intend to be able to predict learning outcomes from students' interaction with representational systems and in the longer term generalise these predictions to support decisions between alternatives. The basic premise for the quantitative interpretation is that systems are considered as variable state machines or automata. The magnitude of states is equivalent to the magnitude of expression possible with a system and so determines an expressivity measure. A system is considered in terms of the relation between how much there is of what is being represented and how much the system is capable of representing. This ratio determines the specificity attribute. The hypothesis is drawn from the theory that information enforcement aids communication and can be succinctly stated — that graphical representation systems which are most specific will provide the best teaching aids. The next section considers this principle with a simple system for showing inhabitants of an office. We call this system WOT? or Who's in the Office Today? and its variants WOT—2, WOT—3 and so on.

2.1 An Office Indicator System

The WOT system might be used for communicating the presence or absence of researchers sharing an office. The graphic could be fixed to the door outside the office and would help to show visitors the state of the situation inside. The representation shows four quadrants of a square where each quadrant can be either shaded or unshaded. The system exhibits no interactivity; rather it provides information by means of a system of conventions, which so long as they are understood will reveal information to the visitor without the need to establish the information in other ways.

In a first variation of WOT called WOT—1 there are four quadrants each of which may be shaded or unshaded. All instances when rotated by 90°, 180°, or 270° are assumed to be rotationally equivalent. There are then exactly six states in which the system can exist and they are depicted in Figure 1. The measure of expressive cap-

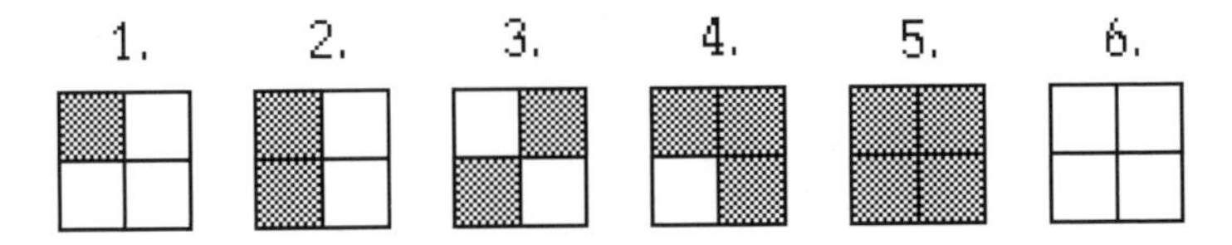

Figure 1: The office indicator system with all semantically distinguishable states assuming rotational equivalence

ability is determined in WOT—1 by the number of states available and in this case, $n=6$.

In the situation represented so far four people may be in the office and assuming they only ever sit at their own desks, 2^4 alternative situations where only 6 are distinguishable using WOT—1. There are: four ways of achieving the first picture, four ways of achieving the second picture, two ways the third, four ways the fourth, and, one way each for the fifth and sixth pictures. A ratio of real-world situations for which there are 16, to graphical system states for which there are 6 leaves a ratio of $16 \div 6$ or 0.375. In short, there are not enough states in the graphical system to show all possible information. If we wanted to know who is in the office then the system does not have enough flexibility to communicate this information. A representational system can be made more expressive by adding conventions which constitute the rules for interpreting it. On removal of the rotational equivalence feature there will be an exactly one-to-one mapping of the domain to the WOT representation. This new variation can be called WOT—2. By removing the rotational equivalence condition there are more states allowable and the ratio value will then be exactly $16 \div 16$ or 1.00 since there are 16 real-world states and 16 states to be represented. Assuming conventions are well understood, an inquirer should easily see who is inside the office.

Increasing the size of the domain to include information about whether the office inhabitants are holding a telephone conversation or are available might yield a scenario like WOT—3. These alternatives may be shown using colour codes, say green for availability and red for non-availability. Numbers of states for the two situations are shown in Table 1. Each graphical system has symbolic tokens used to represent some state in the world which may have different values, say red or green, and can be thought of as variables. In the office system when there is a one-to-one mapping from states to representations there are two tokens.

One is the position of a region and the other the value of the shading — shaded or not shaded. By the combination of each of these two variables information about who is in the office, how many are in the office, whether they sit across from each other or diagonally when there are only two, can all be easily derived. Where rotational equivalence is assumed in WOT—1 there are four ways to show one person in the office and one way could mean any of four situations, that is any of the four office inhabitants could be in the office for the diagram to be a good representation. The third system includes the dimension of availability for meeting as well as being in or out of the office. It is now possible not only to "see" if someone is on the office but

Table 1: Numbers of states represented with model diagrams in the office indicatory system

Diagram	Situations in WOT—1 and —2	Situations in WOT—3 and —4
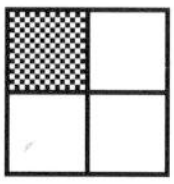	4	$4*2 = 8$
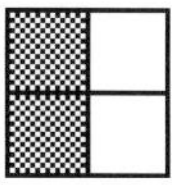	4	$4*2^2 = 16$
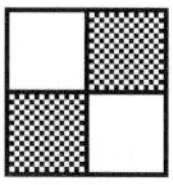	2	$2*2^2 = 8$
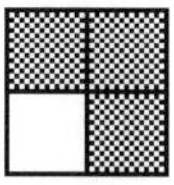	4	$4*2^3 = 32$
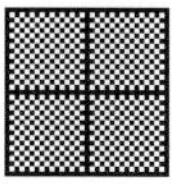	1	$2^4 = 16$
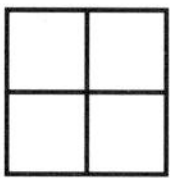	1	1
Total	16	81

Note: Rotational equivalence is assumed for all systems.

also whether they are willing to be disturbed for a meeting. In WOT—3 it is still impossible to know who it is that is in the office apart from some situations where, for example, all are present. By adding a new convention a larger set of variables is established and a larger domain can be represented. We are now beginning to describe a class of graphical systems which are related to each other in that they differ in the scope of allowable values for their tokens.

A representational system can be measured in terms of the information it is able to represent and this measure is called expressivity (Table 2). When the expressivity measure is compared with the substance of the communication, a ratio of the system's capabilities over the use to which it is being put emerges. This does not yet show the effect on learning but the analysis should help to predict the communicability of a domain when used in real teaching and learning environments. The hypothesis is that systems are the most appropriate for a teaching and learning situation when they are most specific. The remainder of this chapter returns to the domain of syllogistic

Table 2: Features of the office indicator system

	Expressivity (e)	**Domain size (d)**	**Specificity (s)**
WOT[1]	6	16	0.375
WOT[2]	16	16	1.000
WOT[3]	16	81	0.195
WOT[4]	81	81	1.000

Note: Specificity is at unit value when the expressivity of the system is the same as the domain size.

reasoning which is chosen for the study since there are a well established set of representations which have been used over time, and because the literature is rich in explanations for misconception and error. These systems have been constructed to be used on paper with a tutor or alone as a student. They are effective generally because of constraints on interpretation which help conclusions fall out more clearly than is possible with linguistic representations. Moves which need to be made with these systems for adding a second premise or reading a conclusion are to some extent automated. All these systems are fully implemented in interactive software.

2.2 Venn Diagrams

The Venn system starts with this basic diagram below (a). Assertions can be represented on the diagram in one of two ways. Universal assertions are depicted by shading out infeasible regions. In diagram (b), shading denotes the fact that All A are B since the part of A which is not also in B is shaded and therefore infeasible. Diagram (c) indicates that No A is B, the intersection being precluded by shading.

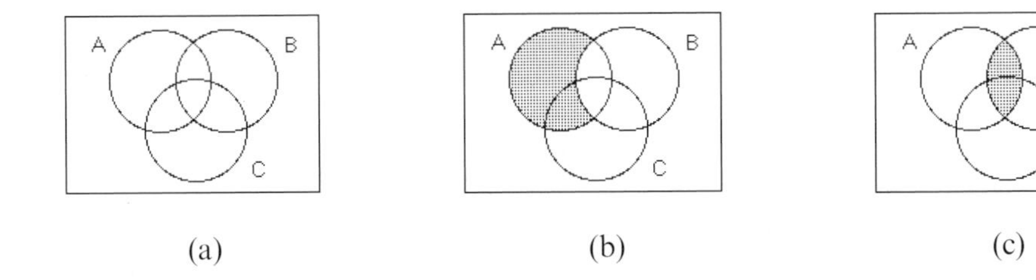

(a) (b) (c)

Existential assertions are represented by marking with an X a region which is not empty. The complementary nature of marking a cross for existence and shading out non-existence mirrors directly the equivalencies: (a) Some A are B Not (No A is B), and (b) Some A are not B Not (All A are B). Thus the following diagrams all indicate that Some A are B.

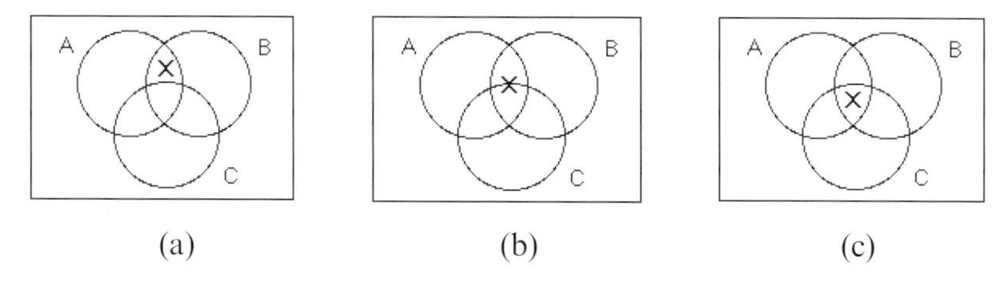

(a) (b) (c)

The difference here is in the additional information concerning the relationship between A and C. In (c) there is a clear assertion that Some A are C while in (a) there is a suggested implication that this is not so. In fact if the rules are adhered to then such an implication is not supported without appropriate shading so that it may be argued that (a) is quite acceptable. Nevertheless we adopt (b) as the conventional way of representing the statement Some A are B without further concern for C. In precisely the same way the following diagrams all signify that Some A are not B.

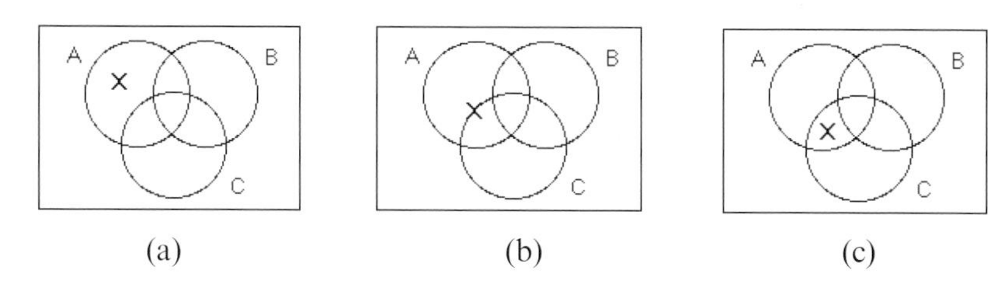

(a) (b) (c)

Here (a) also suggests, but again this is unsupported, that No A is C while (c) actually asserts that Some A is C. We adopt (b) as the conventional representation of Some A is B. In total there are eight forms which the first premise of a syllogism can take.

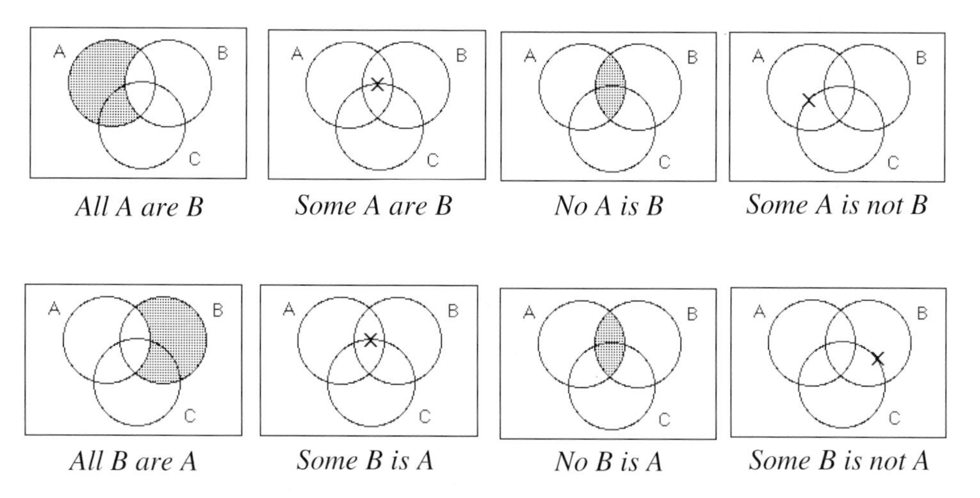

All A are B *Some A are B* *No A is B* *Some A is not B*

All B are A *Some B is A* *No B is A* *Some B is not A*

Logically equivalent premises produce the same diagrams; for example the diagram for Some Bs are As is the same as for Some As are Bs.

Following the counting argument used to show the office indicator system, this section treats the Venn diagram systems in the same way and eventually we show that the total number of distinct diagrams in Venn evaluates to 1,784,593. A superset of distinct diagrams is achieved by the possible combination of "x"s and shadings which the system allows. The count starts by assuming that an "x" may be placed in the interior of any of the eight regions or on any of the 12 arcs that separate those regions. The total number of possibilities is then 2^{20}. Of course any of the eight regions may be shaded, giving a further 2^8 possibilities. In principle these could be combined as independent but earlier remarks concerning inconsistency place a restriction on this and the counting is a little more complicated. When any region is shaded then we no longer permit any of its boundary arcs to carry an "x". The counting is more easily achieved by using a second-order graphical system or intermediary representation. The regions of the Venn diagram are treated as vertices of a cube with contiguous regions joined by edges of the cube. This particular representation system has been used as a method for solving the syllogism but has not been seen in the literature for the purpose made here.

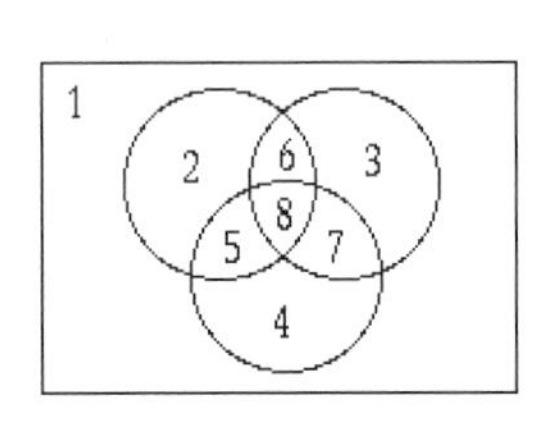

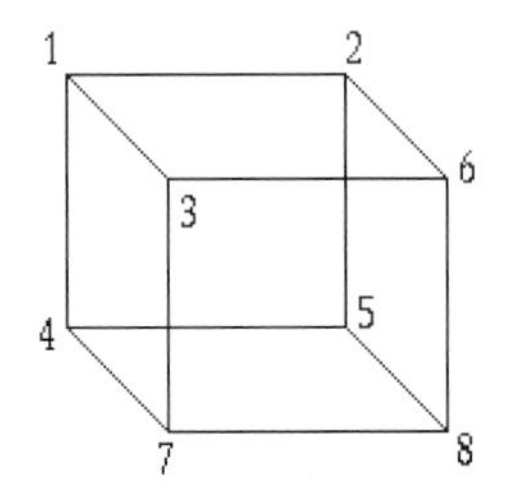

If no regions are shaded then an "x" can appear in any one of them together with any of the 12 arcs which yields 2^{20} possibilities. There are eight ways of shading one region each leaving seven regions and nine arcs and accounting for 2^{16} possibilities accounting for the first two rows in the table below.

Continuing in the same vein using the geometry of the cube to find the number of available arcs given a particular shading, we find the possibilities in Table 3 and the total number of states can be thought of as the total possible expressive capability of the Venn system.

2.3 The Carroll and Euler Systems

Lewis Carroll, whose real name was Charles Dodgson, was, apart from being an author of great children's literature, a teacher of logic. Carroll's graphical system for demonstrating the syllogism is based on a grid system which requires placement of shaded and unshaded counters. This system is demonstrated in chapter 2 of this book. The counting approach was applied to the Carroll system in much the same way as with Venn. The Carroll and Venn systems are actually very similar although

Table 3: A summation of the Venn counting argument showing possible states in Venn system

Shaded regions	States
0	2^{20}
1	8×2^{16}
2	5×2^{15}
3	19×2^{11}
4	237×2^{5}
5	19×2^{6}
6	5×2^{5}
7	16
8	1
Total	1,784,593

there are fewer constraints limiting possible combinations of grammatical items in Carroll and so the language of this system is more expressive. Relative to Venn, Carroll is less specific and so would be expected to be less valuable in communication according to the principle.

The third system was created before the others by a Swiss mathematician named Leonard Euler. Euler's requirement to tutor a Russian princess led him to develop his circle system (1772). The system has been interpreted differently at various stages in the Psychology literature but has recently been described in detail by Stenning and Oberlander (1995).

Table 4: Specificity for the three systems with and without interpretation

System	Expressivity	Specificity
Euler	1,600	25
Venn	1,784,593	27,884
Carroll	3,486,784,401	54,481,006

There is no space here to show the details of the counting for Carroll and for the Euler system, but this can be found in other places (Dobson, forthcoming). Table 4 describes the results of this counting process and shows a measure of expressive capability for each of the three systems. Of all the systems Euler is clearly more specific than either Venn or Carroll.

3 Evaluating Information Enforcement

Since there is now a measure for each graphical system according to the information enforcement principle, it is possible to test the value of the principle as a decision support mechanism for choosing between alternatives. The crucial information required is how well each of these alternatives supports the learner developing the syllogistic reasoning skills they are intended to teach. To get at this information a study was done with 42 students, seven per condition in six conditions. Three software applications representing the graphical systems shown above were used with students from both mathematics and humanities backgrounds. All subjects were matched for prior exam scores. These examinations are internal and used to prepare students for externally moderated ones which are in turn used for university entrance. The results of these tests are known to be a good indicator of success in the entrance exams. None of the subjects has been exposed to prior logic training of any kind. The mathematics students were enrolled in advanced-level mathematics courses and the humanities students in a range of courses including history, geography and English at the same advanced-level.

The study involved a training period, a pre-test, an exploration period and a post-test, and the whole interaction for each student lasted just over 2 hours. Training materials were created based on the individual graphical systems. A range of problems was demonstrated to the students including simple and difficult ones. Difficulty measures are well known in this area and have been tested in cross-cultural comparisons (Johnson-Laird & Byrne, 1991). The 15 pre-test and post-test questions were isomorphic in terms of the structure of the problems administered. The only difference between these tests was the ordering of the questions and the terms used in the premises. Changing the terms could interfere with the difficulty levels of the questions, since it is known that believability has an influence on many aspects of the human reasoning process (Oakhill, Garnham & Johnson-Laird, 1990). To avoid this problem, premises were assessed by a group of peers for believability. The pre- and post-tests asked the student to indicate validity or non-validity of offered conclusions and to draw a correct version of the diagram in the post-test phase. These diagrams led to a generative model of error which could be diagnosed through actions made by learners on an interactive graphic.

3.1 Results

Here we look at the effects of exposing the two discipline conditions, humanities and mathematics students, to the three software systems. In all mathematics conditions students showed a significant improvement in performance using Carroll, Venn or Euler. In the Euler condition, subjects took 50% longer to achieve the improvement and to finish the post-test. This is difficult to explain in terms of the specificity count as outlined above. Harder to explain are the results of the humanities students in both Euler and Venn conditions. Humanities students showed a decline in performance on post-tests for the same problems (Table 5).

In all mathematics conditions where students used the Venn, Euler or Carroll systems (a), (b) and (c) scores increased significantly from pre- to post-test. In the first condition from mean scores of 6.2 to 12.1 (Wilcoxon signed rank test: $T = 3$, $N = 7$, $p < 0.05$). In the second condition, mathematics students used the Euler system

Table 5: Scores, percentages and times for all conditions in specificity evaluation

Test item	Condition					
	Venn $n = 14$		**Euler** $n = 14$		**Carroll** $n = 14$	
Maths						
Pre-test	6.2		6.5		7.3	
Post-test	12.1	(20)	10.4	(30)	12.9	(20)
Mean Pre post-test Δ	5.9		3.9		5.6	
SD	1.41		1.27		1.2	
Humanities						
Pre-test	7.3		5.0		6	
Post-test	5.1	(20)	3.6	(30)	6	(20)
Mean Pre post-test Δ	−2.2		−1.4		0	
SD	0.78		0.21		0	

Note: Values inside brackets indicate time taken to complete post-test phase in minutes.

and scores increased significantly from 6.5 to 10.4 ($T = 0$, $N = 5$, $p < 0.05$). In the third condition scores increased from 7.3 to 12.9 ($T = 2$, $N = 5$, $p < 0.05$). In two of the humanities conditions (d) and (e) those in the Euler and Venn dependent groups a significant decline in performance took place from mean scores of 7.3 to 5.1 ($T = 4$, $N = 5$, $p < 0.05$) in condition (d) and from mean score of 5.0 to 3.6 in condition (e) ($T = 2$, $N = 5$, $p < 0.01$). Differences in outcome between Venn and Euler show that the Venn system proved a significantly better supporting system than did Euler and was tested with the Mann–Whitney U-Test ($U = 8$, $U' = 41$, $p = 0.02$). The most striking difference is that between the two learner groups, mathematics including conditions (a), (b) and (c) vs. humanities including conditions (d), (e) and (f) (Exact 2 — tailed $p = 0.0173$, $U = 38$, $W = 116$), mathematics students showed much higher learning gains over the humanities group while using either Euler or Venn and this result is really quite remarkable. The students had been taken from groups with almost equally matched scores on their previous practice advanced level examinations in their own chosen disciplines and so there is no reason to believe interference of general intellectual differences.

4 Discussion

The work reported in this chapter initially set out to evaluate the information enforcement principle to see how it might contribute to a modality choice algorithm in a teaching system. Previous work (Dobson, forthcoming) had provided a method for interpreting this principle numerically. The experimental design was primarily organised to find any correlation between learning outcomes and the results of that analysis.

The first outcome of the study was that this correlation did not occur. The principle did not predict the learnability of the reasoning skills from the systems. The results showed the opposite of what was predicted from the metric, that the least specific system produced the best learning outcomes. The interpretation of the principle is understood to be fair, and was checked several times with its originator (K. Stenning, pers. comm., 1995). There are a few minor reservations about the way in which the metric was derived from the principle, however, they are not enough to change the rank order of the systems as they are described here. It is more likely that the principle itself, while important, does not have such an overriding effect as was thought. Other variables are likely to have an impact of learnability and several of these were not controlled for in these studies.

There are several features of representational systems which may be important beyond the simple specificity ratio. One problem may be in the way in which the metric was calculated. Complexity can be described in different ways. The gross number of states in a system provides one way to measure that complexity. The gross number of states may have been created from combinations of similar conventions, or the system may have a new convention for each state. In complexity, this difference is described as the dimensionality of the system. Intuitively, it is possible that systems with the same measure of expressivity may have different learnability. If a system is created from a few common and repeated conventions, it is likely to be more easy to learn than several dissimilar ones. Dimensionality describes the way in which numbers of states are created. A system which has 10 possible states may have two variables where each variable may be in one of five states. An example system could have two lamps, each with five levels of brightness. Another system may have 10 completely independent states. The difference may be described in terms of dimension, the first system having two dimensions and the second only one.

Another interfering variable is in what the students bring with them to the study. Students always have some prior knowledge and expectations about how things are going to work. Another study reported in Dobson (forthcoming), looks in detail at how the students' prior expectations affected the way they perceived the meanings of conventions used in the syllogism representations.

The most unexpected result from this study was the drop in humanities students' scores after using the software. Many teaching interactions have no positive effect on outcomes, but few ever manage to decrease student performance. Explanations for this have been hard to infer, and remain speculative. It is possible that, although humanities and mathematics learners score similarly well on pre-tests they may be doing so for different reasons. This difference may be the key to the later changes in performance. Humanities students may be making particular use of belief cues in early stages of problem solving. During exploratory learning with the software and through the instruction, the importance of belief cues may have been undermined. Problems which have believable conclusions turn out to be invalid during this phase.

At a point in the learning process those students will generally degrade in performance, having rejected a partially useful strategy without replacing it with anything better. Believability cues are both useful and counter-productive in real-life reasoning. Making an unbelievable inference especially from data taken to be contingently true, will challenge currently held beliefs and so would cause an affective reaction much stronger than reaching a conclusion which did not challenge beliefs (see for example, Newstead, 1990; Revlin & Leirer, 1978; Wason & Johnson-Laird, 1972). The realisation that such affective cues are not necessarily useful in solving syllogisms is a fact demonstrated by both Euler and Venn applications. This may leave the student for some period without an alternative strategy to replace it. Without more experiment the hypothesis is not supported but this study could have caught these students without a viable strategy.

4.1 Implications for Graphical Interactive Systems in Learning

A key attribute of both interactive graphical representations and microworlds is the use of direct manipulation graphical interfaces to represent the domain material. This chapter has shown that in the context of learning formal reasoning there are a number of attributes that graphical representation systems should have in order to promote effective learning.

Microworlds are said to add value in exposing and testing students' prior conceptions and thus provide conditions favourable to cognitive conflict. Resolution of this conflict requires the learner to make explicit the causal relations between the components of their own view of the domain model (Draper et al., 1992). Direct manipulation of simulations which are representative of the domain are meant to provide the space for conflict and then for resolution. However, it has often been reported that persistent errors in student conceptions, even after semi-structured and guided discovery interaction with the simulations, still persist. This persistence is usually assigned to the inability of students to create a runnable model and thereby guarantees a correct understanding of the principles.

It may be, however, that the representations in the simulations could be improved. Specificity as a general rule of thumb is intuitively useful. It is true that the metric developed here did not turn out to provide accurate predictions at the smallest level of granularity. However very poor results were achieved in another study (Dobson, forthcoming) that used Tarski's World (a far more expressive system). These results can easily be attributed to its relatively broad range of expression. The implication of this could be that the information enforcement metric is useful at predicting between systems with very large differences of expression. At a finer granularity, it might be that other factors, such as prior conceptions and knowledge, begin to interfere. Some simulations only provide concrete representations, others choose to plot changes in variables as a simulation progresses. The work here would suggest that this choice is the most important decision in creating pedagogic interfaces to simulations. One general rule that may help to improve learning from simulations is the development of representations which are complex enough only to show the key relations between variables in a simulation. Cheng argues that translation between his alternative representations of simulations for showing elastic collisions is core to successful learning. His choice of representations seems also to be quite limited in expression. Furthermore, the process of translation between representations effectively reduces expressivity as a whole. Usually a reduced set of meanings will be understandable in each of the representation languages. Translation, by itself, effectively reduces the expressivity of a communication system.

Another implication of this work for interactive teaching designs is the importance of the dynamic aspects of the graphics. As multimedia becomes more "active", there is increasing pressure to add "activity", even while it does not contribute to benefit for the learner. The dynamic nature of the representations of Euler has been underexamined in this work. It may well be that access to the limits of constraint in a simulation by means of graphical manipulation is more important even than formal measures of expressivity. O'Malley (1990) has distinguished between semantic directness and articulatory directness. An interface is semantically indirect if the user has to engage in a lot of planning and problem solving in using the system to perform some task. Articulatory directness refers to the extent to which the user's understanding of the meaning of expressions maps onto the form of expression required by the system. Her comments are directed at interface designs in general. The message applies to interactive diagrams even more strongly. The set of male instructional designers can be represented by a circle within another depicting all instructional designers. But it seems instructive that wherever you move the inner circle, none of its area may extend beyond the boundary of the outer circle. There is value in the meaning of that manipulation since it corresponds to the exploration of a key constraint. The metric is only a starting place along a path towards an ideal micro-media selection model envisaged by, among others, Dijkstra (1997). It has been shown that the specificity principle can contribute to a functional metric for media selection but that more information would need to be available to such a metric to make it reliable. We have shown how to use an interpretation of specificity for systems of representation that can be thought of as finite machines, however, most representations cannot be considered in such a way since the presence of unlimited strings in a system cause an unlimited number of states for these cases. An improvement in the interpretation could include complex state machines augmented transition networks and other devices to analyse non-finite systems. Through the use of these devices other principles might be looked at such as viscosity (Green, 1989). This principle is related to specificity in that it is concerned with the overhead a system causes to a learner but instead of predicting that overhead through a cardinal measure of ability to express domain information, viscosity is more concerned with the difficulty a user has in manipulating symbols to solve a problem. It is not immediately possible to see how this principle might be quantified or formalised, however Green (1996), has indicated that the search for a metric from this principle might be of benefit, both in developing the principle further and in leading towards a model of modality or notational choice.

The design of interactive diagrams may derive benefit from other areas of learning theory. Constraints and their manipulation have been very important areas of investigation in this project. These constraints are similar to the scaffolding that supports the practice of learning to become skilled in any cognitive field (Collins, Brown & Newman, 1989). The presence of these constraints may be treated in the same way as scaffolding. It may be that providing the constraints in early student interaction should be followed by the fading of those constraints later on. This aspect of design for interactive graphics has not been dealt with and should be included in any agenda for better understanding of diagrams in learning. The continued study of interactive graphical representation will prove useful in guiding the development of simulations, multimedia and better teaching.

6

The Role of Prior Qualitative Knowledge in Inductive Learning

Maarten W. van Someren and Huib Tabbers

1 Introduction

People acquire much of their knowledge from experience. Many authors have argued that knowledge that could be taught directly is learned better when the learner explores a domain and learns from experience, in particular if discovery is guided so that the learner encounters the relevant information (e.g. De Jong & Van Joolingen, in press; Mayer, 1987). This means that a learner is placed in the position of a researcher who is assigned the task of discovering the general principles and laws that explain certain phenomena. Just as scientific research, this is a difficult task. Compared with scientific research, the learner's task is simplified because the phenomena that are to be studied are relatively well defined and singled out from the immense range of events that we can perceive in the real world. As in scientific research, several different types of representation play a role. The learner can perform experiments in which aspects of the environment are manipulated and others are measured. The events during an experiment can be observed and the results of experiments can be represented in tables or graphs. From direct observation or from analysing tables and graphs, learners typically make qualitative discoveries ("A seems to have a positive effect on B" or "if C equals 0 then D has no effect on E") and they can also discover numerical laws (e.g. $B = A^2$). In regular teaching (without inductive discovery) this type of qualitative knowledge is normally presented and learned before the quantitative counterpart. Students are first taught, for example, that the speed of a falling object increases during its fall and after that they are taught the laws that describe this phenomenon.

In this chapter we focus on the role of qualitative knowledge in the induction of quantitative regularities (in particular laws of physics) in the context of science domains. Prior studies, especially of problem solving, have shown that qualitative knowledge is very important in these domains. Several authors have shown that, in the context of problem solving, qualitative analysis should precede quantitative modelling and problem solving. For example Ploetzner and Spada (1993, 1998) show that instructing students to perform qualitative analysis improves their problem-solving performance in physics. In the work by Ploetzner and Spada "qualitative analysis" refers to the application of abstract physics concepts to a concrete situation. For example, a balloon is viewed as a closed thermodynamic system with the associated properties. This is not so much a qualitative description as an abstraction: a concrete, observable object is described in very abstract terms from physics.

Less is known about the role of qualitative knowledge in induction. Klahr, Dunbar and Fay (1990) and Klahr, Fay and Dunbar (1993) show that some students tend to perform badly on an inductive discovery task because they do not systematically traverse the space of possible hypotheses. Qin and Simon (1995) report that their subjects are able to construct images from qualitative descriptions of problems and use these images to construct equations. At least for certain problems and persons this is easier than constructing equations directly from text. Schauble et al. (1991) observed a complex set of effects of prior knowledge on discovery learning. Apart from effect on experimentation strategies, which we shall not consider here, subjects with more prior knowledge state more predictions and explanations of phenomena that they observed during experimentation (with simulated electrical circuits), consider more possible experiments and are more persistent in finding explanations.

Although there is therefore substantial evidence that qualitative knowledge plays a role in the induction of quantitative knowledge, the precise role of qualitative knowledge is less clear. In the study by Schauble et al. (1991) the effect of prior knowledge was measured in a very general way and no detailed analysis was made of the role of this knowledge in learning. Interesting enough, the effects observed by Schauble et al. all involve strategic aspects of discovery: having more prior knowledge about a domain seems to lead to different discovery behaviour. In this study, prior knowledge was measured and not manipulated. The results show that across subjects, prior domain knowledge is either correlated with prior strategic knowledge or that prior domain knowledge somehow leads to better discovery strategies. Schauble et al. did not offer a theoretical analysis of the relation between these and these cannot be separated in the empirical data. Van Joolingen and De Jong (1997) showed that learners seem to operate in a search space of hypotheses that is structured from very general qualitative hypotheses ("A is related to B") via more specific hypotheses ("if A is higher then B is lower") to numerical hypotheses ("$A = 1/B^2$").

Why would prior qualitative and conceptual analysis improve problem solving and inductive learning? In the context of problem solving, it was observed that qualitative analysis was beneficial and, perhaps because this seemed consistent with the adagio "think before you act", no detailed explanation was given initially. Chi et al. (1989) give a cognitive explanation of this effect. Typical novice behaviour is to work backwards from what is asked in a problem. If the question at the end is "What is the speed when the ball hits the ground?" then a novice will look for a formula that contains "speed" and that can be used to solve the problem by finding the values of the other variables in the equation or by finding additional formulae containing these variables. This strategy tends to overlook two important aspects of a problem. The first is the "applicability" of a formula. Formulae often hold only under certain conditions ("ideal gas", "no friction") and the backward strategy may overlook this aspect because it focuses on the formula as mathematical objects and overlooks qualitative preconditions. A second and related problem is that the "mathematical" approach that is typical of novices tends to overlook the interpretation of a formula to a complex situation. For example, pumping a tyre is a form of compressing a gas. This suggests that this process can be described by Boyle's Law. But which gas is compressed? The gas inside the tyre? And does this happen with or without heat exchange with the environment?

Under which conditions will qualitative understanding of a domain support inductive learning about the domain? Here we shall explore this question and argue that a necessary condition is that the relation between qualitative and quantitative knowledge is known to the learner. Although for some domains this is relatively simple, there are many domains in which this is rather complicated. To understand the conditions under which particular qualitative knowledge is a useful first step for learning particular quantitative knowledge, we need a more detailed model of the role of qualitative knowledge. In the next section we first describe a rational model of inductive learning. We illustrate this with an induction task that will be used in the experiment described in the sections entitled "The Experiment" and "Results". We consider three models of discovery: a purely inductive search based model, a model that includes general heuristics and a model that includes prior qualitative knowledge about the domain.

2 Task Analysis

Discovering a domain by performing experiments is a complex task. The environment consists of a set of possible situations. Some aspects of the situations can be manipulated and others can be measured. The situation can be presented in many different ways. It can be a practical situation in which objects can be manipulated and visual observations can be made or measurement tools be applied. Usually, a symbolic representation is also used with abstract descriptions of practical situations, for example a table with the values of variables in different experiments. Another aspect is the goal of the task. It is not obvious what the goal of inductive discovery is. In general the task given to learners is to find "the" regularity in the situations. However, the learner cannot easily test if a regularity that he found actually holds for all possible observations. As noted by Popper (1959) there may always be an unobserved counterexample somewhere. There may even be no regularity at all (compare the attempts to discover the regularity in roulette) or there may only be a partial regularity so that many aspects will remain unpredictable and unexplainable.

Our analysis is derived directly from research in machine learning and the SDDS (scientific discovery as dual search) model (e.g. Van Joolingen & De Jong, 1997; Klahr & Dunbar, 1988; Klahr, Dunbar & Fay, 1990; Langley & Zytkow, 1990; Simon & Lea, 1974). One approach to this task, data-driven discovery, was explored by Langley and others (Langley & Zytkow, 1990). The aim of this research was to model scientific discovery behaviour. The methods that follow this paradigm are based on the following basic scheme. The data consist of a set of "experiments". Each experiment consists of values for a set of variables. The learner can generate "hypotheses". A hypothesis is an equation for (some of) the variables in the experiments. An equation is "consistent" with a set of experiments if the equation is true for the values in the experiment. The goal is to find an equation that is consistent with all experiments. There are two subtasks in discovery: finding "good" hypotheses and defining "good" experiments (or other observations). In this chapter we do not address the second aspect and we restrict ourselves to a situation in which the data of a number of experiments are given.

2.1 Induction without Prior Knowledge in Symbolic Domains

We first consider a relatively simple case in which the data of a number of experiments are available and the task is to find the regularity in the data in the form of an equation that holds for all experiments. There is no other prior knowledge about possible equations.

For a given set of experimental data, inducing the underlying law involves searching the space of possible equations. Because the data may be qualitative the equations may include non-numerical terms (for example Boolean functions). A criterion is needed to decide when an acceptable hypothesis is found. A generally used criterion is Occam's razor: the simplest hypothesis that covers the data is the best. If no *a priori* knowledge about the equation is available then induction involves searching all possible equations for the simplest one. The number of possible equations with a given set of variables is of course extremely large. This means two things: not literally all possible equations can be considered and a method is needed to keep track of the hypotheses that are inconsistent. In the literature a number of methods are described (see Langley & Zytkow (1990) for an overview).

2.2 Application of the Analysis to the Experimental Domain

Here we illustrate the analysis above in the context of the induction task that will be used in the experiment described in section 3. This task is formally analogous to the task of discovering the formula of lenses from data of experiments with lenses. The variables in the experiments are b, the distance between lens and image, f, the focal distance of the lens and v, the distance between lens and object. The formula relating these variables is $1/b = 1/f - 1/v$. The induction task is to discover this formula from a set of data of experiments with different lenses (different focal distances) and different object distances.

To prevent subjects from remembering the original formula (which is part of the standard physics curriculum in secondary school) a "cover story" was invented around this formula. This story went as follows:

> "A new weapon was developed, a kind of laser gun. A drawback of this gun is that it can only be fired in a horizontal position, so the beam it produces always goes straight ahead. To be able to hit an object on the ground it must be possible to deflect the beam. This can be done by putting a screen in the way of the beam. The beam then either passes through the screen or it is reflected, depending on the distance between gun and screen. If the beam passes through the screen, normally it is deflected. There is a "breaking point" when the beam passes right through the screen without being deflected. The distance between the gun and the screen at which this happens is called the "breaking distance". This distance is different for dif-

ferent screens. There is a formula that describes the relation between, on the one hand the distance between the gun and the screen (distance A) and the breaking distance (distance B) and on the other hand, the distance at which the beam hits the ground (distance C)."

The actual formula that relates A, B and C is the same as the formula for lenses: $1/C = 1/B - 1/A$. A different way to write this, and one that is perhaps a bit more natural if C is viewed as the effect of A and B is: $C = (A * B) / (A - B)$. The induction task is to discover this relation from the following data from experiments with the new weapon and screens with different "breaking distances".

The instructions accompanying the dataset (Table 1) state that the formula to be discovered must fit exactly on the data, and that there are no noise or measurement errors. Besides, the subjects were asked only to use these five operators "plus", "minus", "times", "division" and possibly some numbers as constants.

In this case it seems natural to try to write C as a function of A and B. This is already a choice and there are other possibilities but the cover stories suggest a causal relation of this form.

Consider the search space without heuristics. If we allow only the operators *, /, + and – we have 10 possible equations of A and B (division and minus can be applied in two ways and * and + can meaningfully be applied to a single argument). In general we have a search space with a branching factor of 10. At the second step we have $10 * 10 = 100$ equations and at the third step we have $10^3 = 1000$ equations. The actual figure of possible hypotheses is somewhat lower because some equations are

Table 1: Dataset of the induction task

Number experiment	**A (distance from gunman to screen)**	**B (breaking distance)**	**C (distance to where beam hits the ground)**
1.	2	5	–3.333...
2.	3	5	–7.5
3.	4	5	–20
4.	5	5	infinite[1]
5.	7	5	17.5
6.	10	5	10
7.	5	1	1.25
8.	5	2.5	5
9.	5	4	20
10.	5	6	–30
11.	5	10	–10
12.	5	15	–7.5

[1]Infinite means a division by zero in the formula.

equivalent. The correct equation is at this level. The complete search space is of course much larger than the approximately 1000 possible equations because much more complex equations can be constructed. The actual search space will even be considerably larger than this because we made several simplifying assumptions that may not hold for the actual task:

- We decided to write C as a function of A and B. Allowing other forms of equations increases the search space significantly.
- We did not consider including constants and considering these has an exponential effect on the size of the search space. Also we assumed that C can be written as a function of A and B. If this assumption is not made then there are many more possible equations. The exact size of the search space depends on which constants are allowed and if any restrictions are used in introducing constants. Each operation that allows the introduction of a constant to be included in a hypothesis increases the branching factor of the search space and results in an exponentially growing search space with a higher factor.
- We excluded threshold functions ("if Var1 > 0 then Var2 = Var3 * Var4"), other well-known mathematical functions such as logarithmic and trigonometric functions and constants. We shall not consider these more complex equations here (see Langley & Zytkow, 1990).

Even this basic search space is thus quite large and hard to manage. It is likely, however, that people can do better than more or less "blind" search by using heuristics. The nature and effect of these heuristics is discussed in the next section.

2.3 General Heuristics for Inductive Search

If it is decided in advance that only a subset of all possible equations will be considered, it is possible to reduce the search space and to define heuristics that guide the search. The result is that fewer possible equations need to be considered. Here we consider the heuristics that were used in the BACON family of discovery systems (see Langley & Zytkow, 1990). These systems rely on the following method. They are presented as a table with results of experiments. Each entry in the table consists of the values of certain variables that are observed in one experiment. The method also uses a relation between patterns in the data and functions of two variables. The method operates as follows:

```
For each pair of variables do:
   Test if a pattern exists in these data for all
      or some of the experiments
   If a pattern is found then construct a new variable
      that is defined as a function of the two variables
until a variable is constructed that is constant over all
experiments or no patterns are found
```

The heuristic knowledge consists of patterns associated with functions. Examples of such heuristics are "*if* the values of variables V_1 and V_2 are positively correlated *then* construct the new variable $V_i = V_1/V_2$" or "*if* the values of variables V_1 and V_2 are negatively correlated *then* construct the new variable $V_i = V_1 * V_2$". The advice leads to the construction of a new term that is intended to become part of the final equation. The new term can be viewed as a function of prior variables and it can be evaluated for the experiments. For example, when "V_1/V_2" is constructed then the value of this new term can be computed from the values of V_1 and V_2. This new value is added to the data and used for further induction. Because the new variable can be correlated with other variables, including new constructed variables and including its own input variables, this process can gradually construct more complex terms. For example if V_i (= $V_1 * V_2$) has a negative correlation with V_1, the heuristic suggests to construct

$$V_j = V_1 * V_i = V_1^2 * V_2$$

A similar heuristic (explored by Gerwin, 1974; Gerwin & Newsted, 1977) is known as the least-squares induction method, see for example Langley (1996). Experiments with these heuristics show that they reduce the effective size of the hypothesis search space and still make it possible to find a number of important physics laws.

2.4 The Role of General Discovery Heuristics in the Experimental Induction Task

To understand the effect of this on the induction task we reimplemented a version of the BACON method.[2] We built the heuristics from the BACON system into our model and gave the model the data in Table 1. The result was that the system did not solve the problem because its heuristics did not find any patterns or promising functions in the first step (and we did not include a weak search strategy as a default). Consider Table 2 to see why this is the case. In the table we list the value of some functions applied to the variables A and B that should explain C. BACON exploits the idea that these functions should be a better approximation of C than the original variables. However, the table shows that the values of these functions do not correlate strongly positively or negatively with C. Our original BACON implementation therefore abandons them. We adapted the system to search deeper in a breadth-first way but the system was not able to reduce search effectively over its breadth-first strategy. This shows that the basic heuristics of BACON are not strong enough to solve this induction task effectively. Human subjects behaving in the style of BACON are likely to give up on partial solutions because these do not seem promising.

This analysis shows that the inductive approach without heuristics results in a large search space and that the set of simple heuristics that was used for a machine dis-

[2]This implementation was built by Irma Houtzager and Wouter Hanegraaff and extended by Philip Lafeber, based on Langley and Zytkow (1990).

Table 2: Some simple functions of A and B

Number experiment	A – B	A + B	A * B	A/B	C
1.	–3	7	10	0.4	–3.333...
2.	–2	8	15	0.6	–7.5
3.	–1	9	20	0.8	–20
4.	0	10	25	1	infinite*
5.	2	12	35	1.4	17.5
6.	5	15	50	2	10
7.	4	6	5	5	1.25
8.	2.5	7.5	12.5	2	5
9.	1	9	20	1.2	20
10.	–1	11	30	0.84	–30
11.	–5	15	50	0.5	–10
12.	–10	20	75	0.33	–7.5

covery system is not adequate for this domain. The heuristics do not guide the search but only leave a "plateau" in the search space.

In the induction task that was used in the experiment, subjects received a little, almost explicit help in the form of the comment that the value "infinite" means a substitution in the formula that evaluates to 0 in the denominator. This suggests that the denominator is O – F or F – O. This gives away one of the three steps deep to be taken. In that case the size of the basic search space (no heuristics, only basic functions as operators) is reduced from $6^3 = 258$ to $6^2 = 36$! The other search spaces (with constants and other types of equations) are also reduced by a factor but remain rather large.

Next we consider the role that qualitative knowledge plays in inductive learning.

2.5 The Role of Qualitative Knowledge in Inductive Learning

What does qualitative, conceptual analysis mean and which role can it play in induction? First we note that there are different types of knowledge under the heading of qualitative knowledge. In physics practical work, learners use real world observations but also pictures, diagrams and qualitative symbolic knowledge. Here we focus on the notion of qualitative knowledge. By qualitative knowledge we mean qualitative symbolic descriptions of situations and general symbolic equations (or inequalities) that describe a set of situations (in the same way as a set of quantitative equations). Consider for example the law of falling bodies. Suppose that we have performed experiments with falling bodies. In each experiment an object is dropped from a tower at differing heights. Data were collected for "falling time", mass and height (Table 3).

Table 3: Example data for falling object experiment

Height	Time	Mass
1	9.8	1
2	0.5	2
4	0.64	2

Using bottom-up induction the law height = 9.8 * $time^2$ could be discovered from these data. However, suppose that we have the following prior knowledge:

1. *The mass of a falling object does not influence its falling time*
2. *The force that acts on a falling object is the force of gravity. This force is constant and it results in a constant increase in falling speed. If an object moves at constant speed then the time for moving over a distance d is a linear function of d. A constantly increasing distance will result in a quadratic function.*

In this case, prior qualitative knowledge can be used to infer the "form of function" for the hypothesis. This reduces the search space for this problem from the large initial space to equations of the form height = X * $time^2$ which makes it easy to find the actual law. More general, prior knowledge can be used to reduce the general hypothesis space to a subspace that is characterised by a class of equations. However, this requires knowledge of classes of equations and about their meaning in terms of observations, as in the example above. We argue that in general qualitative knowledge reduces the search space for inductive learning. Examples of particular techniques for this are the work by Clark and Matwin (1993) and Bredeweg et al. (1992). The idea behind these techniques is to filter candidate hypotheses by testing whether they are consistent with a known qualitative model. Experiments with these techniques show that this leads to more accurate and more comprehensible hypotheses than pure inductive techniques. Nordhausen and Langley (1990) also use the qualitative model in a generative way in the IDS system. The model is used to determine which variables of a situation can influence other variables. Only these variables are then used to define the search space for quantitative induction. In both approaches qualitative analysis reduces the "naïve" search space of quantitative relations between observable variables.

2.6 The Role of Qualitative Knowledge in the Context of the Experimental Domain

Let us return to our example of inductive discovery of the relation between three variables. Qualitative knowledge about this domain means that something is known about the relation between A, B and C but in qualitative rather than quantitative terms. The qualitative knowledge is formulated in terms of the story given in section 2.3. It can be summarised as follows. The meaning of the variables A, B and C was explained in diagrams like Figure 1.

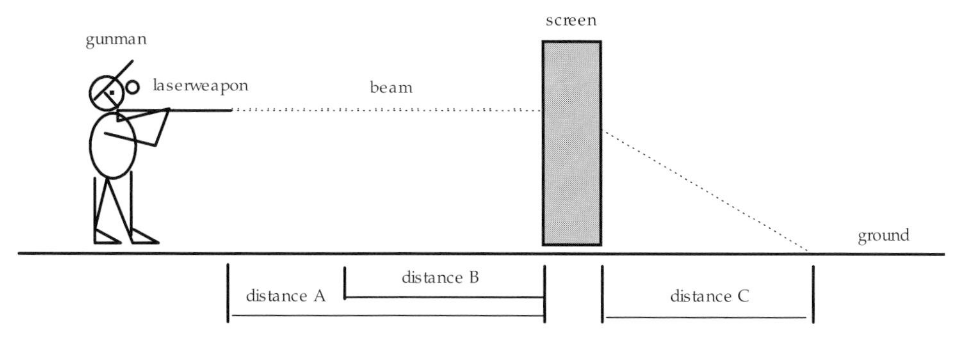

Figure 1: Diagram of a laser beam experiment

Qualitative knowledge consists of a set of qualitative relations between A, B and C. These relations were given in verbal form in the context of diagrams like Figure 1. Subjects were taught the following qualitative relations:

1. *if* $A < B$, *then the beam is reflected* ($C < 0$)
2. *if* $A > B$, *then the beam passes the screen* ($C > 0$)
3. *if* $A = B$, *then the beam passes the screen straight ahead* ($C = \infty$)
4. *if* $|A - B|$ *decreases, then* C *increases*
5. *if* $A > B$ *and* $|A - B|$ *increases, then* $C - B$ *decreases*
6. *if* $B > A$ *and* $|A - B|$ *increases, then* $C - A$ *decreases*
7. *if* $A = 2B$ *then* $C = A$
8. *if* $B = 2A$ *then* $C = B$

Following our analysis above, this qualitative knowledge should now be used in two ways: to characterise a class of possible equations (and only search this hypothesis set) or to use this to evaluate (partial) hypotheses that are generated by applying functions to variables. However, as the reader will note, in this case this reduction of the hypothesis space requires quite some mathematical knowledge, which most people will not have.[3] To the mathematically sophisticated, qualitative facts 3 and 4 suggest that a kind of hyperbolic function is involved and that a possible form of the function is

$$\frac{<\text{term}>}{A - B} = C \quad \text{or} \quad \frac{<\text{term}>}{B - A} = C$$

If we combine this with facts 1 and 2 we note that if $A < B$ then $A - B$ becomes negative. This suggests that we need a positive term at that point for <term>. How can we guarantee a positive term here? Possibilities with the operators that we have are $A + B$, $A * B$, A/B and also positive constants and the given functions plus a positive constant. This narrows down the set of functions considerably.

[3] We have not been able to specify how this qualitative knowledge can be used to characterise the set of possible (numerical) equations and we challenge the reader to do this!

This analysis implies that qualitative knowledge will only support induction if it can be used to reduce the hypothesis space. This may require quite some mathematical sophistication. Empirically, it implies that the qualitative knowledge above will not help the learners in discovering the law above. To test this implication, we conducted the experiment below.

3 The Experiment

The experiment uses the induction task introduced in section 2.2. To avoid large variations between subjects in the data used for induction, subjects were not allowed to select and perform experiments but they were all given the same set of data with the task to discover the regularity. Only students with some affinity with maths and natural science were asked to enrol for the experiment. Sixteen first-grade psychology students participated for course credits. Thirteen other students, most of them from science subjects, participated on a voluntary basis. Data of one person were discarded because this subject misunderstood the instruction. The experiment was run in small groups (three or four people), which were randomly assigned to the experimental or control condition. Table 4 gives an overview of the experiment.

Table 4: Overview of the experiment

Qualitative instruction **(n = 13; duration ± 40 min)**	***Control*** **(n = 16; duration ± 60 min)**
1. Introduction	1. Introduction
2. Qualitative instruction (10 min)	2. Induction task (20 min)
3. Studying qualitative model (10 min)	3. Qualitative reasoning test (5 min)
4. Induction task (20 min)	4. Exit questionnaire
5. Qualitative reasoning test (5 min)	
6. Exit questionnaire	

Half of the subjects (the control condition) received the dataset given in Table 1 and started with the induction task. The other half received qualitative instruction in the form of 12 pictures of experiments with the new gun. With each picture, a qualitative description was given. In these experiments the distance between gun and screen and the "breaking distance" of the screen were varied and the effect was shown on where the beam hits the ground. The following illustrates the qualitative instruction. Two sample figures are given with corresponding instruction text and questions:

1.

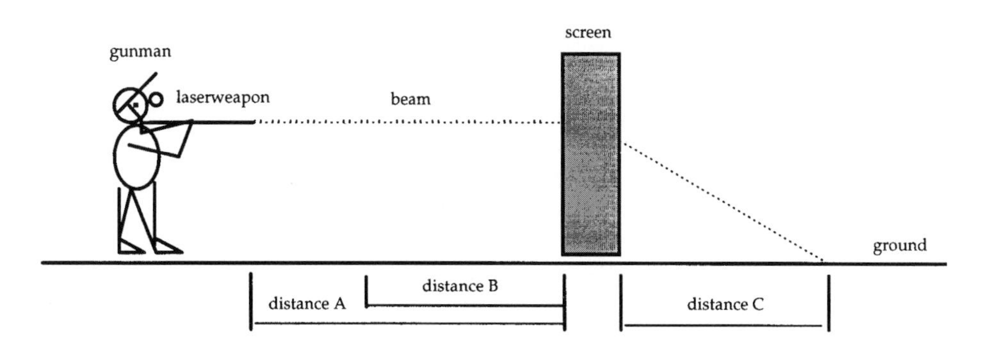

2.

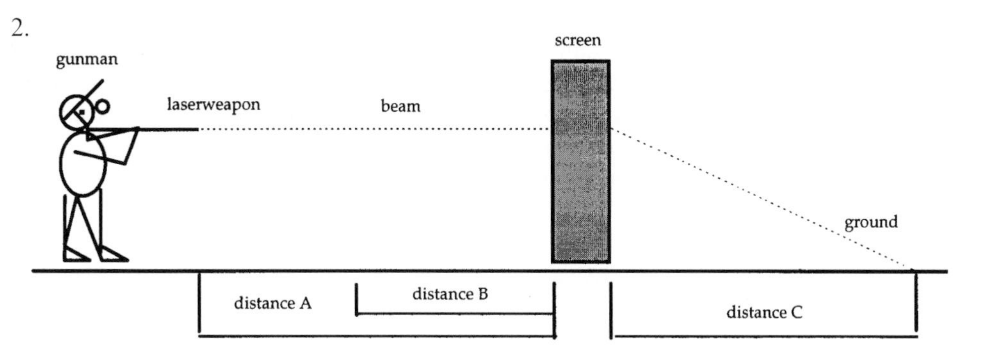

Sample figures:
A > B; B remains constant; A increases

Sample instruction text:
In these experiments the screen and therefore the breaking distance is the same. In the second picture the man with the gun is further from the screen. In both pictures the man with the gun is more than B away from the screen. As you can see, the distance between the screen and the point where the beam hits the floor is also increasing.

Example questions:

1. *If A is larger than B then the beam hits the ground . . . of the screen*
- *right*
- *left*

2. *If A increases B then C will . . .*
- *increase*
- *decrease*
- *remain constant*

Each page in the qualitative instruction contained three figures of "laser weapon" experiments that illustrate a particular effect (the effects of distance between gunman and screen and of different screens). From each set of three figures of the laser weapon experiments, subjects had to answer three questions that could be answered from the diagram. The purpose of these questions was to make certain that subjects would pay attention to the important aspects of the figures and that they would learn the qualitative relations that are illustrated in the figures. For this they had 10 min. The results showed that subjects answered these qualitative questions almost perfectly (average 14.5 correct on a scale that ranged between 5 and 15; the lowest score was 13 correct). Then they got another 10 min to study an overview of the qualitative knowledge that reviewed all the qualitative relations between the variables. This summarised the qualitative relations given in Table 2. Only after this did they receive the dataset. All subjects were given calculators, pen and paper. At the end of the dataset a remark was added, explaining that "infinite" means a division by zero in the formula. As soon as a subject had discovered the formula, he or she wrote it down on the answering form and the experimenter recorded the time. There was a limit to the time that the subjects could search for the formula (20 min), to prevent subjects from getting stuck. When someone discovered the formula in less than 20 min the time was recorded and he or she was asked to relax until the 20 min were over. In case a subject had not discovered the right formula in time, he or she was asked to write down any idea they had formed about the formula's shape.

After the induction task all subjects were given a qualitative reasoning test, in which they had to make predictions about new experimental situations. The goal of this test was to measure qualitative knowledge about this domain. It consisted of nine multiple-choice questions about experimental situations that were not in the qualitative instruction. There were three such situations and three questions about each situation. These were phrased in the same style as those in the qualitative instruction. Subjects had five minutes to answer these questions.

Finally, all subjects were asked to fill in an exit questionnaire. In this questionnaire subjects were asked about their maths level and their experience with induction tasks.

4 Results

One subject was removed from the sample, because the exit questionnaire showed that she had not understood the instructions of the experiment. The total number of experimental subjects is 28 with 15 in the experimental and 13 in the control condition. The exit questionnaire showed that there was no difference between experimental condition and control condition in science background. The results of the induction task are depicted in Table 5.

The results clearly show that the experimental condition with qualitative instruction does not perform better in discovering the law. In fact the control condition does slightly better but the difference is small and not statistically significant (difference between proportions is -0.215; with a normal approximation this gives $P(Z < -1.13) = 0.13$). The exit questionnaire showed that seven subjects, four in the experimental

Table 5: Observed frequencies on induction task

	Found correct formula	Wrong/no formula	Total
Experimental condition	$n_{expc}=6$	$n_{expw}=9$	$n_{exp}=15$
Control condition	$n_{conc}=8$	$n_{conw}=5$	$n_{con}=13$
Total	$n_c=14$	$n_w=14$	$n=28$

and three in the control condition, had recognised the lens formula before or while making the induction task. When we remove their results from the table of frequencies, the difference becomes 3/11 – 5/10 = –0.227 (which also is not significant). The same pattern emerges when we look at the time. It took the subjects in the experimental condition on the average 12.7 min (SD = 7.5) to discover the formula, while the subjects in the control condition only needed 6.2 min (SD = 2.4).

Table 6 shows the results of the qualitative knowledge test that was administered to all subjects at the end of the experiment. The table gives means scores and standard deviations for various groups of subjects.

Table 6: Results of qualitative test

	Experimental	Control	Total
Found correct formula	6.0 (1.1)	5.5 (1.2)	5.7 (1.1)
Found incorrect formula or no formula	5.2 (1.9)	3.2 (0.8)	4.5 (1.7)
	5.5 (1.6)	4.6 (1.6)	

When we compare the scores, we see, however, that in the experimental condition (mean 5.5) the score is higher than in the control condition (4.6). This difference is not significant (two-sided *t*-test gives $p=0.14$). Inspection of the table shows that the subjects who did not receive qualitative instruction and who did not discover the formula performed poorly on the qualitative reasoning test. The subjects who only did the inductive learning task and who did discover the correct formula performed as well as the subjects who also received qualitative instruction. There are several possible alternative explanations of these findings. One is that the qualitative instruction did not work in the sense that subjects did not actually acquire qualitative knowledge about the domain. However, this explanation is not plausible because they answered the qualitative questions during the instruction almost perfectly. A second possibility is that in spite of random assignment to conditions the subjects differed widely on relevant characteristics. In the exit interviews subjects indicated which relevant science subjects they studied at secondary school and also their experience with induction tasks. We defined a simple scale with two levels to measure the amount and level of science and maths which students studied at school (A did less mathematics and physics than B). There was a significant relation between this scale of "science subjects at school" and performance on the induction task (Table 7).

Table 7: Association between science level and discovery of correct formula

Science subjects at school	A	B
Found correct formula	5	9
Incorrect or no formula	10	4

Table 8: Distribution of science level over conditions

Science subjects at school	A	B
Experimental condition	8	7
Control condition	7	6

A chi-square test of Table 7 gives $P = 0.06$. However, the distribution of levels A and B over the two conditions in the experiment was about equal (Table 8).

Although this factor is relevant it does not explain the difference between the two conditions because the distributions in the two conditions are about equal. We also asked subjects about their experience with similar tasks. The results show the same pattern as above (Tables 9 and 10).

Here again there is a substantial association between prior experience with induction tasks but there is no substantial difference between the conditions. Note that these results should be interpreted with some care because the answers can be influenced by subjects' experience in the experiment.

We conclude that the qualitative knowledge gathered in the qualitative analysis has had no positive effect on the performance on a quantitative induction task and that subjects who only performed the induction task and found the correct formula performed at least as well as the group that received qualitative instruction *and* performed the induction task.

Table 9: Association between experience with induction and discovery of correct formula

Experience with induction	No	Some	Much
Found correct formula	3	7	4
Found no or incorrect formula	2	11	1

Table 10: Association between experience with similar tasks and discovery of correct formula

Experience with induction	No	Some	Much
Experimental condition	2	13	0
Control condition	3	5	5

5 Discussion

The experiment shows that qualitative knowledge about the domain does not always improve performance on an inductive discovery task in the same domain. According to our analysis in section 2, subjects should be able to apply qualitative knowledge to reduce the set of possible equations that are considered during induction. However, as we pointed out, this is extremely difficult for this particular domain and we may assume that it is beyond the capabilities of our subjects. Acquiring the qualitative knowledge might even have a negative effect, because subjects may spend time during induction trying to use their qualitative understanding. Directly starting off to search in the space of possible equations will result in shorter solution times and better performance, especially if the clue can be used that the value *infinite* means a division by zero.

This interpretation that this is due to the lack of knowledge which relates qualitative understanding to the form of the equation receives some support from other data that were collected in the experiment. Six of the 15 subjects in the qualitative condition indicated that they had no idea about the form of the equation to be discovered. One subject in this condition thought that there would be a logarithm in it and four remembered the lens formula from their physics education. Three subjects inferred that the formula should contain a division by a difference between A and B. Only two subjects reported the qualitative rules as an answer to the question which information had been most helpful in finding the formula.

The difference on the test of qualitative knowledge shows that inductive discovery of the regularity in the data does not take these subjects up to the level of the qualitative knowledge group with respect to performance on qualitative knowledge questions, although the qualitative predications follow from the (correct) equation. The results show that in particular subjects in the condition without qualitative knowledge who did not discover the correct formula did badly on the test. Those who did discover the formula performed as well on this test as the qualitative knowledge group.

Qualitative understanding can indicate which variables are irrelevant and say something about the form of the relation (the type of function) which in turn simplifies the learning task. However, this will only work if a person has the knowledge to relate qualitative or visual understanding to the mathematical formula. This requires a certain mathematical knowledge. For example, the notion of "constant acceleration due to constant force of gravity" suggests that velocity of a falling object will grow linearly with time and that the position of the object will change with the square of time. In this case, the relation between the concepts force, position, velocity and acceleration is relatively simple as compared with the behaviour of light in lenses. It is almost impossible to derive the form of the formula from qualitative knowledge about object distance, focal distance and image distance. Unlike other studies that demonstrate more in general the difficulty of relating different representations to each other (e.g. Schijf and Simon, Ainsworth, Bibby and Wood, and Boshuizen and Van de Wiel, all this volume), in this case we can point out precisely which knowledge is required for this mapping, which gives a detailed explanation of that fact that instruction does not really work here. Without this knowledge qualitative understanding will not reduce the search space for the formula and that trying to do this will only cost time, leading to longer inductive learning times than simply trying out possibilities. Even worse, it is unlikely that discovering the law of lenses from the data will have little effect on qualitative and visual understanding.

The phrase "qualitative knowledge" has several different meanings. In our analysis and in the experiment we used this phrase in its original meaning, in contrast to quantitative knowledge. One can argue that the experimental group was taught knowledge that is not only qualitative but also "general" and "visual" or "pictorial" and that it does not include any abstract physics concepts. Semantically, the qualitative knowledge presented with the figures in the experiment represents generalisations of the pictures. It states general versions of relation of which the pictures are instances. An interesting question that deserves further exploration is if people reason about this domain using the symbolic, qualitative knowledge or using the visual knowledge (or a combination). The qualitative physics knowledge in our experiment does not involve any abstract physics concepts (or, more precisely, it does not involve the relation between abstract concepts and concrete situations). For example, in the domain of optics of which our discovery task was derived, the concept "focal distance" is an important abstract concept that is defined in terms of the underlying theory of the behaviour of light. In our experiment the concept focal distance appears only as a property of objects that is not further defined. We believe that this is a suitable way to study inductive learning but it does not address the aspect of acquiring abstract concepts.

An issue that we did not address here but that is important from a practical viewpoint is the role of experimentation. The data of laser weapon experiments that we gave our subjects in the induction task corresponded to those of sensible experimentation. One variable was kept constant while the other was varied. As we noted in section 1, there is substantial evidence that prior knowledge plays an important role in the task of defining informative experiments. However, even very informative experiments will not help if a person cannot apply qualitative knowledge to the induction task and we expect that allowing people to generate experiments will result in larger variation between subjects but not in a strong effect of qualitative instruction on performance on the induction task. However, we suspect that guiding experimentation must also be based on a structure imposed on the hypothesis space and this structure will have to rely in part on mathematical concepts, in particular, classes of equations and their properties. Since we expect that our subjects lack the mathematical concepts to structure the hypothesis space, we expect that allowing subjects free experimentation will not change the effects that we found in this experiment.

5.1 Implications for Education

From an educational viewpoint discovering a law is not the main goal of inductive discovery learning. This form of learning is viewed as a means to teach something else. Our analysis suggests that, in terms of the types of knowledge that we distinguish, learners should acquire at least three types of knowledge: (a) the general formal, mathematical model (in this case the equation and its mathematical properties), (b) general qualitative scientific knowledge (e.g. "focal distance", "acceleration") and (c) the mapping between properties of the formal model and the more intuitive qualitative model. If the knowledge to construct this mapping is not available, students will inevitably acquire separate formal and intuitive qualitative models, even when discovery learning is used for teaching. Our analysis suggests that mathematical concepts (in particular properties of equations) are needed to construct this mapping. In general, as argued by De Jong and Van Joolingen (in press), this requires a rather detailed analysis of the knowledge involved.

Acknowledgements

The authors want to thank Ton de Jong and Hans Spada for their comments on an earlier version of this chapter.

7

Analysing the Costs and Benefits of Multi-Representational Learning Environments

Shaaron E. Ainsworth, Peter A. Bibby and David J. Wood

1 Introduction

Software that employs multiple external representations (MERs) has become increasingly available at all levels of education. Common packages such as spreadsheets employ tables, equations and many forms of graphs. Learning environments such as Geometry Inventor (LOGAL / Tangible Math) allow tables and graphs to be dynamically linked to geometrical figures. Function Probe (Confrey, 1992) provides graphs, tables, algebra and calculator keystroke actions and allows students to act upon any of these representations. One of the biggest areas of expansion in educational software is with multi-media technologies. By definition, these systems involve MERs, often including video and spoken text. Even traditional classroom uses of MERs, such as using an equation to produce a table of values which can then be plotted as a graph, have been significantly altered by the introduction of graphical calculators.

Given this growth of multi-representational software, it is appropriate to ask what evidence is there that providing learners with MERs facilitates understanding. Although research suggests that MERs are advantageous when learning, much less is known about the conditions under which MERs are beneficial. Consequently, designers and educators have few principles to guide their use of MERs.

We propose that one significant factor hampering the development of generalised principles for learning with MERs has been the failure to recognise that MERs are used for quite distinct purposes. Consequently, while there has been much research on individual examples and some theoretical explanation, an integrative framework has been slow to develop.

In this chapter, three different ways in which MERs can be used to support learning are considered. In addition, we consider how these benefits of employing MERs can be counteracted by their associated costs. Two multi-representational learning environments are presented and we discuss how they have been evaluated in order to examine under what conditions MERs aid learning.

2 Uses of Multiple Representations

We propose that there are three distinct classes of use of MERs: (a) that MERs support different ideas and processes, (b) that MERs constrain interpretations and, (c) that MERs promote a deeper understanding of the domain.

2.1 Different Ideas and Processes

A common use of MERs is when the information varies between the representations in the multi-representational system, so each representation serves a distinct purpose. For example, "MoLE" (Oliver & O'Shea, 1996) is a learning environment to teach modal logic. MERs are used to express different information about a domain which ultimately must be described in sentential form. One representation is used to express the relation between different modal worlds, and another to illustrate each world's content. In this case, the two representations describe different information. This approach to using MERs is ideal if a single representation is insufficient to carry all the information needed about the domain or a representation which, if it did so, would be too complicated for people to interpret.

A second example of this use of MERs is to support different processes. This follows from the fact that representations that are informationally equivalent may still differ in their computational properties (Larkin & Simon, 1987). For example, they proposed that diagrams exploit perceptual processes, by grouping together relevant information, and hence make processes such as search and recognition easier. Further research has shown that other common representations differ in their inferential power (Cox & Brna, 1995; Kaput, 1989). For example, tables tend to make information explicit, emphasise empty cells (thus directing attention to unexplored alternatives) and highlight patterns and regularities. The quantitative relationship that is compactly expressed by the equation "$y = x^2 + 5$" fails to make explicit the variation which is evident in an (informationally) equivalent graph. Therefore, by using MERs we can obtain the different computational properties provided by each of the individual representations in the system.

That MERs can support different information and processes provides learners with a number of advantages to facilitate their achievement of learning goals. These advantages can be seen in task, strategy and learner variables.

Learning objectives often require multiple tasks to be undertaken. However, a single representation is rarely appropriate for all of these tasks. Gilmore and Green (1984) proposed the match–mismatch conjecture — that performance will be facilitated when the form of information required by the problem matches the form provided by the notation. Evidence for this proposal has been obtained in a number of domains. For example, Bibby and Payne (1993) examined how different, informationally equivalent, representations (table, procedure, diagram) supported acquisition of various aspects of device knowledge. Looking at performance on a simple control panel device, they found cross over effects. Subjects given tables and diagrams identified faulty components faster. However, those given procedures were faster at deciding which switches were mispositioned.

Research that examined the relation between different representations and strategies has also provided support for the use of MERs. Tabachneck, Koedinger and Nathan (1994) showed that the different representations used to solve algebra word problems were associated with different strategies. No single strategy was more effective than any other, but the use of multiple strategies was about twice as effective as any strategy used alone. As each strategy had inherent weaknesses, switching between strategies made problem solving more successful by compensating for this. Cox (1996) observed a similar effect when students solved analytical reasoning problems. He found that subjects tended to switch between representations at impasses and on difficult problems.

Finally, it is often proposed that there are individual differences in representational and strategic preference. Thus, if MERs are provided, users could act upon the representation of their choice. Research examining the impact of various personality or cognitive factors in relation to learning with external representations has proposed, differential effects of IQ, spatial reasoning, locus of control, field dependence, verbal ability, vocabulary, gender and age (see Winn, 1987). A common (although by no means invariant finding) is that learners defined as showing less aptitude in the domain benefit from graphical representations of the task (see Cronbach & Snow, 1977; Snow & Yalow, 1982).

2.2 Constraints on Interpretation

A second use of MERs is to help learners develop a better understanding of a domain by constraining interpretation that can be made about other representations and the domain to be learnt.

An additional representation may be employed to support the interpretation of a more complicated, abstract or unfamiliar representation. The second representation can provide support for a learner's missing or erroneous knowledge. For example, microworlds such as DM^3 (Henessey et al., 1995) provide a simulation of a skater alongside a velocity–time graph (among other representations). Two misconceptions common to children learning Newtonian mechanics are that a horizontal line on a velocity–time graph must represent a stationary object and that negative gradient must entail negative direction. These misinterpretations of the line-graph are not possible, however, when the simulation shows the skater still moving forward. A further example is that of Yerushalmy (1989) who describes a multi-representational learning environment for teaching algebraic transformations. It presents users with an algebraic window where they transform algebraic expressions. It also provides three graphs: the first displays a graph of the original expression; the second displays the current transformed expressions and; the third describes any difference between the two expressions. Consequently, learners are encouraged to check that their transformations are correct as graphs should not change if a transformation was legal.

A second use of MERs to constrain interpretation is when one of the representations enforces a different interpretation of the situation. For example, some representations permit less expression of abstraction. To use an example based on Johnson-Laird's research (e.g. Ehrlich & Johnson-Laird, 1982), the ambiguity in the propositional representation "the knife is beside the fork" is completely permissible. However, an equivalent image would have to picture the fork as either to the left or to the right of the knife. Thus, when these two representations are presented as a multi-representational system, interpretation of the first representation must be constrained by the second when the representational system is considered as a whole.

2.3 Deeper Understanding of the Domain

Kaput (1989) proposed that MERs may allow learners to perceive complex ideas in a new way and to apply them more effectively. By providing a rich source of representations of a domain, learners can be provided with opportunities to build references across these representations. Such knowledge can be used to expose the underlying structure of the domain represented. On this view, mathematics knowledge can be characterised as the ability to construct and map across different representations. Similarly, Resnick and Omanson (1987) suggested that mapping between representations plays an important role in developing a more abstract representation that encompasses both. When they describe the process of abstracting over Dienes blocks and written numerals, it is the quantities that both representations express that permit mapping.

Schwartz (1995) provides interesting converging evidence that multiple representations can generate more abstract understanding. In this case, the multiple representations are provided by different members of a collaborating pair. With a number of tasks, he showed that the representations that emerge with collaborating peers are more abstract than those created by individuals. One explanation of these results is that the abstracted representation emerged as a consequence of requiring a shared representation that could bridge both individuals' representations.

Therefore, although research with this aspect of MERs seems more speculative than research on the first two purposes of MERs, evidence from both individuals and pairs suggest that an abstracted understanding can result from working with MERs.

As we have seen, there are many different reasons why MERs should be beneficial for learning. Furthermore, MERs used in a single learning environment may fulfil two or more of these purposes simultaneously. For example, representations used to describe different aspects of a domain may also encourage abstraction if learners can map over them. However, employing MERs in a learning environment does not come without associated costs. In the next section, we will consider what additional learning demands may be involved in using MERs.

3 Costs of Multiple Representations

Learners are faced with three tasks when they are presented with MERs. First, they must learn the format and operators of each representation. Secondly, learners must come to understand the relation between the representation and the domain it represents. Finally, and uniquely to MERs, learners must come to understand how the representations relate to each other.

To use a representation successfully, learners must understand how a representation encodes and presents information (the "format"). In the case of a graph, this would be attributes such as lines, labels and axes. They must also learn what the "operators" are for a given representation. For a graph, operators to be learnt include how to find the gradients of lines, maxima and minima, intercepts, etc. At least initially, such learning demands will be great. For example, Preece (1983) reports that 14–15-year-old children had problems using graphs due to basic problems with applying and understanding the format and operators. For example, some pupils experienced difficulties with reading and plotting points, they interpreted intervals as points, confused gradients with maxima and minima, etc.

Learners must also come to understand the relation between the representation and the domain it is representing. This task will be particularly difficult for learning with MERs as opposed to problem solving or reasoning, as learners will also have incomplete domain knowledge. To return to the graph example, children must learn when it is appropriate to examine the slope of a line, the height or the area under a line. For example, when attempting to read the velocity of an object from a distance–time graph, children often examine the height of a line, rather than the gradient. Additionally, the operators of one representation are often used inappropriately to interpret a different representation. A representation of graph may be interpreted using the operators for pictures. This behaviour is seen when children are given a velocity–time graph of a cyclist travelling over a hill. Children should select a U shaped graph, yet they show a preference for graphs with a hill shaped curve (e.g. Kaput, 1989).

The final learning demand, unique to multi-representational situations, is that when MERs are presented together, learners must come to understand how representations relate to each other. Without abstraction across representations, the invariances of the domain may remain hidden. Some multi-representational software has been designed to teach such relations. For example, Green Globs (Dugdale, 1982) provides opportunities for learners to relate graphs to equations. A computer displays coordinate axes and 13 "green globs". Students must generate equations that hit as many of these points as possible. This type of learning environment is common when the relation between representations is difficult. Grapher (Schoenfeld, Smith & Arcavi, 1993) consists of three micro-worlds: (a) Black Globs (similar to Green Globs described above), (b) Point Grapher which allows students to define function (e.g. "$y = 2x + 3$") and produces tables and graphs and, (c) Dynamic Grapher, where families of function (e.g. "$y = mx + b$") can be explored graphically. The instructional goal of the micro-world is essentially to develop the complex set of mappings that describe the relation between graphs and algebraic expressions ("the Cartesian Connection" in Schoenfeld's terms).

Other environments have been designed to exploit translation to some other instructional end. One example for the primary classroom is that of the Blocks World (Thompson, 1992) which combines Dienes blocks with numerical information. Users act in one notation (such as the blocks) and see the results of their actions in another (numbers). Thompson found that average and above average students developed better understanding of the number system structure and algorithms than students who had used a non-computerised version.

However, a number of researchers have noted the problems that novices have in learning the relation between representations. Tabachneck, Leonardo and Simon (1994) report that novices learning with MERs in economics did not attempt to integrate information between line graphs and written information. Students' performance on quantitative problems, where answers could be read off from graphs, was good, but it was poor on problems requiring explanation and justification. A similar pattern of results was found for graph generation as well as interpretation. This contrasted with expert performance where graphical and verbal explanations were tied closely together. Similarly, Yerushamly (1991) examined 35 14-year-olds' understanding of functions after an intensive 3-month course with multi-representational software. In total, he found that only 12% of students gave answers which involved both visual and numerical considerations.

Consequently, we can see that for users to benefit from multi-representational software, they must meet a number of very complicated learning demands. In particular, it seems that learners may find translating between representations particularly difficult.

We will now describe two systems that employ MERs within primary mathematics and discuss studies which have examined whether the advantages of MERs can be obtained without being outweighed by the costs.

4 Empirical Studies

Both of the systems used for experiments described in this chapter teach aspects of primary mathematics. Recent approaches to mathematics instruction in the primary classroom emphasise mathematics as flexible, insightful problem solving. This approach requires understanding that mathematics involves pattern seeking, experimentation, hypothesis testing and active seeking of solutions. However, research shows that children hold very different beliefs about the nature of mathematics. For example, Baroody (1987) asserts that due to an overemphasis on "the right answer", children commonly believe that all problems must have a correct answer, that there is only one correct way to solve a problem and that inexact answers or procedures (such as estimates) are undesirable. Our systems aim to address and challenge these beliefs. The first system we discuss examines the idea that there is only one correct way to solve a problem and the second system supports children's learning about estimated solutions.

4.1 Multiple Solutions to Mathematical Problems

COPPERS teaches children to provide multiple solutions to coin problems. This apparently simple task is difficult for primary school children (Ainsworth, Wood & O'Malley, 1998). COPPERS design is based on aspects of Lampert's classroom teaching (e.g. Lampert, 1986). COPPERS operates in the familiar and mathematically meaningful domain of money problems. Users are posed problems such as "what is $3 \times 20p + 4 \times 10p$?" They answer these questions by pressing buttons on a "coin calculator" whose buttons are representations of British coins. The coin calculator is designed to provide a simple way of interacting with the system. It is both easy and fun to use and acts to reduce the burden of number facts. So to answer the above problem, a user may select "20p + 20p + 10p + 50p" or "10p + 2p + 2p + 1p +

5p + 10p + 10p + 10p + 50p". Either answer is equally acceptable as there is no notion of "best" answer in the system. Users do not progress to new problems until they have produced a number of different solutions.

COPPERS provides MERs of learners' answers. They reveal whether an answer was correct and describe the elements of the solution in detail. This information is displayed in two ways and, by means of highlighting the key elements, the system encourages students to map between the different representations. The first representation (RHS Figure 1) is a common one in the primary classroom, and will be referred to as the place value representation. The user is reminded of how many of each type of coin they used. The operations of multiplication and addition needed to produce the total are made very explicit in the place value representation, making the arithmetical operations one of the most salient aspects of the representation.

The second representation (LHS Figure 1) is tabular and similar to the one described by Lampert. In contrast to the place value representation, the summary table is less familiar to the children and the arithmetical operations are implicit. To understand and make use of the information, children must decide what processes are involved and perform them for themselves, hence practising their multiplication and addition skills. The table also displays previous answers to the question. This allows students to compare their answers with those already given and, it is hypothesised, prompt pattern seeking and reflection.

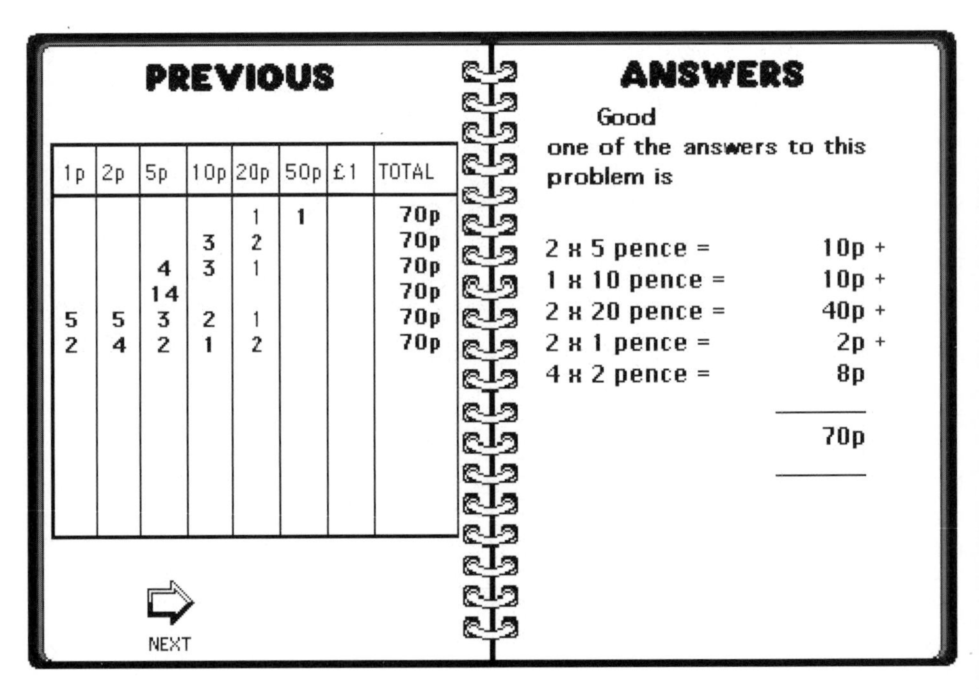

Figure 1: COPPERS — place value feedback and the summary table

These representations are primarily used to make different information salient. In addition, if learners can translate between the representations then they can constrain interpretation. First, interpretation of the tabular representation can be supported by children's understanding of the first representation. The familiarity of the place value representation can constrain the possible misinterpretations of the unfamiliar table representation by indicating the appropriate format and operators for the table representation. Secondly, the two representations can constrain interpretation by exploiting differences in the inferences supported by the representations. Coin problems such as "5p, 10p, 5p, 10p" and "5p, 5p 10p, 10p" may appear very different to a young child if they do not understand commutativity. The tabular representation of coin values used in COPPERS does not express ordering information. Therefore, if children translate between the representations, the equivalence of the two different orderings in the place value representation is more likely to be recognised.

Learning demands were intended to be kept to a minimum with these representations. One representation is familiar (even if the place value concept it embodies is not) to children of the intended age range for COPPERS. The second representation is likely to be less familiar, but learning demands of interpreting the table may be supported by presenting it alongside the familiar representation. Signalling how the representations relate to each other should hopefully reduce the third translation demand.

However, the advantages proposed for including an additional tabular representation to produce a multi-representational system will only occur if children can meet the learning demands imposed by the extra representation. Moreover, it is known that young children have difficulty in working with tabular representations and often fail to use them successfully (e.g. Underwood & Underwood, 1987). Hence, an experiment with COPPERS explored whether including an additional tabular representation improved learning outcomes for children of primary school age.

The study was conducted with 40 6–7-year-old children in a local primary school. Initially, children were pre-tested with multiple solutions coin problems and required to draw coins to make up the total value of the sums. As predicted, initial performance was surprisingly low for such an apparently simple task. Although, there are literally hundreds of different correct answers for each of these problems, children produced on average less than one correct answer per problem.

An intervention phase followed where children used COPPERS twice. Half the children used a version with only row and column representations and half with both the place value and tabular representations. A further factor controlled whether children gave one or many answers per question. Hence in some conditions, the table displayed a trace of several answers and in some only one (see Ainsworth, in press). At post-test, we found that all children had improved their scores significantly, producing over four times as many correct solutions per question. However, children who had seen a table in addition to the place value representation produced significantly more, and more, correct solutions compared with those who had not. Hence, children as young as 6 can learn more effectively to produce multiple solutions when using MERs in comparison to a single representation.

This study demonstrated that appropriate combinations of representations can lead to increased learning outcomes. It was proposed that this result showed that the design aim of keeping the learning demands of the representations low had been met successfully, particularly the translation learning demand. In the next experiment, we examined further what properties of representational systems affect how learners translate across representations.

4.2 Computational Estimation

CENTS (see Figure 2) has been designed to help children learn the basic knowledge and skills required to successfully perform computational estimation (e.g. Reys et al., 1982; Sowder & Wheeler, 1989). In particular, it aims to teach some basic estimation strategies and to support the development of the conceptual knowledge that underpins successful performance in estimation. A central goal is to encourage understanding of how transforming numbers to produce estimations affects the accuracy of the

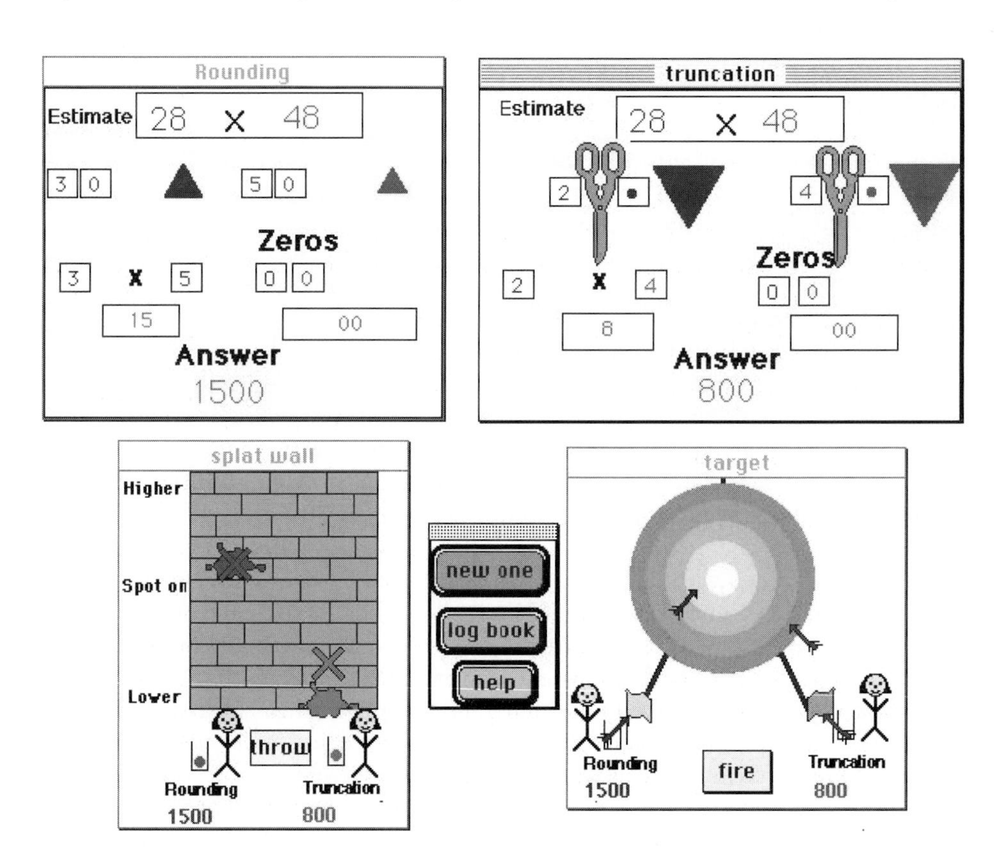

Figure 2: A completed problem with CENTS using pictorial representations

solution. This objective is supported by the use of MERs of estimation accuracy. CENTS employs a hypothesis testing strategy to encourage this understanding. Users make predictions about the accuracy of an estimate, then produce an estimate and finally compare the results to their predictions. To help them learn about accuracy and to address conceptual issues, users produce at least two different estimates for each problem.

CENTS supports this process by guiding the learner's selection of an appropriate intermediate solution, supporting place value correction, and by monitoring number facts, providing help where appropriate. After each problem, children log the results of the different estimation strategies in an on-line work book. At the end of a session, children are encouraged to review the log book to investigate patterns in their estimates.

The prediction and comparison stages are supported by MERs of estimation accuracy which are used for both action and display. The primary purpose of these MERs is simply to support different ideas and processes by allowing a variety of different views on the task to be displayed. However, they may also help children to come to understand the domain more fully if children can translate between or abstract over them.

All representations are based on the percentage deviation of the estimate from the exact answer [(exact answer − estimate / exact answer) × 100]. No matter how the nature of surface features of the representations differ, the deep structure is always based on this relationship. The system can display this information in a variety of ways. Representations can be mathematical ones such as numerical displays or histograms or they can be pictorial representations (e.g. an archery target). It is also possible to present combinations of pictorial and mathematical representations.

Further dimensions that can be manipulated include the resolution of presented information (in these cases it was either categories of 10% deviations or continuous representations accurate to within 1%) and the amount of information per representation (e.g. representations which either display direction or magnitude separately or display both dimensions simultaneously). This is particularly interesting when considering multi-representation systems because it allows for different levels of (informational) redundancy across representations.

These dimensions have been manipulated to address how features such as similarity of format and redundancy between representations affects ease of translation between representations. This allowed us to examine predictions about the effectiveness of different combinations of representations. For example, in Ainsworth, Wood and Bibby (1996) we examined pictorial, mathematical or mixed representations that were partially redundant. Unlike the COPPERS experiment reported above, automatic translation was not provided by the system as the experiment aimed to analyse how difficult children would find mapping across representations that differed in format.

A series of predictions concerning the effectiveness of the representational systems was made. For example, pictures should be beneficial for initial stages of learning and also for lower performing children. By the same token, mathematical representations will take longer to learn, but will ultimately prove to be a more effective representation. Mixed representations may offer the best situation, in that the pictorial representations can be used to bridge understanding to the more symbolic ones. However, this can only occur if children see how the two representations relate to each other.

These predictions were tested with 48 10–11-year-old children in a local primary school. At pre-test, we measured the children's ability to perform estimation and the insight they had into the accuracy of these estimates. Pre-test performance was very low on both dimensions. For example, estimates were an average of 96% away from the correct answer. Two periods of computer training were followed by further tests. We found that all experimental children improved significantly at performing estimation, becoming more accurate and using more appropriate strategies; 23% of answers were generated using a recognised strategy at pre-test compared to 60% at post-test. The (non-intervention) control group did not improve. However, the insight into the accuracy of estimates (the skill most directly supported by the representations) improved in only the pictures and mathematics groups.

In order to provide an explanation of these learning outcomes, analysis of children's use of the representations during the intervention sessions was performed. We examined how accurate they were at predicting the accuracy of their estimates. This provides information about children's developing domain knowledge and about the first two learning demands associated with using a representation (the format and operators of a representation and the relation between the representation and the domain). It was clear that the poorer performance of the mixed group was present during their computer use as well as at post-test. However, each representation used unsuccessfully in the mixed condition was used effectively in the pictorial or mathematical conditions. Therefore, we proposed that the explanation rests in the learning demands of translating between the representations.

To examine this third learning demand we proposed that if learners understand the relation between the representations their actions should be identical over both representations, even if their prediction is wrong with respect to the domain. Hence, the children's predictions on the two representations were correlated to give a similarity judgement (representational coordination). It was shown that over time, the behaviour of mathematical and pictures group did become more coordinated across the representations but that children in the mixed group did not show this behaviour. We believe that the mixed group did not successfully map between the pictorial and mathematical representations and this failure in translation led directly to the reduced learning outcomes.

By examining the properties of each representational system in turn, we can explain why translation occurs between the pictorial and mathematical representations, but not between the mixed representations. The pictorial representations are based on the same metaphor, are acted upon in the same way (clicking at some distance from the goal) and read off in similar manners (accuracy as distance from centre). Hence, both the format and the operators for these representations are almost identical. In addition to these similarities, pictures are "ambient symbol systems" (Kaput, 1987). We know that expertise is needed to use diagrams successfully and children of this age will have had considerable opportunity to interpret language and pictures, but relatively little experience with other representations. Translation between the different mathematical representations also occurs successfully. Although action and read off involve very different processes and one representation is graphical and one is propositional, mapping between the representations may be facilitated as both representations use numbers. DuFour-Janvier, Bednarz and Belanger (1987) suggested that children only believed that two representations were of the same thing if they both used the same numbers. The mixed representations differ in terms of modality (propositional and graphical) and reading and action procedures. In addition they also mix mathematical and non-mathematical representations. Failure of overlap therefore occurs at all levels. Given this analysis, children's inability to translate over representations is completely understandable.

5 Representational Coordination

These two experiments have shown that in the right circumstances, children can learn with MERs. In COPPERS, MERs are used for displaying feedback about solutions and were associated with better performance. However, the system was carefully designed so that the new representation was supported by a familiar complementary representation. In CENTS, this issue was examined further by manipulating the learning demands of mapping between representations. When this cost was too high, children seemed unable to coordinate the MERs and failed to learn the domain material. In subsequent experiments, the results have been replicated and extended. Failure to coordinate representations has been observed over long periods suggesting that the problem is not simply one of cognitive load due to the excessive burden of initial task demands. We have also shown that even when representations are fully redundant such that the information in one representation is completely derivable from the other, coordination is still difficult (Ainsworth, Wood & Bibby, 1997).

Failure to coordinate seems a persistent property of learning MERs emerging for many combinations of representations we have studied and consistent with the research reported earlier which described how learners rarely answered using MERs (e.g. Yerushalmy, 1991). Consequently, we need to understand why learners find translating between representations difficult and to consider what factors affect translation across representations.

Why should learners find it difficult to map across the mixed representations? Superficially, this task appears simple. Learners were given a large amount of experience (in the experiments discussed above between 64 and 128 problems were solved) and were provided with feedback on each attempt. Moreover, the information needed to be mapped between the representations used in CENTS is not particularly complicated consisting of a maximum of two dimensions (direction and magnitude). In contrast consider the mapping between representations such as equations, graphs and tables (see Schoenfeld, Smith & Arcavi, 1993 for details on the complexity of mapping between representations). However, there is abundant evidence that mapping between different situations can be difficult for learners. Research on analogical reasoning has consistently shown that people find it difficult to recognise the similarity between problems (e.g. Gentner & Toupon, 1986; Gick & Holyoak, 1980). They become misled by the surface dissimilarity and fail to recognise that the deep structure is the same. This is particularly true for novices who tend to categorise problems by their surface features (e.g. Chi, Feltovich & Glaser, 1981). As this is the case and by definition learners are novices in the domain to be learned, it would seem important to identify what aspects of learning with MERs makes mapping between representations difficult.

Translating between representations will be mediated by a number of variables: the nature of the representations, task and domain issues, and learner characteristics. Current research is not sufficiently advanced to allow a definitive statement of what these factors are and how they interact. For example, predictions about the representations factors should ideally be based upon an integrative taxonomy of representations. However, based upon current evidence, candidates for affecting ease of translation include:

- the modality of the representations — propositional vs. graphical;
- including representations that differ in levels of abstraction;
- the degree of redundancy across representations;
- whether the representations encourage different strategies;
- differences in labelling and symbols.

When considering learner characteristics, a number of factors that have already been shown to influence understanding of single representations may also be important for learning to coordinate MERs. For example, if learners are already familiar with the domain or the representations, then learning to translate across the representations should occur more rapidly. Cognitive style and aptitude have often been related to performance with particular types of representations and, in addition, Oberlander et al. (1996) argue that one distinguishing characteristic of people who were classified as diagrammatic reasoners may be their ability to translate information across representations more successfully.

To predict how easy it will be for someone to understand the relation between a set of presented representations for a given task, both the individual and the representation's characteristics will need to be considered.

6 Design Implications

Designers of multi-representational learning environments face complex demands if such software is to be effective. The different benefits of MERs discussed above will not be achieved if learners are overwhelmed by the costs of working with MERs. Moreover, these different benefits of MERs allow distinct predictions to be made about how the learning goals should be supported.

The first use of MERs proposed above is that they can be used to support different ideas and processes. This use is ideal if just one representation would be too complex if it included all of the necessary information. It is also advantageous when the computational properties of alternative representations support and focus on different aspects of the domain. In this case, learners may have no need to understand the relation between the representations. Hence, for a designer the main consideration must be to determine whether they have selected the appropriate representations for the tasks and whether the proposed learners are familiar with the format and operators of the representations and how these relate to the domain to be learnt. The need to learn to translate between representations should be kept to a minimum. For example, the learning environment could perform any necessary translation or representations could be presented sequentially.

The second category of use of MERs is to constrain interpretation. The first way that this can be achieved is when a known representation supports the interpretation of an unfamiliar abstract representation. For example, simulation environments often present concrete illustrations alongside more abstract representations such as graphs or equations. This is ideal as it reduces the learning demands for the familiar constraining representation. But, in addition and in contrast to the first use of MERs, designers also need to ensure that learners can translate between the representations. Selecting complementary representations or supporting the mapping between representations is therefore desirable. For example, in COPPERS the relation between the place value and table representations is highlighted to help translation.

A more complicated case is when constraint is achieved because of the inherent properties associated with one of the representations — for example when one representation permits less expression of abstraction. In this case, it may not be possible for the constraining interpretation to be more familiar with less associated learning demands. Once more translation is required for this goal to be fulfilled so that supporting co-ordination between representations is crucial.

The third category of use of multiple representations is when MERs are designed to promote a deeper understanding of a domain by providing alternative views which must be abstracted across to reveal its underlying structure. This goal provides designers with hard choices. If users fail to translate across representations, then abstraction cannot occur. It was demonstrated above that learners find translating over representations which are superficially dissimilar to be difficult. However, in contrast to the other cases where translation between representations is desired, in this case translation between representations should not be made too easy. If alternative representations do not provide sufficiently different views on a domain, then abstraction of invariances can not occur. Research that addresses which properties affect coordination seems crucial to answer this question. Designing for successful abstraction will depend upon the trade off between the benefits of using MERs to support abstraction and the cost of the learning demands.

7 Conclusion

In this chapter we have presented evidence that learning with MERs, although often successful, can be disrupted by the additional costs associated with using more than one representation. Furthermore, we have shown that certain combinations of representations make translation between these representations particularly difficult and shown how this can affect learning outcomes. By identifying particular uses to which

MERs can be put, we proposed a set of design principles aimed at maximising the effectiveness of multi-representational learning environments by adjusting their learning demands.

Acknowledgements

The work was supported by the ESRC at the ESRC Centre for Research in Development Instruction and Training. The authors would like to acknowledge all the other members of the centre for providing a stimulating and supportive atmosphere in which to conduct research.

PART 2

Problem Solving and Learning with Multiple Representations

8

Problem Solving with Multiple Representations by Multiple and Single Agents: An Analysis of the Issues Involved

Henny P. A. Boshuizen and Hermina J. M. (Tabachneck-)Schijf

As young children people have already learned that they have five senses: their eyesight, hearing, smell, taste and sense of touch. The receptors of all these senses transform information into a format that can be dealt with by the human brain. People use these different kinds of information in a coordinated way; e.g. we know a person from his or her appearance, sound of voice and movements such as footsteps or coughs and sneezes, smell, sometimes touch and maybe even taste. These different "parts" together form the full "description" of this person. They are one whole and people are very surprised when different parts do not match. For instance, suppose you hear your colleague Frank coming up to your room (footsteps, coughs, knock on the door), but instead of Frank, Francy comes in. What would you think? Probably something like "Where did you leave Frank?", "Am I becoming senile?" or "Are they playing a trick on me?"

In Western societies people grow up with the idea that multiple representations in the form of texts with illustrations are helpful for building up an adequate understanding of the topic at hand. At least, that is how school books and books for young children are designed. The Frank/Francy example shows how integrated these different representations are. Normally the use of multiple representations feels like a very smooth process. The same applies to the process of building up multiple representations of new objects or events. Only in extreme situations may we experience that this process can be very difficult. The following quote illustrates this difficulty. It has been taken from a chapter by Oliver Sachs (1995) in which he describes the "adventure" of a middle-aged man, Virgil, who became blind at a very young age, but whose eyesight was partly restored after a glaucoma operation.

> As we settled down, Virgil's cat and dog bounded in to greet and check us — and Virgil, we noted, had some difficulty in telling which was which. This comic and embarrassing problem had persisted after he returned home from surgery: both animals, as it happened, were black and white, and he kept confusing them — to their annoyance — until he touched them, too. Sometimes, Amy (his wife, HB) said, she would see him examining the cat carefully, looking at its head, its ears, its paws, its tail and touching each part gently as he did so. I observed myself the next day — Virgil feeling and looking at Tibbles with extraordinary intentness, *correlating* the cat. (Sachs, 1995: 121–122, italics added)

This man not only has problems with the visual information as such, but also with the integration and translating back and forth of the visual and the tactile information. Another patient who became blind during adolescence and recovered after about 10 years reported similar problems.

Examples like Virgil's have been known for quite some time, but so far little is known about the everyday-life difficulties of learning and problem solving with multiple representations. We do not know if the naive belief that illustrations "help" with building an understanding is right. The fact that normally no effort is felt does not mean that such is indeed the case. This is exactly the leading thread of the present chapter and the others in this section. Here, the assumption that multiple representations help and hence that coordinating or integrating multiple representations is more or less effortless, is questioned.

In this part of the book several chapters are brought together which explore this rather contradictory situation in different domains, asking the question if even under less extreme circumstances integration of multiple representations and solving problems with multiple representations requires an extra effort and if so what the extra effort is. As an extra problem the integration of multiple representations held by different people (multiple representations in multiple agents) is investigated. In the present chapter a formal analysis of the different ways representations can be integrated is given, followed by an extrapolation to the multiple agent situation.

1 Some Essentials of Representations[1]

A *representation* is a format for recording, storing and presenting information together with a set of operators for modifying the information (Tabachneck & Simon, 1995). Neither component, *information format* nor *operators*, alone defines a representation. A representational format can be almost anything. Data can be words, pictures, numbers, parts of an X-ray, etc. Operators modify the information stored in the data. "+" and "−" are well-known arithmetic operators. "Scan" and "zoom" would work on pictures. For example, Roman numerals (M, D, C, L, X, V, I), are data pieces that combine in strings that represent cardinal numbers. They are meaningless without definitions of the individual symbols (format) and rules (operators) for performing arithmetic operations on them (e.g. XXXXX → L; XX + XX → XL). The effectiveness of a representation depends both on its operators and on the speed with which the operations can be performed. The latter depends also on familiarity with the representation: expert diagnosticians will draw conclusions more rapidly and accurately than novices from X-rays.

Both the data structures and the operations in distinct representations will be very different. However, inferential goals may be achievable in both representations. One can for instance find the solution of a set of two simple linear equations by solving

[1]Parts of this section have been adapted from Tabachneck and Simon, 1995.

them or by drawing them and reading off the coordinates of the intersection of the two lines. The fact that some goals will be much easier to achieve in one representation than in the other provides the rationale for using the superior representation: it is more *efficient* than the other. The effectiveness of representations for communicating and instructing depends first and foremost upon their computational efficiencies. Consequently we must distinguish the informational from the computational equivalence of representations.

Two representations that are *informationally equivalent* (the information in the one can be obtained from the other and vice versa) will usually not be *computationally equivalent* (will not obtain the information with the same amount of computation), for their efficacy depends on the operators that act on them. Two sets of operators may differ in their capabilities for recognising patterns, in the inferences they can carry out and in their strategies for controlling search. Diagrammatic and sentential representations, for example, employ operators that differ in all of these respects (Larkin & Simon, 1987). Efficacy is not necessarily a fixed attribute of the representation format. Ways can be found to do the operations faster — for instance, long division on Roman numerals with a calculator for that purpose would be faster than long division on Arabic numerals by hand. Efficacy of a representation can alter drastically when ways are found to perform the operations more rapidly. These ways can be found through exercise, invention, or "help".

A special type of operator is the *inference* operator. Inference operators help us get information out of data that is implicitly, but not explicitly present. The process of going from a question to an answer is an inference. Inference operators define a kind of logic on the database — although not, in general, formal logic. There are several types of inferences. For example, going from "1+1 = ?" to "2" using knowledge about arithmetic is a deductive inference. Going from seeing 20 geese, all white, to "all geese are white" is an inductive inference. Inferring cancer from a small blob on an X-ray requires experiential inference — the diagnostician has learned, from books but more so from being exposed to many, many X-rays, that a certain type of blob can mean cancer. He can, of course, be wrong — the blob may merely be an innocent growth. Unlike the tautological inference rules of logic (e.g. syllogism, modus ponens), which remain the same in every domain of reasoning, inference rules that incorporate matters of fact must be specially tailored for each knowledge domain, and will usually not remain invariant over time. Usually, it is worth the cost to acquire a substantial set of inference operators tailored to any domain in which we wish to reason and solve problems frequently, because such operators may accelerate enormously the speed with which we can extract knowledge from a database.

An example from medicine is the following. Suppose we have a patient with pain in the upper abdomen. The medical student knows, from his book learning, that this type of pain can originate in many organs in that area: the liver, gall bladder, stomach, pancreas, etc. and from many different processes, like a tumour, inflammation, stone or trauma. While it takes a medical student quite a while to come to a diagnosis, experienced diagnosticians know what to look for in order to come to a likely correct diagnosis. They will not only look at physical symptoms, like degree of pain, and place of pain, but also at the patient's background. Confronted with a 45-year-old male complaining about severe continuous pain in the upper abdomen, many possibilities exist. However, when questioning reveals him to be an alcoholic, then chances that he has an inflammation of the pancreas or the liver are high. If a 10-year-old child came in with the same complaint, the doctor would most probably not even think of the pancreas, simply because it is a known fact that children of that age do not get that type of disease. Inference rules incorporating age of the patient and lifestyles are tailored to diagnostics. But such rules must change with altering circumstances. For example, 10 years ago, a general practitioner would question a patient with a peptic ulcer extensively about stress and might even refer this person for psychiatric treatment. Nowadays, it is known that peptic ulcers are in the first place a result of bacterial infection with *Helicobacter pyloris* (first identified in 1983), alcohol and aspirins. Medical practice has changed accordingly.

1.1 Interactions Between Different Kinds of Representations

Modes of representation are aligned with natural human input modes — hearing, seeing, touch, smell and taste. We now know that seeing constitutes at least two modes — spatial and objects modes, which are handled by different areas of the brain, and are eventually reunited again (for an overview of the research, see Farah, 1990). For this chapter, we will concentrate on hearing (verbal communication, be it spoken or written) and seeing. We realise that touch and smell are also very important senses, especially in medical diagnostic tasks, but will for the moment ignore these. We hypothesise that each mode of representation has its own format and its own operations, and that one mode cannot directly modify a representation of another mode. However, one mode *can* activate, or connect to, a representation of another mode (for a description of a computational model embodying these principles, see Tabachneck, Leonardo, & Simon, 1997).

Although different representations cannot alter one another directly, it is possible to translate one representation into another. One representation may be computationally efficient for dealing with one part of a problem-solving or reasoning task, while another representation may be more advantageous for another task. For instance, a graph of a set of experimental outcomes may be quite fast in conveying that there is an interaction between two dependent measures — the lines cross. However, finding out whether this interaction is significant is much easier using statistical mathematics. Translating from one representation of data to an informationally equivalent one may be a non-trivial task. For example, it took several top-flight physicists a number of months to show that Heisenberg's and Schroedinger's independently discovered representations of quantum mechanics were informationally equivalent. Both representations continue to be used in physics today (nearly 70 years later), because each has computational advantages over the other in solving particular classes of problems.

1.2 Natural and Artificial Learning Settings

We have already suggested that in humans, dealing with multiple representations is very common. In effect, multiple representations probably seem the rule rather than the exception, especially for everyday learning. On eating a pancake, we see the characteristic shape and colour, we smell the typical pancake smell, and we taste the typical taste. All three representations are stored. While all three together offer the largest likelihood of recognising that we are dealing with a pancake, each one separately also makes a good chance. How things are learned and which representations are involved, is linked to the input senses, to the task a person has, to his or her expertise and to the characteristics of the communication setting. However, learning in more artificial settings, like a classroom, can sometimes, surprisingly, lack representation forms that can turn out to be very important for future performance. For instance, medical students are initially taught from books, and do not come in contact with the patients. The pictures in these books do not afford smelling and touching, or hearing a patient wheeze or groan. Furthermore, textbooks show typical signs of a single illness; patients often have more than one disease at a time. When medical students first start applying their book knowledge to patients, or, rather, when medical students first have to lay connections between their book knowledge and the patients' signs and symptoms they have to translate the new representations into the better known format. For instance, they must recognise that the specific sound heard, is a "crepitation" that according to the books should sound as footsteps in the snow: *correlating* the patient, one might say, just like Virgil.

Although multiple representations seem the rule in humans, there are several complications that will be highlighted in this chapter. The next two paragraphs are meant to investigate these complications by analysing the boundary conditions that affect "cooperation" between different representations in multiple agents (e.g. teams of specialists) and multiple representations in a single agent. We start with the multiple agents, because this domain magnifies certain problems that are also inherent, but less easily detected in single agents. In this approach multiple representations in multiple agents are not only dealt with for their own sake; the multiple agents are also, so to speak, a metaphor for the single agent situation, with all the pitfalls a metaphor can have.

2 Distributed or Multiple Representations in Multiple Agents

"Distributed or multiple representations in multiple agents" is defined here as the circumstance where multiple human or artificial agents have dissimilar representations about one object, person, interaction or situation. These representations may be dissimilar in terms of their data. This is very often the case, for instance when people think differently about an object, or when they have different perspectives on an object or situation. An example of the latter can be found in research by Baum and Jonides (1979), who showed that inhabitants of one part of the country have more detailed representations of their own part than of other parts farther off. Hence inhabitants of the same country can have very dissimilar representations of locations and distances between cities depending on their own place of living. Representations can also be different in terms of the formats, e.g. when one representation is propositional and the other visual. And finally they may differ regarding the operators that are applied, e.g. logical vs. experience based. In everyday life the different contents of representations are most eye-catching and can be the source of many misunderstandings. For instance, just the other day a colleague, Jim, told us about showing up in The Hague for a meeting, while the person he was supposed to meet was in Groningen. They had agreed to meet in John's office. Unfortunately, John had changed offices half a year before, and was now stationed in Groningen, 2 hours from The Hague by train. Both men confidently assumed that the meeting place was well known. This was a large difference in representation content, but small differences are an everyday occurrence. Sending a child for "a loaf of bread" may result in anything from white to brown, to raisinbread, depending on the child's representation of "a loaf of bread". Different operators can appear very unexpectedly and can lead to real quarrels, because one person thinks that the other reasons "unfair". An example of the use of different sets of operators is using a legal approach toward solving a misunderstanding between people vs. using more common sense ideas about it.

A typical situation where people work together and have different representations is the team of specialists who work together on one task. The following is an example. It has been taken from the domain of surgery where teams of specialists with very different backgrounds work together.

2.1 Advantages and Disadvantages of Multiple Agents in a Team

Teamwork has many advantages over the situation where one person has to do all the work. A single person does not need to have all knowledge and all skills necessary to perform the task. This makes the learning process before the task can be done at the required level much shorter than when every expert would have to do all sub-tasks. In some cases, it is even impossible for one person to meet the standards set for all participating specialisations. Also, one person cannot be at several places simultaneously, which might be a task requirement as well.

If knowledge were completely different in each agent, cooperation would not be possible. They would not know what the other requires. Experts working together must share at least part of their representations. This may require that they extend their own representations. For example, the surgeon in the operating theatre team must have basic understanding of anaesthetic procedures and problems and the anaesthesiologist must understand the needs and problems of the surgeon. Both must know which part of the representations is shared and which they do not share. But even that does not guarantee that they always understand each other immediately; communication is not automatically smooth and spotless, especially in cases of uncommon situations or procedures. For example, a 4-month-old child with sickle cell anaemia (a very uncommon condition in The Netherlands where this happened) needs a hernia operation. From the perspective of the surgeon this is a standard operation that only requires some extra blood cells because of the child's anaemia; a "piece of cake that can be done on a Friday afternoon". For the anaesthesiologist it is a very complicated situation, not only because the time span in which blood has to be given is very short, but especially because problems with the blood circulation may occur afterward and require extra precaution. Hence she refuses to do the procedure in a local hospital that does not have the equipment to keep the child properly monitored after the procedure.

An example from another domain is the different representations that a team of people working on new product design will have of the product. The engineer will look at the product aspects that make the product work smoothly; the designer will focus on whether the product looks and feels pleasing; the marketing person will look at the product in the light of how it should look and act in order to sell better; the human factors specialist at those aspects that have to do with how the customer will interact with it. All are looking at the same product; all have a different representation of it, with different data sets, and different operators. It is no wonder that team members sometimes have a hard time talking to each other — in a way, they really *are* thinking about a completely different product. Similarly, the surgeon and anaesthesiologist are dealing with different (aspects of) patients and draw different conclusions. The misunderstandings that may result from such differences in perspective are a serious disadvantage that can only be solved through communication. This, unfortunately, takes time, another disadvantage. On the other hand, time can be saved as well. The product design team can complete the assignment in less time than one single person can. A disadvantage can be that a team member sometimes has to wait until the other has completed his or her share in the task.

This tentative analysis of the advantages and disadvantages of people working together in teams and hence of distributed representations in multiple agents reveals several factors that can affect this cooperation: (a) (dis)similarity of representations, (b) the (lack of) coincidence of individual tasks and goals and the common goal of the team, (c) the necessity of a good timing and coordination of the activities. Another factor that has not yet come up in the analysis but that may largely affect the success of a team is (d) the responsibilities of the team members and the social structure that coordinates these individual responsibilities. We see this very often in medical teams responsible for the care of multi-problem patients. The specialities in such teams are not organised hierarchically (each has its own responsibilities that cannot be overruled by the others). Sometimes it is possible to find a way out of a dilemma by taking extra measures, but sometimes, when evaluation procedures fail to come up with one definitive solution, the team may end up in a situation where there is no good solution. They are caught in an approach–avoidance dilemma that people try to prevent as much as possible. An example of such a situation is the discussion whether to operate or not on a very important person, e.g. the Pope or the president of the former Soviet Union. In the case of Boris Yeltsin in the mid-1990s, who was said to have local and generalised problems with the heart and blood vessels, long and serious discussions took place, because not only the health or even life of the Russian President was at stake, but also the professional status and prestige of many doctors.

The heart surgeon said that he could not take the responsibility for longer postponement of the operation; the anaesthesiologist and the physician said that the man's condition was so poor (he had problems with the kidneys and liver, and generalised atherosclerosis) that his chances of surviving the operation were minimal. Finally they consulted a very famous and old American heart specialist to come to a mutual agreement. In the following paragraphs each of the factors that could affect cooperation will be analysed.

2.2 (Dis)similarities of Representations

Representations can be dissimilar in the data, the format and the rules that operate on them. Experts working together must at least share some of their representations. How far this similarity must go, is unknown. In daily practice it is impossible, yes even undesirable to have complete similarity of representations. It would annihilate the advantage of saving learning time. In practice we see a combination of shared representations and trust in the validity, feasibility, etc. of the other's representations. An operating theatre team (partly) shares their representation of the standard procedure applied; their representations of the patient, however, largely deviate since both specialities take different aspects into account. The anaesthesiologist primarily focuses on the vital functions of the patient with the help of hi-tech equipment monitors, while the surgeon focuses on the patient's anatomy, a part the anaesthesiologist often even hardly sees because of his or her position in the room.

Two people can "cooperate" to make data from their representation available to the other person. We discuss two different ways in which this can occur: coordination and integration.

2.2.1 Coordination Between Two Representations

What does it mean for two people to coordinate between representations? Anaesthesiology and surgery require completely different skills, and therefore operators and databases. So different, that although both are needed for a successful task execution, one can do his or her job nearly independent of the other. The part that is *not* independent, however, is crucial and coordination is a necessity. The surgeon and anaesthesiologist are fortunate in that there are only a relatively small number of pieces of information that one needs from the other during the operation. Mostly, *data exchange* suffices. With the term *data exchange*, we mean the furnishing of pieces of information from one representation to another, which the other representation can use. Pure data exchange does not, *a priori*, require translation of the data; although most often, for data exchange to be successful, one of the partners will have to translate in order for the data to fit the other's representation. For the director of a business, it is crucial to know what profits the company made. The accountant gives him or her a summary of the company's operation, translated to terms the director has learned to understand. The director does not need to learn accounting, only to learn to understand the summary. The accountant does not need to know how to run the business, just how to compose the summary and how to provide other information needed for the director to make sound decisions. In most cooperations, there are clear rules as to which data is needed and in what format, and who will do the translation.

For successful data exchange, therefore, the cooperating partners must know for what reason which data need to be exchanged, and the format in which they need to be exchanged. Some data need to be exchanged because the other person needs to know that to perform his own procedures in the right way. In the surgical team that might be the information from the anaesthesiologist that the patient is anaesthetised at a proper level so that the surgeon can start his or her procedure. Some data need to be communicated to the other team members because they indicate an emergency coming up, for example cardiac fibrillation that requires a resuscitation. The pieces of data that need to be exchanged between the team members form the "links" in the *team* process; they are needed to continue it. In terms of representations, these data are the places where their representations touch each other — where they coordinate.

Operator exchange is another form of representation coordination. However, giving someone an operator is only useful if the other has the correct data structure. Let us say you receive this operator: "If a lower number appears before a higher number, the two numbers form a unit and you have to subtract the lower number from the higher number to get the correct number." If you do not know what Roman numerals are, and what value the basic numbers have, then you could be given the string XLIX, but you cannot apply the operator. Hence, operator exchange is only useful if there is a data structure to apply it to.

Also, non-trivially, people need to know that a cooperation needs to exist. The anaesthesiologist and the surgeon have both gone to medical school, and have both been exposed to basic issues in anaesthesiology and surgery. Through acquiring partly the same representations, they have learned that they need each other, what for, and probably what the rules of data exchange are. If this initial exposure to other representations does not occur, people may have no idea that they need someone else to successfully complete a task. For instance, consider computer programming — especially the user-friendliness aspect. In the early years of computer programming, and unfortunately even often today, programmers were rarely exposed to human factor, social or cognitive psychology issues relevant to software usability. Consequently, they constructed software that is easy to use for them, but not for the (non-programmer) people who have to use it. Most often users of software do not share the representations and knowledge of the software designers and programmers. Trying out the software on one's colleagues does of course not help. The software designers are not to blame though — they have no way to even know that they are constructing non-user-friendly software, or of realising that cooperation with human–computer interaction (HCI) experts may dramatically improve their product. Cooperation was not stimulated at the organisational level; it was not part of the programmers' formal training experience.

2.2.2 Integration of Two Representations

Data exchange, and an understanding of each other's needs by (often very) limited shared representations is generally all that is needed for cooperative partners to do the task. When someone has to take over another's task, or continue another's work, integration of representations has to occur: the whole or partial takeover of a representation so that newly incoming information can be interpreted similarly. One can only integrate representations, i.e. make it possible to freely exchange operators and data, if representations have the same data structures and operators (they do not need identical contents; but the more overlap of content, the easier the exchange). The more complex the task, the more difficult this is. Academic students, in general, need at least 10 years of preparatory work to be able to continue parts of the work of their professor. In work situations, passing on "experience" is often very difficult. If this knowledge is considered critical, a knowledge engineer can be called in to record the representation of an expert. This is difficult and time-consuming work, and requires many iterations with the expert to ensure that the representation has been recorded correctly. Presenting the knowledge in such a way to the next person so that it can be acquired is another problem again, but as volumes upon volumes have been written about education, one we will leave to others.

2.3 Other Factors that Affect Cooperation Between Multiple Agents

2.3.1 Task and Goal Structure

The previous paragraphs have shown that coordinating and integrating representations can be very difficult and time consuming. How important it is that representations are coordinated and integrated, and how and when this has to be done depends to a large extent on the task characteristics.

Let us return to our team of medical specialists concerned with the preparation and performance of surgery on a patient. This team has one common, overarching task to which every team member contributes in his or her own way. Everyone is responsible for his or her own sub-tasks in the whole, plus exactly knowing when and what data to exchange in what format. In such a task situation these coordinations should not be done on the job. It could even be disastrous. A substantial part of the success of the team lies in preparation and in the prevention of problems (Gaba, 1992). This can be done by testing procedures and the communication lines in the team by running simulations of emergencies. An important aspect of doing surgery is that the job must be done in a specified period. The shorter the better. Problems will always lead to a longer operation and more anaesthetic, and more stress to the patient. Problem prevention is to be preferred over problem solving. Surgery is not unique in this respect. The same applies to aviation. While in the air it is impossible to park the aeroplane and perform trouble shooting.

Domains like these have their own dynamics and timing. Deadlines are very, very real, as someone will literally be dead, unlike with the project "deadlines" that a design team is confronted with. Those might be merely expensive. Another difference is that the normal task must go on during trouble shooting, however difficult it may be. The aeroplane must stay in the air, and the patient must be kept alive.

2.3.2 Working Procedures

Although cooperation in surgical and aviation teams is heavily practised, other collaborative tasks have less clearly worked-out structures and standard procedures. An example is the team of road accident analysis in chapter 9 by Alpay, Giboin and Dieng. But even new groups, working on tasks the individual members have never done before, could do better by using a more or less standard way of project management to organise their activities, manage their agenda and allocate resources. Many innovation projects have about five phases (Cumming & Worley, 1993). In the first phase an exploration of the problem or task and of the situation takes place, also including the environment that can dictate boundary conditions for future solutions. This soon narrows down the field to a limited set of possible solutions. The analysis then follows; pros and cons of the solutions are sorted out, and possibilities are estimated. As a next step, a small-scale try-out can be initiated. After this phase has been successfully completed, larger scale implementations can take place. The final phase involves an evaluation of the innovation.

Working procedures need to describe what has to be done, when, by whom, who is responsible for what, who takes which decision, who communicates with whom, etc. Clearly, standard procedures and scripts dictate the communication between participants and how their representations should be tuned to the task.

2.3.3 Expertise as a Team

Multi-disciplinary teams often consist of experts with different backgrounds. We have seen several examples in this chapter: the surgeon, anaesthesiologist, cardiologist and physician who are responsible for the management of a multiple-problem patient or the product development team consisting of a designer, an engineer, a human factor specialist and a marketing specialist. All these people have to learn to work together, not only by coordinating and/or integrating their representations, but also by learning how the work of the other is done and how it affects their own work.

An example is found in the work by Gaba (1992) who investigated the cooperation of operating room teams. In his opinion every time an anaesthetic is given, it should be perceived as a disaster waiting to happen; good teams prevent disasters as much as possible. They do this by intensive investments in team training, by doing pre-use checkout of the equipment, preoperative evaluation of the patient and by monitoring the care and skill of the team members. The better they know each other's strong and weak points and idiosyncratic peculiarities (e.g. a left-handed surgeon) and how these affect their own task, the better they can prevent eventual problems. Prevention of these problems is to be preferred over emergencies that have to be dealt with on the spot.

3 Distributed or Multiple Representations in Multiple Agents Revisited

All the examples given and analysis performed were meant to serve one goal: the analysis of the essence and of the factors affecting the cooperation of multiple agents having multiple representations. Looking back, a few important points seem to have emerged in our analysis and examples:

- Most prominent is of course the ways multiple representations can be dealt with: coordination between representation, including data and operator exchange, and real integration consisting of (partially) handing on knowledge and experience.
- Boundary conditions are the recognition by one agent of which data or operators are needed by the other agent, or the other way around when the agent him- or herself needs which data from the other agent. This rather metacognitive aspect of the cooperation is made much easier when all participating agents have a common understanding of the task to be performed and the procedures that have to be followed. These standard procedures and protocols coordinate the tasks and responsibilities of the different team members; they also dictate and monitor the timing of the task performance. Since these procedures are common knowledge, they also regulate the interaction between the agents. Expertise built up by the team in running these procedures will result in faster and smoother communication and in prevention of problems rather than solving problems after the fact.

Two chapters in this part of the book explore aspects that are fundamental in the cooperation between multiple agents. Bromme and Nückles (chapter 10) investigate what determines the problems professionals of different backgrounds have when working together, i.e. doctors and nurses working together in a paediatric oncology ward. They suggest the notion of "perspectives" to refer to knowledge which is distributed among different persons and *shaped* by different (professional) approaches. Relevant differences seem to be: knowledge about the task demands of the other, representations of the presumed perspectives of the other, responsibility for effective interprofessional communication, power differences, different languages used by differently trained professionals and the existence of one dominant language, i.e. the doctors' language. Alpay, Giboin and Dieng (chapter 9) basically address the same phenomenon: cooperating experts with different professional training, i.e. different specialists in a team that analyses road accidents. Their chapter describes a quest after the dimensions along which the representations of different specialists vary and how coordination of these representations can be best described. Ostwald's model of representations for mutual understanding, but as Alpay et al. also found for mutual agreement, turned out to be very helpful in explaining the differences and indicating directions for improvement. Their work also shows that unclear task structures magnify the problems due to a lack of knowledge about the information needs of other team members. Although approaching from very different angles these two chapters allow similar conclusions concerning the factors that jeopardise shared cooperation in a team due to non-coordinated multiple representations.

4 Again Multiple Representations in One Agent

From the above analyses, it is clear that it takes work and effort to coordinate different representations in multiple agents; it should come as no surprise that we assume that coordinating different representations in a single agent is no less effortful though less "visible". In one agent, representations can also be coordinated through connections via data, or, more rarely, operators, and representations can be integrated. Coordination occurs by connecting representations through a piece of data, which then becomes common to both representations. Integration occurs by interconnecting two similar but separate representations into one. An example of this process in a single agent is the following. Suppose that you have learned to do statistics in the context of a mathematics class. Now, you take a few psychology classes, and after doing a few experiments, you collect a set of data. You can then learn to apply the statistics learned in the mathematics class to the psychology data, which have been translated to a form that now fit the statistical formulas. In this way, you can integrate the psychological data structures with your statistical representations.

In textbooks of nearly all disciplines, pictures, graphs, text and equations are often mixed — each, hopefully, being used to most efficiently convey knowledge to the reader. When reasoning *between* representations is required, i.e. when one must connect knowledge from different representations into one body of knowledge, one must map knowledge from one (the base) onto another (the target), preserving among the target objects a system of relations that holds among the base objects. This can be quite difficult for novices (Tabachneck, 1992). In case of between-representation mapping, the same domain knowledge is mapped from one representation to a different one. For instance, in a *graph* of an economic market, demand is represented by a downward-sloping line and supply by an upward-sloping line, with price on the y-axis and quantity on the x-axis. The equilibrium price and quantity are found by noting the coordinates of the intersection of the demand and supply lines. The graphical information can also be expressed verbally, for example, the supply information by the *statement* that the quantity of a commodity that consumers buy will be smaller, the higher the price. A successful representational mapping does not only add flexibility to the reasoning process; it may also offer new insights into problem structure and new computational possibilities. In one study of a physics and a geometry problem, diagrammatic and verbal representations facilitated different search control strategies, different recognition patterns, and different inferential processes (Larkin & Simon, 1987).

Novices' difficulties in within-domain mapping from one representation into another have been well documented both in basic research (e.g. isomorphs of the Tower of Hanoi puzzle (Hayes & Simon, 1974–1976) and the mutilated checkerboard (Kaplan & Simon, 1990)), and in applied research. In economics, the base representation is verbal, and students come in with little or imprecise knowledge or with misconceptions. They do have some general knowledge about graphs, but the graph elements are meaningless until mapped onto the particular concepts they represent in economics. The operations required to interpret graphs include both general domain-independent (perceptual) operations and special operations that are meaningful only in the domain under consideration. A similar distinction applies to other representations, including the verbal (e.g. knowledge of English syntax vs. knowledge of terms used in economics). In medicine, it is hard to say which is the base representation: verbal, spatial, mental models, etc. To really grasp medical knowledge all these representations must be coordinated. This process goes on for many, many years. The diagnostic and management tasks medical students are trained for require that medical events are analysed using all these representations: how a tumour or inflammation, or other process, affects the spatial relations between organs, how they affect their functioning, and as a result the functioning of other organs, how the body will detect these disturbances and give alarm signals, and the other way round, how to interpret these alarm signals and signs of malfunction, and/or spatial changes, and integrate that into a clear image of what is wrong. Evidently the diagnostic task requires that students can apply their knowledge starting from the signs and symptoms, but medical education most often starts from the other end, describing the disease, requiring extra coordination effort from the student.

Four chapters in this part of the book deal with the problems of the coordination and integration of multiple representations in one single agent. All three do this from the perspective of expertise building, clearly signalling that they see this change as an important part of a learning process that can take very long. Savelsbergh, De Jong and Fergusson-Hessler (chapter 13) show how the problem representations generated by poor and proficient novices and experts, and hence the knowledge structures these representations are based on, show remarkable differences especially in structure and flexibility. Expert knowledge is more coherent and experts have better knowledge about properties of situations that make that knowledge applicable or not. Furthermore, expert knowledge allows for multiple redundant representations, which is a powerful instrument to recover from errors and to find alternative solutions, once a first solution has failed. Expert knowledge also seems better linked to application situations. Both Tabachneck-Schijf and Simon's and Boshuizen and Van de Wiel's chapters address the question how this process takes place. The chapter by Tabachneck-Schijf and Simon (chapter 11) gives a detailed example of how novices in economics learn when confronted with materials in different representations and how little success these learners had. The chapter by Boshuizen and Van de Wiel (chapter 12) also shows how difficult this integration process is. It shows how a more advanced medical student struggles with the coordination of different verbal and spatial representations when solving a problem, while the expert uses both representations flexibly, in a coordinated way. The student only happened to shift from a verbal, clinical knowledge base to a spatial, non-clinical, biomedical one when she got stuck in her problem solving, trying to support her verbal reasoning with spatial knowledge. Van Someren, Torasso and Surma (chapter 14) discuss the problem of optimising knowledge that can be represented in different representational forms. The context is diagnosis that can be performed by reasoning from a causal model or by reasoning from a memory of cases that were solved before. The case memory is "functionally subsumed" by the causal model but in general retrieving a case is more efficient. However, this is not always true. The problem now is to optimise the case memory.

No chapter, either dealing with multiple or single agents, focuses the importance of procedures, strategies, protocols and scripts that appeared crucial to good task performance in multiple agent teams. In a real paradigmatic shift in research, this aspect has slipped out of the focus of psychological research into the acquisition of expertise and advanced knowledge acquisition. Problem-solving research in the 1960s and 1970s investigating the use of algorithms and heuristics has been replaced by research into the use of knowledge in problem solving and the restructuring that is required for it. Only a few researchers (e.g. Schraagen, 1994) have investigated the role of more "generic" strategies in expert performance, indicating that experts integrate their domain knowledge in generic procedures that are also used by professionals who do not have knowledge of and experience with the specific domain. These procedures that are packed with or linking relevant knowledge in the expert largely guide their actions. Hence, the following chapters unravel aspects of working with multiple representations in single and multiple agents by making the obvious non-obvious. This is a very important contribution to the field. However, we do not hesitate to say that this is only part of the work to be done. Tasks, task structures, procedures, and maybe even the old research findings on algorithms and heuristics, have to be given a place too, later.

9

Accidentology: An Example of Problem Solving by Multiple Agents with Multiple Representations

Laurence Alpay, Alain Giboin[1] and Rose Dieng

1 Introduction

"When we interact with one another", Norman (1993: pp. 81–82) writes, "we have to transform our thoughts [let's say also our 'mental representations'] into surface representations [e.g. words, gestures, objects] so that others can have access to them." Norman illustrates this idea by giving a virtual example of a person — Henri — who had a car accident, and who relates it to his friends like this:

> " 'Here,' Henri might say, putting a pencil on the tabletop, 'this is my car coming up to a traffic light. The light is green, so I go through the intersection. Suddenly, out of nowhere, this dog comes running across the street.' With this statement, Henri places a paper clip on the table in front of the car to represent the dog. 'I jam on my brakes, which makes me skid into this other car coming from the other direction. We don't hit hard, but we both sit there stunned.' [...]" (Norman, 1993: pp. 47–48.)

In Henri's relation, the tabletop, pencils and paper clip (among others) are artefacts used to represent real objects such as the street, the cars and the dog. These surface representations, we will say, allow one to make oneself understood, that is, they are "representations for self-understanding". The same representations, Norman adds, are also "a tool for social communication: Several different people can share the tabletop and the story at the same time, perhaps suggesting alternative courses of action". For example:

> " 'Look,' Marie might say, picking up one of the pencils, 'when you saw the dog, you should have gone like this.' 'Ah, but I couldn't,' Henri might respond, 'because there was another car there,' and he puts yet another pencil on the tabletop.' " (Norman, 1993: p. 48.)

In this collective relation, the tabletop becomes "a shared workspace with shared representations of the event." In this situation, the surface representations, we will say, are "representations for mutual understanding", i.e. representations providing "a shared and explicit ground for communication" (Ostwald, 1996).

[1]The theoretical framework presented in this chapter was proposed and developed by the second author Alain Giboin, i.e. the application of Ostwald's model of "representations for mutual understanding" leading to the notion of "representations for mutual agreement".

Although hypothetical and simple, Norman's situations of an accident report give an idea of multiple representations (internal and external) used in groups. Our aim in this chapter is to explore further multiple representations in groups, by considering real and complex situations of accident reports. The situations considered here are those where car accident specialists, or *accidentologists*, analyse accidents cooperatively to explain them, and to produce an accident report (or accident folder). Our analysis of the multiple representations used by the accidentologist's teams is based on the model of "representations for mutual understanding" proposed by Ostwald (1996) and which accounts for software designers' activity. To some extent, the accidentologists' activity can be considered as a design activity, a "retrodesign" activity, i.e. reconstruction of the accident. This activity can in turn be considered as a kind of problem solving. That is, accidentologists have to reconstruct some event which occurred in the past. We aim in this chapter to assess the validity of Ostwald's model for the accidentology situations, and consequently to establish a basis for a future model of collective representation management in accidentology.

We first present the domain of accidentology, the method and theoretical framework used to study the representations and their management in the accidentology teams. We then describe a number of dimensions and attributes of the multiple representations used in the accidentologist's teams that emerged from our analyses. Furthermore, we will look at some aspects of multiple representation management in the accidentology teams, including cooperation and collaboration among accidentologists. To conclude, we discuss some of the consequences of our results on several issues of multiple representations evoked in this volume.

2 Domain, Method and Theoretical Framework

2.1 The Domain of Accidentology

The accidentologist's teams that we studied belong to the Department of Accident Mechanism Analysis of INRETS (the French National Institute for Transport and Safety Research). These teams are multi-disciplinary, and they associate researchers and investigators from different specialities covering the components of the driver–vehicle-infrastructure (DVI) system, i.e. infrastructure engineers, vehicle engineers, psychologists (who could be called "driver engineers"). The DVI system helps the accidentologists in their analysis and understanding of the accidents (see section 3.1).

The tasks of the accidentologist's teams are:

1. to analyse the malfunctioning mechanisms that occur in the DVI system, and which generate road accidents;
2. from those analyses, to elaborate diagnoses as a mean for improving transport safety;
3. from those diagnoses, to help in designing infrastructure and vehicles, to train infrastructure and vehicle designers, road planners and users.

Teams of investigators perform the first task. The investigators operate at the site of the accident along with the emergency and police units. Usually, when there is an accident, INRETS is called immediately, having been informed by the fire crew. In general, a team of two investigators is formed: one is responsible for interviewing the drivers involved in an accident, while the other is responsible for gathering information on the accident itself (e.g. photos of the cars, records of the tracks left by the vehicles). They then exchange their first impressions of what happened and their first hypotheses. They determine what information is missing which will lead to an additional brief. The investigators then write a synthesis report. Based on these data, a kinematics analysis is performed (Lechner et al., 1986; Lechner & Ferrandez, 1990). This phase aims at identifying the movements of the vehicles involved in the accident, e.g. their positions, speed and acceleration. The final brief includes the synthesis of the accident, ending the pre-analysis (Ferrandez et al., 1995).

The second and third tasks are performed by researchers. Researchers use the briefs elaborated by the investigators to perform these tasks. The second task is also called thematic analysis (Van Elslande, 1992). An example of a thematic analysis is that of drivers of small fast cars of the GTI type (Girard & Michel, 1991a, b).

Investigators have to cooperate to interpret specific road accidents and produce the related accident folders. Researchers have to cooperate with investigators so that investigators can produce relevant accident folders and so that researchers can exploit these accident folders for their thematic analyses.

2.2 Method

2.2.1 Data Collection

Accidentology data were collected from three kinds of source constituted initially for a study aiming at designing a computer system to support the accident analysis task (see Alpay et al., 1996; Alpay, 1996; Dieng et al., 1996). The sources are:

(a) INRETS reports, papers and books, e.g. on models of accidents (Fleury, 1990; Fleury, Fline & Peytavin, 1991).
(b) Interviews with a limited number of accidentologists. The purpose of the interviews was to get a description of how accidentologists view their tasks.
(c) Transcriptions of experimental accident case analyses performed by pairs and trios of accidentologists from the selected set. The purpose of the case analyses was to get an account of how accidentologists actually perform their tasks, especially how they produce the scenario of a specific accident, and how they identify the factors that determined this specific accident.

The accidentologists who were interviewed and observed were mainly researchers, and more precisely: two psychologists (E-psy1 and E-psy2), two vehicle engineers (E-veh1 and E-veh2), and three infrastructure engineers (E-infra1, E-infra2 and E-infra3). Note that the majority of these researchers had been investigators in the past. Both psychologists had conducted interviews of the drivers involved in an accident, just after the accident, and analysis of such interviews through discussions with the vehicle engineers or the infrastructure engineers who had recorded the vehicle tracks and taken photos of the infrastructure. The vehicle engineer E-veh1 had focused on the record and the analysis of the tracks left by the vehicles on the pavement. The two infrastructure engineers, E-infra2 and E-infra3, had been investigators, but only E-infra3 was still practising. E-infra1 had never worked at the scene of the accident.

In this chapter, we will make frequent references to one of the accident case analyses we observed. This analysis involved E-psy1, E-veh2 and E-infra3 who studied the following accident case:

Case 003 (see Figure 1). The accident happened on a national road, with three lanes; the central lane being non-allocated, i.e. cars coming from both directions can use the central lane. The accident occurred at night between three cars: a Toyota Sunny, a Renault 21 and a Nissan Micra. The driver of the Sunny was driving on the right lane. He urgently needed to get gasoline. When he saw a petrol station on the left side of the road, he crossed the three lanes to get to the station. The driver of the Renault 21, who was at that moment overtaking a car and thus was on the central lane, crashed into the Sunny. The driver of the Micra, who was coming from the opposite side of the road, also crashed into the Sunny.

2.2.2 Data Analysis

The analysis of the collected data consisted of identifying the multiple representations and the processes of multiple representations management, using the theoretical framework described in the next section.

2.3 Theoretical Framework

Our analysis of multiple representations in the accidentology domain is based on the work of Ostwald (1995, 1996) on supporting collaborative system design with representations for mutual understanding. According to Ostwald, in collaborative design, a shared understanding between designers is necessary to coordinate work efforts. A special case of collaboration occurs when designers and users design together. In this case, the necessary shared understanding must be created in spite of great communication difficulties.

To address the problem of shared understanding between designers and users, Ostwald proposed an approach that emphasises the construction of representations to facilitate communication among partners, or representations for mutual understanding. Representations for mutual understanding are artefacts for constructing individual and shared understandings. That is, they are external representations, i.e. some "explicit expression (e.g. verbal utterance, diagram, computer code) of some idea".

In this model, representations are said to have meaning "only in the sense that they are interpreted by someone, or something (such as a computer)". Agents interpret

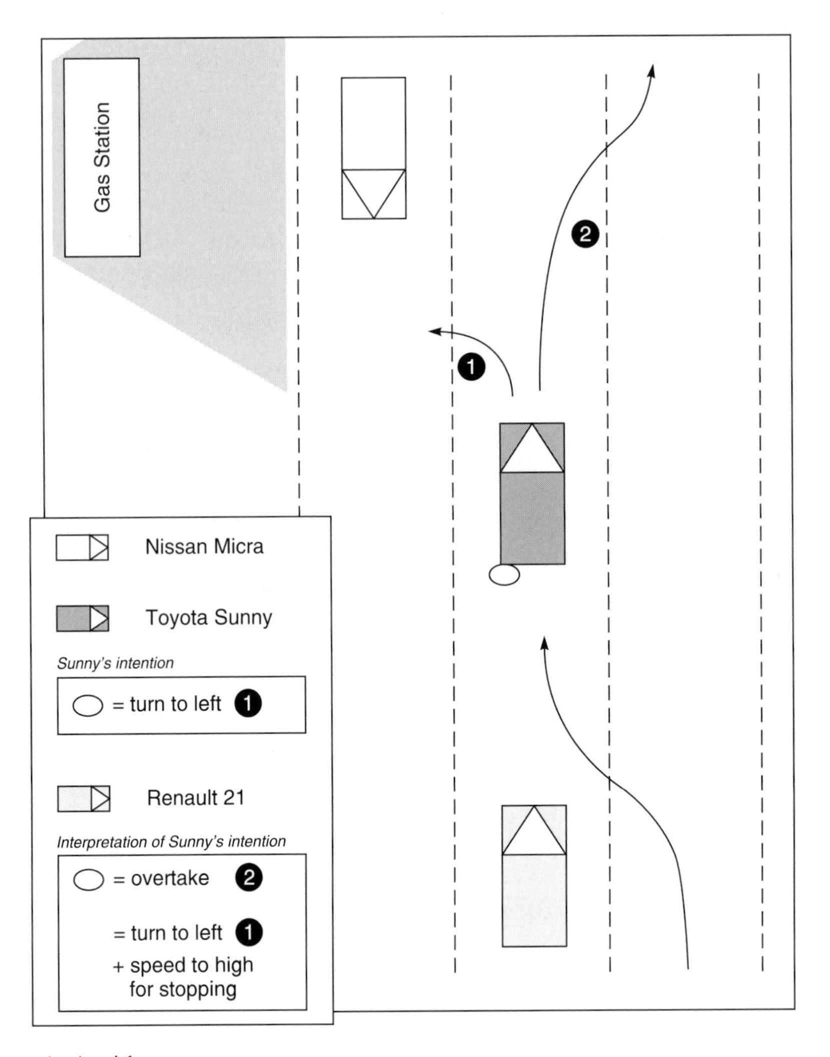

Figure 1: Accident case

representations "within a social context and against their individual background". Between members of a common culture, representations are produced (by the so-called "speakers") and understood (by the so-called "listeners") "against a rich background of shared experiences and circumstances". When speakers and listeners have a different culture, they have little shared context, or they have a different context (see Figure 2a). Such non-shared context is largely tacit and cannot be completely described or expressed. This leads to communication breakdowns. On the contrary, when the speaker and the listener share (i.e. intersection) more context (see Figure 2b), they lessen the risk of mismatch between the speaker's meaning and assigned meaning by the listener.

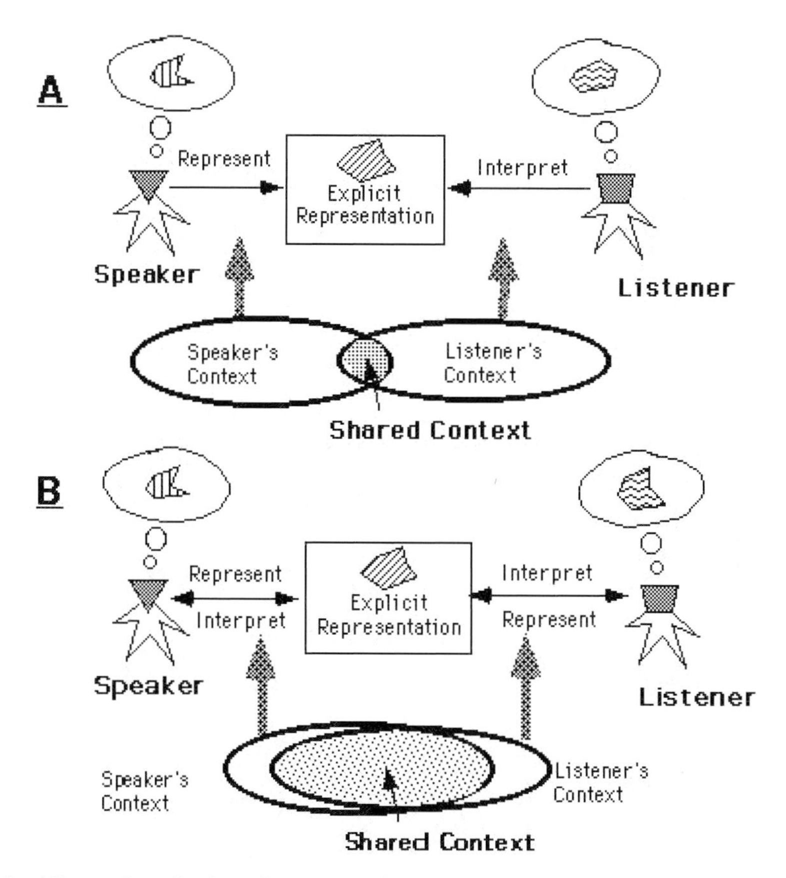

Figure 2: The role of shared context in communication (Ostwald, 1996). (A) Problematic situation: breakdowns in communication occur when there is little shared context between speaker and listener. (B) Desired situation: a shared context ground communication between speaker and listener

According to Ostwald, external design representations can help to establish such a shared context for communication. These representations provide referential anchoring (Clark & Brennan 1991; see also Bromme & Nückles, chapter 10) as an object that can be pointed to and named, helping partners to make sure they are talking about the same thing. Grounding communication with external representations helps to identify breakdowns and serves as a resource for repairing them.

Finally, representations for mutual understanding support three processes. (1) They support the *activation* of tacit and distributed knowledge by providing an object that may be reacted to. For example, descriptions of current work practice can activate domain knowledge when developers and users make the descriptions the focus of discussion. (2) They provide a shared and explicit ground for *communication*. As representations are created and discussed this common ground accumulates. (3)

They support *envisioning* of future work practices by communicating developer's ideas in a form that can be experienced by users, rather than merely imagined.

As we will see in the remainder of this chapter, Ostwald's model needs to be complemented or modified to account for the multiple representations used by the accidentologist teams. Thus, we used other existing theoretical elements, in particular:

(a) the *Common Ground* model of Clark (1992), which is referred to by Ostwald through Clark and Brennan (1991), and on which is based the chapter by Bromme and Nückles (chapter 10);
(b) the *Computational Model of Multiple Representations* (CaMeRa) of Tabachneck, Leonardo and Simon (1997), on which is based the chapter by Tabachneck(-Schijf) and Simon (chapter 11);
(c) Sumner's (1995) description of *Multiple Design Representations*. Sumner, a colleague of Ostwald, observed that a major part of a designer's job is to create and evolve external representations of the design being constructed, to facilitate communication and collaboration with each partner. Sumner called these representations, "multiple design representations". She noticed that these representations are tailored to the special needs of each of the major partners of the design team.

3 Types of Multiple Representations Used in Accidentology Teams

In the context of cooperative software design, Ostwald (1996) proposed a spectrum of (external) representations for mutual understanding, including text and graphics, scenarios (describing what a system should be, e.g. what tasks it should support and what steps are necessary to accomplish a task), simulation games and prototypes (determining how the system tasks can be performed). This spectrum can help to account for accidentology representations (e.g. accidentologists use text and graphics, scenarios and so on). However, it is not enough, and a more refined typology is necessary. At present, we have only some possible dimensions and features of such typology.

3.1 Dimensions of Multiple Representations

Several dimensions can be elicited from our analyses to describe the multiple representations used in the situations of accidentology we studied. Actual representations combine these dimensions. Some examples of such dimensions are given below.

3.1.1 Internal/External

A first dimension is the distinction between external and internal representations, or "between information in the environment and information in the brain" (Tabachnek et al., 1997), or between "Knowledge in the head — what we can remember, what we know how to do" and "Knowledge in the world — representations we create to help visualise, understand and remember things" (Norman, 1993). Accidentology situations involve both external and internal representations of information, e.g. the "DVI system" (which can be used both as an external and internal representation of information).

DVI system (Figure 3): The DVI system has three interacting components: (1) D: the driver, (2) V: his/her vehicle, and (3) I: the infrastructure (the road and its surroundings) within which D and V circulate. In some accidents more than one driver and more than one vehicle may be involved, and are thus included in the DVI system. The different components are interrelated in their actions. For instance, the driver takes information from the dynamic environment, or the driver behaves in a certain way affecting the vehicle directly and indirectly. Furthermore different vehicles may respond in different ways to the conditions of the road. In a normal situation, the three components of the DVI system, the driver, the vehicle and the infrastructure interact in accordance. However, an accident which occurs is the result of the malfunctioning between these elements of the DVI system. Aspects such as the driver's behaviour, the state of the car and the setting of the infrastructure have to be taken into account.

3.1.2 Abstract/Concrete

In the domain of software design, Sumner (1995) showed that design partners use representations at various levels of abstractions. The abstract/concrete dimension refers to the distinction between concrete and abstract representations. In accidentol-

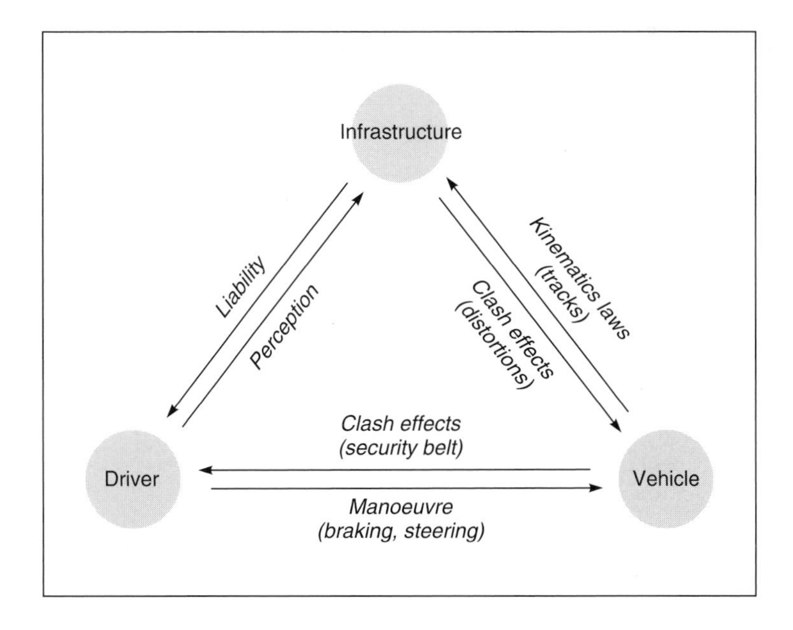

Figure 3: The DVI system

ogy, *concrete representations* are, for example, photos of crashed vehicles, photos of car crash tests (see Figure 4), photos of car tracks, textual transcriptions of drivers' interviews. *Abstract representations* are, for example, the DVI model (see Figure 3) and the functional model of the driver (see Figure 6).

Linked to this dimension is the notion of forming abstractions. In accidentology, experts in their problem-solving task have to handle various forms of information, be it "raw" data, e.g. night (the accident happened at night), or "subjective" information such as the views of the investigator on the accident. As in many other problem-solving domains (e.g. medical problem solving), in accidentology, the experts have to transform problem data supplied in the accident folder into a set of factors which explain the accident. Furthermore, owing to the interdisciplinary nature of the accidentology domain, the experts have to combine knowledge about various aspects of the accident (e.g. the car, the infrastructure and the driver). The expert's speciality will of course determine which aspect is more developed.

3.1.3 Permanent/Temporary

Permanent representations are not tied up with a specific accident case and they are reused for processing several accident cases. They are already part of the specialist's expertise. Examples of such representations are: the DVI system, generic scenarios constructed from past analyses and the model of decomposing the accident into phases.

Figure 4: Photo of a car crash test

The model of decomposing the accident into phases (see Figure 5) includes different situations such as: (i) the driving situation before the accident; (ii) the accident situation, usually created by an unexpected element; (iii) the emergency situation which occurs just seconds before the crash point, and can only be solved by avoidance manoeuvres; and (iv) the actual crash point and its consequences. Some experts personalise this model by introducing an approach situation and a pre-accident situation.

In contrast, *temporary representations* are representations which are built dynamically in the task of analysing a specific accident. In this situation, the expert progressively constructs the course of the accident using the clues which are salient, and so on. Among such representations, some may be instantiations of a permanent representation. Others may be built specifically for the case at hand. Table 1 gives an example of the content of such temporary representations.

3.1.4 Shared/Non-shared

The shared/non-shared dimension is the most crucial dimension to account for multiple representations in groups. *Shared representations* are representations that:

(i) are used by several agents;
(ii) seem to characterise accidentology independently of any discipline aspect and any given accident;
(iii) overlap across the agents' specialities;
(iv) are similar enough to be considered as variants of the same representation;
(v) are more or less agreed among the agents. Some domain models are instances of shared representations. For example:

The functional model of the driver (see Figure 6). This model is aimed at describing, from the point of view of the driver, the mechanisms that probably caused the accident and at explaining the possible driver's malfunctionings. The model decomposes the functioning of the driver into four main activity phases: information acquisition, information processing, decision of action and action.

It is worth noting that even with common shared representations, communication between experts is not automatically smooth and spotless. For example, the experts have different terminology viewpoints on the notions of scenario and factor (see section 4.2).

The opposite of shared representations, some domain models are specific to a discipline and can be viewed as *non-shared representations*. For example:

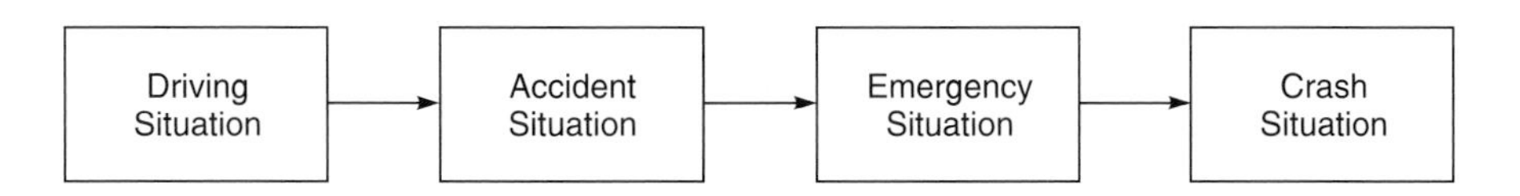

Figure 5: The permanent model of decomposing the accident into phases

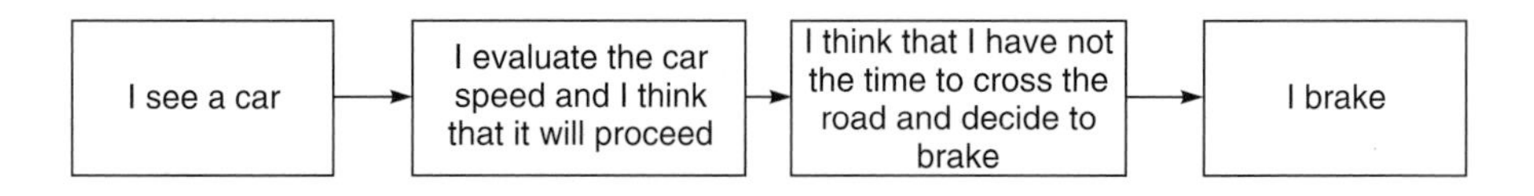

Figure 6: The shared functional model of the driver

- Only the vehicle engineers made explicit a model of vehicle's mechanical defaults and a model of kinematics sequences.
- Car tracks are mainly exploited by the infrastructure engineers and by the vehicle engineers.
- The cognitive models of the driver are typical to the psychologists and to most of the infrastructure engineers (e.g. regarding the influence of the infrastructure on the road user's behaviour). Within a given discipline, we can also take into account the specific models acquired by an expert thanks to his thematic research. For example, E-psy1 has a model of drivers' malfunctioning (see Figure 7) and a model of help to driving while the other psychologist, E-psy2, has a model of the cross-road driver and of the GTI vehicle driver.
- The detailed models of infrastructure are specific to the infrastructure engineers. Incidentally, one of the psychologists, E-psy2, has an expertise due to his thematic analyses on the drivers in cross-roads, and this expertise appears through his deep

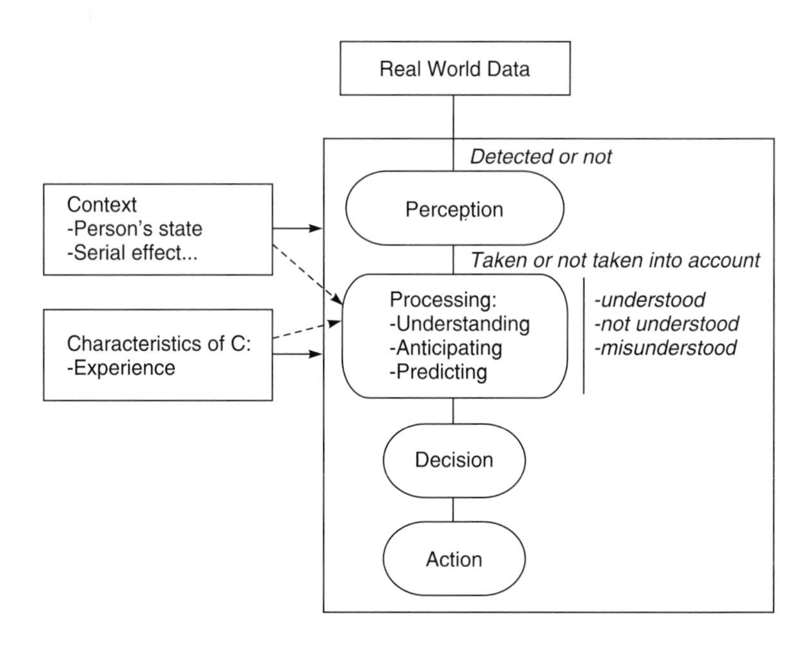

Figure 7: The non-shared functional model of the driver of E-psy1

knowledge of cross-roads. The models built by E-psy2 are specific in that they describe the infrastructure (cross-roads) as it is used by drivers.

- Models of reasoning strategies (including strategies for searching an hypothesis, for filtering an hypothesis, for testing an hypothesis) were elicited from three experts (two psychologists, E-psy1 and E-psy2, and one infrastructure engineer, E-infra3) (Alpay, 1996). Each model of how the reasoning strategies were applied, reflected the specificity of the individual expert. For example, for E-psy2 and E-infra1, the hypothesis generation was found to be a two-step process: an explication of mechanisms and an explication of factors. However, E-psy1 did not focus on the factor generation, as for him a mechanism concerned hypotheses (on the malfunctioning) and the factors which led to them. In spite of certain differences, preliminary results also show that all the experts applied the set of given strategies.

3.2 *Features of Collective Multiple Representations*

As previously stated, the shared/non-shared dimension is crucial to account for multiple representations within groups. It characterises the "collectiveness" of representations. Thus, a key issue is to determine more specifically what features make a representation collective. One way of determining this is to explain the reason for sharing representations.

According to Ostwald (1996), the function of multiple representations is to achieve *mutual understanding*. For designers and users, Ostwald argues that to collaborate in design means to come to a mutual understanding of a design solution. Our analysis of accidentology situations showed us that the function of multiple representations is to achieve both mutual understanding and *mutual agreement*. Partners in accidentology have not only to understand why an accident happened, but also to agree on a mutual interpretation of the accident; they have not only to explain/understand interpretations of accidents, but also to justify/accept these interpretations. In other words, to be said to be "collective", multiple representations have to "possess" understandability features (e.g. clarity, perspicuity, accuracy) and acceptability features (e.g. soundness, plausibility, consistency, relevance) (see Giboin, 1995).

4 Multiple Representations Management in Accidentology Teams

Describing multiple representations management is answering questions such as: Which goals do agents want to achieve in this management? Which operations do they perform, with which tools? Which factors determine management? Answers provided by Ostwald (1996) in the context of cooperative software design can help account for accidentology "retrodesign", i.e reconstruction. However, these answers must be adapted or complemented to take into account the specificities of the accidentology domain.

4.1 Goals of Multiple Representation Management

At a higher level, the management of multiple representations is aimed at supporting cooperation and collaboration among accidentologists. At a lower level, as previously stated, representation management is aimed at coming to a mutual understanding *and* to a mutual agreement on the interpretation of a specific accident (and sometimes on a thematic analysis of several accidents).

Case 003 illustrates that accidentologists also manage representations to come to an agreement on the interpretation of an accident. Accidentologists want to convince their partners of the soundness of their interpretation, or they want to be convinced. For instance, arguments are exchanged to decide which factors to include in the synthesis of the accident case. Table 1 presents the contents of the individual and collective representations of Case 003 factors mentioned by the accidentologists. Columns "E-psy1" and "E-infra3" refer to the individual representations of experts E-psy 1 and E-infra3 at the beginning of the task of elaborating a common list of accident factors (the individual representation of E-veh2 is not included because this representation was provided by E-veh2). Column "Trio" refers to the collective (shared temporary) representation resulting from the discussion between the three experts E-psy 1, E-veh2 and E-infra3, at the end of the same task. It can be noticed that this final shared temporary representation is very close to the initial non-shared representation of E-psy1.

4.2 Operations and Tools of Representation Management

To achieve mutual understanding and mutual agreement, agents perform various operations of representation management with different tools. In the context of software design teams, Ostwald indicates that designers and users perform operations such as *creation*, *accumulation*, *structuration* and *discussion* of representations. As components of the three knowledge construction processes, Ostwald focuses on activation of existing knowledge (for making explicit some tacit knowledge), communication (for accumulating and updating common ground) and envisioning (for deciding what ought to be).

In her study of a voice dialogue design team, Sumner (1995) reports on more specific operations, for example maintaining consistency across (or managing the co-evolution of) the various design representations. This operation is important because: (a) there are complex relationships between the representations; (b) this interdependency evolves as the design progresses; and (c) many design errors, as shown by empirical studies (e.g. Guindon, Kramer & Curtis, 1987), result from designers' cognitive limitations when managing the dependencies across representations.

With regards to the tools used by agents to perform the operations mentioned above, Sumner reports that designers assemble a collection of software tools, referred to as the "toolbelt", and use them to create different design representations, e.g. a

Table 1: Individual (E-psy1 and E-infra3) and collective (Trio) representations of case 003 accident factors

Factors	Representation Content Elements	Representations of *E-psy1*	*E-infra3*	*Trio*
Infrastructure	*Nissan Sunny and R21*			
	Three-lane road without direction of allocation for the middle lane ⟹ conflict between overtaking and turning left (use of the central lane).	x		x
	Blurred procedure of use for left turn.			x
	Three lanes — easy layout — straight lines ⟹ high speeds possible.		x	
	R21			
	Rectilinear infrastructure ⟹ prompts to drive fast.	x		x
	Cut in zone ⟹ prompts too close overtaking.	x		
Driver	*Nissan Sunny*			
	Focus of the activity is to reach the gas station ⟹ strong constraint: possible breakdown.	x		x
	Focus of attention on the traffic across the road.	x		x
	Problem of evaluating the speed of the R21 as it gets closer **(speed differential)** combined **with slowing-down speed of the R21 (speed regulation).**	x		x
	Little knowledge of the itinerary ⟹ no knowledge of the next possible exit for a gas station (a few gas stations further away).		x	
	R21			
	Errors of interpretation of the intention of the Nissan.	x		x
	When sees the indicator, thinks that the person wants to overtake ⟹ influence on the regulation of the speed. Thinks that he will let the car overtake.	x		x
	Knowledge of the cut in zone ⟹ the overtaking goal temporarily takes priority over the security goal.	x		x
Vehicle	*Nissan Sunny*			
	Under-inflated tyres.		x	
	R21			
	Brake disks completely bare.		x	

word processor to create text documents, a flowcharting tool to create flowcharts, a database to create tables.

As we will see, the operations and tools described by Ostwald and Sumner can be applied to the accidentology teams to a certain extent, and thus some adaptation and completion are necessary. With regards to the tools, we observe that different software tools are also used by accidentologists, for example to create, modify and exchange external representations, e.g. *CorelDraw* for creating plans of the accident site, *MS Word* to create texts and tables, *Paradox* for creating databases, *Anac2D* for elaborating simulated trajectories of vehicles, before, during and after the crash (*Anac2D* also allows adjustments of various vehicle parameters, and visualisation of the vehicle movements) (see Lechner *et al.*, 1986; Lechner & Ferrandez, 1990).

As for the operations, it can be said that accidentologists, for example, create, accumulate, structure and discuss representations. More specifically, maintaining consistency between representations is a major characteristic of accidentologists' activity. For instance, vehicle engineers explicitly search for coherence of accident scenarios when performing kinematics reconstruction. Coherence of each "individual scenario" (a scenario for each vehicle involved in the accident) and coherence of the "global scenario" (a scenario including all the vehicles involved in the accident). However, Ostwald's framework must be adapted and complemented to account for the operations performed by the accidentologists: some operations need to be specified, while others need to be added. Some directions to adapt and complement the description of operations are suggested below.

Direction 1. To refer to operations described elsewhere, e.g. to the knowledge construction operations proposed by Norman (1982) in his cognitive model of learning: *accretion* — to acquire new information, *restructuring* — to integrate the acquired information into agents' existing knowledge structure, and *tuning* — to assure a "smooth operation" of the knowledge structure, since any increase in knowledge is a change that has rippling effects through the knowledge structure (see also Rumelhart & Norman, 1981). Note that these operations only concern internal representations but they can be extended to external representations. In accidentology, tuning can take the form of "popularisation", i.e. rendering a specialised representation intelligible to non-specialist agents, and consequently making the representation shared.

Direction 2. To specify what, in the representations, is the object of operations. For example, if representations are said to be composed of data (Tabachneck et al., 1997), operations can be data acquisition (to infer internal representations of the others), and data exchange (to make internal representations explicit). For instance, some accidentologists could make their view explicit on the terminology of other experts (in particular, on the notions of scenario and of factor). In the collective analysis of Case 003, we have an example of the psychologist explicitly requesting from the infrastructure engineer information or explanations on the meaning of some infrastructure terms.

Direction 3. To distinguish operation descriptions in terms of the representation type they handle. In particular, a distinction could be made between *topic representations* (or representations) and *control representations* (or *meta-representations*). Topic representations are those representations that are under discussion, e.g. the scenario of the accident currently analysed. Control representations are those representations that guide operations on topic representations, e.g. the DVI system model, the functional model and the model of decomposition into phases to help construct the accident scenario. Thus, the maintenance of the notion of consistency across representations could be split into maintaining consistency of topic representations and maintaining the consistency of control representations. An example of the latter is the maintenance of the notional consistency. We will illustrate this by some uses of the notions of "scenario" and "factor".

- *Scenario*. E-psy1 makes a distinction between the spatio-temporal scenario of the accident and the typical scenario of the accident. E-psy2 distinguishes between the scenario and the process of the accident. E-veh1 distinguishes the typical scenario and the typical accident, and makes references to the complete kinematics scenario. E-veh2 refers to the generic scenario and the global scenario. Both vehicle engineers use the notions of first scenario and individual scenario. E-infra1 distinguishes the typical scenario (defined as a class of accidents), the permanent scenario and the *ad hoc* scenario. E-infra3 makes a difference between the family scenario and the history scenario.
- *Factor*. E-psy1 classifies factors as potential, terminal, aggravating and triggering. To the more classic notion of causal factor, he prefers the notion of initiator or explanatory element. E-psy2 emphasises the distinction between the factor (i.e. a static notion) concerning the components of the DVI system, and the failure (i.e. a dynamic notion) which takes into account the interactions between at least two of the components. Finally, E-infra1 makes a distinction between the factor and the mechanism of the accident.

Direction 4. To specify the operations in terms of the characteristics of the representations that are operated, and especially in terms of the important characteristics, i.e. those which are necessary to achieve the representations management goals. Maintaining consistency is an example of such a specification. Another example is maintaining the accuracy of representations. Sometimes the data collected by trainees or non-experienced investigators are not accurate enough for the experts to use them in their thematic studies. Thus, there is a need to train investigators to fill in the checklists better and more efficiently. That would help, not only the investigators themselves, but also the experts. Furthermore, it also means that a better communication between investigators and researchers would result in a greater closeness of representations among these two groups.

Another example is maintaining the plausibility of representations. For example in the first phase of one accident case study, one accidentologist had an hypothesis upon which he had built his reasoning. In phase two, discussions with the other accidentologist took place, and this other accidentologist used the kinematics reconstitution to demonstrate that this hypothesis was in fact wrong.

4.3 Factors Determining Multiple Representation Management

To account for the management of multiple representations in accidentology situations, we also need to identify the factors which determine it.

4.3.1 Agents' Goals

A first factor has been already mentioned. It is the goals pursued by the agents, in particular the understanding and agreement goals. These goals can be considered as meta-representations. Since the goals are multi-levelled and distributed among agents, a difficulty in multiple agent problem-solving situations is that of the matching of the agents' goals.

When constructing specific accident folders, investigators or researchers as investigators had different goals in mind. For example, in Case 003, the common goal of the researchers was to make a synthesis of the given accident. They had a goal of integration, that is, to find common elements to make this synthesis. Researchers also had individual goals such as to hold on one's own point of view, or to take minutes of the session and so on.

Other goals to be considered are problem-solving and learning goals. When an expert analyses an accident (either collectively or individually), he is usually in a problem-solving mode. In the data collected in accidentology, we cannot talk about learning goals as such. Learning had mainly been incidental, or by doing (George, 1983), resulting in new permanent representations. By doing, i.e. by performing some accident analysis action, and by evaluating the consequences of the action, the experts got feedback that led them to learn incidentally about the accident analysis task. Learning occurred when, for example, experts from different specialities have to analyse a case together, e.g. when the expert E-psy2 learns about the coding system of the infrastructure file. In another example, the infrastructure engineer E-infra3 who often works in real life with the psychologist E-psy2 was influenced by this psychologist and made for instance references to cognitive models of the driver, models adapted for his own use.

The issue of the agents' goals as a determinant of representation management could be also studied through Clark's (1992) grounding criterion, which states that speakers, to establish common ground, produce a representation allowing the partners to understand what is meant "to a criterion sufficient for current purposes".

4.3.2 Agents' Background Knowledge

In section 2, we have seen that agents produce and interpret representations "within a social context and against their individual background" (Ostwald, 1996). Background refers to the knowledge of the agents, especially their speciality knowledge. The agents' background may determine the type of representations that can be produced and understood. For example:

- Psychologist E-psy1 has been working on a typology of the drivers' errors and his analysis of traffic accidents was based upon it. E-psy2 has performed thematic studies on specific types of drivers such as old people and drivers on cross-roads. In addition, he was very interested in vehicle mechanics. Both psychologists have been living in the region for years and know the characteristics of the region infrastructure very well.
- Vehicle engineer E-veh2 has a more theoretical background and developed *Anac2D*, the software tool for kinematics reconstitution. E-veh1 uses *Anac2D*. Both engineers also know the features of the roads in the region very well.
- E-infra1 has a very deep theoretical background and has worked on a theoretical model of accidents. As he lives in another French region (Eure-et-Loire), he does not know the features of the region infrastructure as well as the other accidentologists who have been living in the region for years. E-infra2 and E-infra3 are both very interested in mechanics and vehicles. One of them, E-infra3 is also interested in the psychological analysis of the drivers, even though it is not his discipline.
- In addition to the expertise of their own speciality, the accidentologists have gained a tremendous know-how from their past experiences related to: (a) their active participation as investigators, in the past or even currently (for the psychologists, for some of the vehicle engineers and some of the infrastructure engineers); (b) their past thematic analyses as mentioned above.

4.3.3 Agents' Status

Besides the individual background is the social context. The social context refers, for example, to the agents' *status*. *Status* can be defined as the permanent or circumstantial situation of an agent within the group. Here are some examples of the agents' *status* as determinants of representations management:

- Psychologists have the permanent *status* of driver specialists. Thus, the driver-related representations they use can be considered as the most relevant ones to explain drivers' behaviour.
- Vehicle engineers have the strong permanent status of "pillars" of accident analysis: the kinematic reconstruction they perform, and the *Anac2D* tool they use (which helps them reconstruct the kinematics of the accident, thus providing *scientific* data), are the "guarantors" of the plausibility of the accident interpretation.
- *Circumstantial status* can sometimes prevail on the permanent status. For example, in the Case 003 situation, we found that the result of the integration of representations of the list of factors was predominantly guided by a "leading" member of the group, namely by the psychologist E-psy1 (see Table 1).
- A special case of the role of the agents' status is worth mentioning here: it is the status of the drivers as partners of accidentologists in accident analysis. We can distinguish partners face-to-face and in real time (in the case of the drivers with the investigators interviewing them), and distant partners and at a later point in time

(in the case of drivers whose interviews are analysed by the researchers later on after the accident has occurred). Drivers provide their representations of the accident; e.g. they verbally report their view of the accident. Accidentologists, especially psychologists, often question the reliability of the drivers' reports, because they have often to face drivers who lie, or at least transform the reality (about their driving speed for example), to minimise their responsibility. Psychologists then feel the need for a kinematics reconstitution to test the reliability of the drivers' statements.

4.3.4 Agents' Styles and Perspectives

Perspectives have been defined by Bromme (1997) as the "epistemic styles typical for a discipline or a domain of research activities". As indicated above, some researchers use abstract, formalised representations, while others use more concrete and down-to-earth representations. This results in having, at one end of the spectrum, very formalised representations and, at the other end, more concrete and down-to-earth representations. For example, in the Case 003 situation, the psychologist E-psy1 used a formalised functional model of the driver. In the transcript of this particular session, the theoretical thinking process of E-psy1 was clearly displayed.

4.3.5 Agents' Preferences

Agents' interests can determine representation management. This interest can be triggered by the use of some representation by some other agent. For example, in a case study where the psychologist E-psy2 worked collaboratively with an infrastructure engineer who used photos intensively to analyse the accident, we observed that E-psy2 came to use more photos to obtain cues explaining the accident, although he previously searched for these cues in the checklists.

4.3.6 Agents' Terminologies

Within the context of the multiple representations, the experts use various vocabularies and terminologies. Indeed, each speciality has its own terminology (see the examples given in section 4.2). Experts might share terms with other experts (from another speciality or from the same one) specifically at the abstract level. However, their terminologies become more specific and specialised at the detailed levels. With respect to the different types of terminology divergences studied (Gaines & Shaw 1989; Shaw & Gaines 1989), we only found examples of "terminology conflict", i.e. several experts giving different meanings for the same term.

4.3.7 Cooperation

In our studies in the accidentology domain, we can distinguish different types of cooperation.

- "Immediacy" of the agents' cooperation which incorporates direct collaboration and indirect collaboration proposed by Ostwald (1996).
- Indirect collaboration occurs when knowledge or products are shared through some persistent medium, such as a database or other repository. Indirect collaboration is required when direct (face-to-face) collaboration is not possible or impractical. Long-term collaboration takes place over arbitrary time frames, and also requires some persistent medium in which knowledge or products can be stored.

In accidentology, the case of indirect collaboration between investigators and researchers is worth considering. For their thematic analysis, the researchers analyse the brief elaborated by the investigators. Often the brief does not describe everything explicitly, so that researchers lack some information important for their work, e.g. information about some specific contextual element of the accident, or information about the investigators' reasoning for elaborating the conclusions of the brief. When the investigators who elaborated the brief are still present in the Department of Accident Mechanisms, they can be consulted by the researchers. The problem arises when the investigators have left the department and are no longer available; the information is then lost. To prevent such information loss, researchers are currently looking for means of having the investigators collect information that might be needed in the future.

Another case of indirect cooperation is what can be called "understanding a third party". For example, the cognitive models of the driver explicitly used by the psychologists and by the infrastructure engineers could serve as a framework for interpreting the analyses carried out by the other accidentologists about the component and its interactions with the other components, even though these other accidentologists do not explicitly evoke these models.

The issue of indirect cooperation can also be studied in terms of agents' responsibility, as defined by Clark (1992) in the context of conversations:

Principle of responsibility. In a conversation, each of the persons involved is responsible for keeping track of what is said, and for enabling the other persons of keeping track of what is said.

When writers or speakers are distant from their addressees in place, time or both, they are assumed to adhere to the principle of distant reponsibility:

Principle of distant responsibility. The speaker or writer tries to make sure, roughly by the initiation of each new contribution, that the addressees should be able to understand his meaning, in the last utterance to a criterion sufficient for current purposes.

We could say, for example, that to make their indirect cooperation with researchers successful, investigators must adhere to the principle of distant responsibility.

- Intra- and inter-cooperation: another way to look at cooperation in accidentology teams is in terms of a cooperation intra-role — between investigators or between experts; and a cooperation inter-role — between experts and investigators.

Cooperation between investigators. Investigators work at the site of the accident, and do work in a cooperative manner. For example, (i) they share the tasks, i.e. one carries out the interviews, and the other one collects information about the vehicles and the infrastructure; and (ii) they exchange information and points of view. The result of their cooperation is in a way implicitly recorded in the accident brief they have to prepare.

Cooperation between the researchers. During the analysis of the accident, all the experts share a common focus, i.e. to understand how the accident happened and to identify the accident-related factors. However, each expert has a responsibility to analyse the causes of a given accident in his own speciality, and thus will have specific sub-tasks (corresponding to his speciality) in order to bring his contribution to the common goal. Furthermore, experts will have different goals when they examine accident cases from thematic perspectives.

The cooperation in the accidentology team can be characterised as "they need each other". Typically, an expert researcher will work individually on a given accident, and will exchange information with his colleagues (from the same speciality or from another one) when he needs some data. For instance, the expert psychologist will ask the vehicle engineer about the speed of the car at the time of the accident in order to confirm or disconfirm an hypothesis. Although the expert psychologist may already have guessed at the speed, the vehicle engineer will provide accurate data.

Cooperation between the researchers and the investigators. A cooperation between experts and investigators can happen when for example an expert seeks additional information about a given accident. To do so, he has to contact the persons who wrote the report.

5 Conclusion

In this chapter, we have explored multiple representations in groups which can be found in real and complex situations of accident reports involving accidentologists from INRETS. As a theoretical framework we chose Ostwald's model of "representations for mutual understanding". Ostwald proposes an approach to cooperative software design that emphasises the construction of representations to facilitate communication among partners, or artefacts for constructing shared understandings.

From our analyses, a number of dual dimensions of multiple representations emerged such as representations which are internal/external, abstract/concrete, permanent/temporary and shared/non-shared. This latter dimension seems to be the most crucial one to account for multiple representations in groups. Furthermore, it was also found that representations used in the accidentology teams are not only for mutual understanding, but also for mutual agreement. "Speakers" use representations to explain (some accident) or to convince (about some accident interpretation); while "addressees" use them to understand or to be convinced.

Given these various types of multiple representations used in accidentology teams, we described the management of multiple representations which aims at supporting collaboration and cooperation among accidentologists. To achieve mutual understanding and mutual agreement, agents perform various operations of representations management with different tools. One of the major operations we found is that of maintaining consistency between representations. To account for the management of multiple representations, we have also identified a number of factors such as the agents' goals, background, status, styles and perspectives, terminologies and preferences.

Our results from the study of multiple representations in accidentology have consequences on several issues of multiple representations evoked in this volume. To close this chapter, we will briefly present two of these consequences.

5.1 *Dissimilarity of Representations*

The notion of multiple (or distributed) representations has been defined by Boshuisen and Tabachneck-Schijf (chapter 8) as "the circumstance where multiple human or artificial agents have dissimilar representations about one object, person, interaction or situation." Boshuizen and Tabachneck-Schijf notice that this dissimilarity (defined in terms of representations data, formats, or operators) "can be the source of many misunderstandings." Our results also suggest that we can describe the effects of representation dissimilarity in terms of "disagreements" — unless the term "misunderstanding" is used *largo sensu*, as referring also to "the act of disagreeing" (on-line *Webster's Dictionary*).

5.2 *Representations and Common Ground*

The notion of "common ground" (Clark, 1992; Clark & Brennan, 1991), used for example by Bromme and Nückles (chapter 10) to account for multiple representations in multiple agents' communication, deals with understanding aspects, not with agreement aspects of multiple agents' communication — unless the term "understanding" is used *largo sensu*, as referring also to "a mutual agreement not formally entered into but in some degree binding on each side" (on-line *Webster's Dictionary*). Our results suggest that the notion of common ground could deal with agreement aspects as well.

Acknowledgements

The authors would like to thank Els (Henny P.A.) Boshuizen, Maarten van Someren, Eileen Scanlon, Karen Littleton and Richard Joiner for their comments on earlier versions of this chapter. The first author's contribution to this work has been partly supported by COTRAO (Communauté du Travail des Alpes Occidentales, France) which funded L. Alpay's research visit at INRIA in 1995. The second and third authors' contributions have been partly supported by the French "Ministère de l'Enseignement Supérieur et de la Recherche" (Grant Number 92 C075), and the French "Ministère de l'Equipement, des Transports et du Tourisme" (Grant Number 93.0033). The authors thank the accidentologists of INRETS-Salon de Provence (Francis Ferrandez, Dominique Fleury, Yves Girard, Jean-Louis Jourdan, Daniel Lechner, Jean-Emmanuel Michel, Pierre Van Elslande) for their cooperativeness.

10

Perspective-Taking Between Medical Doctors and Nurses: A Study on Multiple Representations of Different Experts with Common Tasks

Rainer Bromme and Matthias Nückles

1 Introduction

The cognitive psychology of expertise has yielded important insights into the nature of expertise, such as the relevance of a rich and differentiated knowledge base for successful problem solving (Boshuizen et al., 1995; Bromme, 1992; Chi, Glaser & Farr, 1988; Ericsson & Smith, 1991; Evans & Patel, 1992; Hoffmann, 1992). The concept of "multiple representation (MR)" is well suited to describe the richness of expert knowledge, but MR itself has multiple aspects (Van Someren & Reimann, 1995; De Jong et al., chapter 2). According to the domains in which the experts had studied (chess, physics, medicine, architecture), and according to how well the problems on which the novices and experts had to work were defined, multiplicity of representation is manifested in different ways. Experts distinguish themselves from novices, for instance, by having several alternative strategies of problem solving at their disposal (inductive vs. deductive strategies, cf. Larkin et al., 1980). The appropriate strategy is chosen depending on the difficulty of the problem formulation. This corresponds to what Boshuizen and (Tabachneck-)Schijf (chapter 11) designate as multiple representations of operators.

Understanding the problem, i.e. developing an adequate problem representation, is essential for choosing an efficient problem-solving strategy in any domain (Larkin, 1983). In some domains the point is to establish multiple problem representations, and to be able to relate them and select between them adequately. For example in mathematics and the natural sciences often a combination of qualitative and formal representations is helpful and strategies for flexible decisions about the interplay of both representational formats are necessary. While the main difficulty encountered by novices is in relating the different types of representations, this kind of flexibility is typical for experts (see Boshuizen & Van de Wiel, chapter 12).

In case of relatively unstructured, open problems of the kind typical for practical working life situations, the point is to "read" the environment's demands and affordances in multiple ways and to construct alternative problem representations on this basis (Schön, 1983). In these cases, problem representation is constructive to a higher degree, that is different problems can be defined on the basis of a given set of demands and affordances. If, for instance, a patient consults the doctor with unclear symptoms, the latter can either develop a problem representation which aims at selecting the appropriate diagnostic procedures, or he/she can define the problem as one of selecting another specialist to whom the patient is referred.

Traditionally, in the cognitive psychology of expertise, experimental tasks require a subject to solve a problem in an isolated fashion. This is certainly true for laboratory studies in expertise, e.g. problem-solving strategies in chess playing, but it also holds for field studies in which authentic problems of professional practice have been investigated. The expert is treated as a "lonely problem solver", not only in terms of experimental designs where this may be necessary due to methodological reasons, but also in the way richness of expert knowledge is understood conceptually.

Outside the context of a psychological laboratory, experts usually do not solve their problems in an individualised fashion. Hence, the lonely problem solver is a little bit unrealistic. In many working places, experts typically need to cooperate with others: With colleagues who are educated in the same field as well as with persons having different expertise and whose work is based on a different domain of knowledge, e.g. managers and engineers in factories, psychologists and medical doctors, nurses and medical doctors. Therefore, communication and cooperation between experts is a necessary element of professional competence. To better understand this aspect of professional expertise, it is thus crucial to take into account the cognitive structures and processes which are required for cooperation and communication among experts.

The availability of multiple strategies, multiple representations of problem structures, as well as the flexibility to choose between and to combine multiple representations and strategies, are examples of what is meant by the "richness of knowledge" typical for experts. As we intend to relieve the problem solver of traditional expert research from his/her loneliness and to put him/her in a context of cooperation and communication with other experts, the question arises of how far the theoretical concepts concerning the richness of expert knowledge must be extended. Now, the concept of multiple representation takes on a slightly different meaning: now the point is the social distribution of information between individuals.

Just as the notion of multiple representations (MR) within an individual problem solver refers to different aspects of knowledge (modality, granularity, specificity of knowledge, cf. De Jong et al., chapter 2), MR within a context of cooperation entails at least two aspects (see also Boshuizen and (Tabachneck-)Schijf, chapter 8): MR as parts of a jigsaw puzzle and MR as multiple perspectives.

Experts of the same profession may cooperate because the workload is too high to do the work alone or the work has to be done continually and is distributed between shifts, e.g. in a hospital ward. In this case, MR refers to the distribution of the same type of information between cooperating team members. As any team member can only collect some pieces of information, these pieces have to be combined to form the whole picture, e.g. a patient's actual condition. Such a puzzle is composed of pieces of information whose types (modality, granularity, etc.) do not vary systematically between persons. Nurses of a ward team, for example, have access to similar information, they collect similar data on their patients and their observations are based on a similar system of professional knowledge.

Many teams include different professions and the team members' different specialisations, as well as their professional training, are the very reason of their cooperation. Doctors and nurses in a hospital have different training and different social status but within a hospital they have to cooperate. Another example can be found in the specialisations among oncologists, surgeons, anaesthesists, etc., i.e. subdisciplines within medicine. The transition from intradisciplinary to interdisciplinary cooperation is of course fluid. The members of these different subdisciplines have to cooperate for diagnostic, as well as for treatment, decisions. Their contribution to the whole picture of the patient's illness or of feasible ways of treatment are also examples of MR, but these representations have been developed from different professional perspectives. In this case, MR held by different professions and disciplines vary systematically.

How does the multiplicity of representations influence communication? This question must be answered differentially for both aspects of MR (jigsaw puzzle vs. perspectives). Although the information held by individual team members fits together like a puzzle, communication may become difficult if the interlocutors have different access to relevant information, for instance, different shifts in a hospital ward. Since the new shift cannot know what happened while the previous one was on duty, they may lack certain data. The puzzle of MR is not completed and therefore mutual understanding is threatened. Such differences of information, however, can be harmonised easily, and there are specific routines and media in all institutions with working shifts to convey information at shift change (e.g. patient sheets).

Mutual understanding becomes more difficult if the partners in communication have had different training and if they cooperate on different sub-tasks. What is the impact of interpersonal differences of specialised disciplinary knowledge on communication and cooperation? Why is interprofessional communication frequently experienced as dissatisfactory in many work fields? Typically, such questions have as yet not been closely examined from a cognitive psychology point of view. Therefore, we have set up a research programme to investigate cognitive aspects of communication among experts with different disciplinary and professional background (see http://wwwpsy.uni-muenster.de/inst3/AEbromme for an overview). In our research, we concentrate in particular on the analysis of experts' ability to take another's perspective in communication (Bromme, in press). This methodological focus relies on the theoretical assumption that successful perspective taking is an important prerequisite to mutual understanding in communication (Mead, 1934). In this context, we define the notion of perspective as the task-related cognitive component of the expertise which has been acquired during the expert's professional life and education (Bromme & Tillema, 1995). Here, "perspective" does not only comprise special methods or concepts but also the epistemological style typical for a profession or for a discipline, the so-called "Denkstil" (Fleck, 1979). We suggest use of the differences in perspectives between different groups of experts as a point of departure for the psychological analysis of interdisciplinary communication.

Our interest in the ability of perspective-taking is inspired especially by Herbert Clark's theory of grounding in communication (Clark, 1992, 1996; see also Alpay, Giboin & Dieng, chapter 9), as well as by Krauss and Fussell (1990, 1991) who used a more experimentally oriented approach to investigate the relation between perspective taking and listener-adapted communication (Keysar, 1994). These theories, however, have been developed for everyday interactions and not for professional communication between experts. Furthermore, both Clark and Krauss and Fussell used rather simple communication tasks which only required subjects to use their everyday knowledge, with no need to draw upon specialised expert knowledge (see Johnson-Laird, 1981, for a critique of Clark's theory). Therefore, we want to test whether the concepts of perspective and mutual perspective taking can provide a fruitful heuristic for a better understanding of those MR which are important for communication and cooperation between experts.

It is known from the analysis of everyday communication that a certain stock of shared knowledge, the so-called "common ground", is indispensable in achieving mutual understanding. "Two people's common ground is, in effect, the sum of their mutual, common, or joint knowledge, beliefs, and suppositions" (Clark, 1996: 93). Almost all exchanges in everyday communication are based on assumptions about what need not be said because it is known by the interaction partner. As a rule, a speaker presupposes that the listener speaks the language he uses, understanding terms in more or less the same way he meant them. Speakers and listeners in everyday communication will realise the extent of the common ground they have assumed only when misunderstandings arise. As a necessary prerequisite for communication the partners involved must develop a certain stock of knowledge about the respective partner's knowledge and about the common understanding. This also includes assumptions about the interlocutor's situation and views. When a doctor tells one of his colleagues that "The condition of the patient in 7 is stable again", this is based on the speaker's assumption that his colleague knows which patient he speaks of, what organisational unit is meant by "in 7", and which critical event is alluded to with "stable again". Besides, the speaker will assume that his colleague knows what to do with this information, and why he was told.

One's own assumptions, on which the conversation is based, are designated as **one's own perspective** and that of the other person as the **perspective of the other**. The common ground also contains a reflexive element, i.e. we know that we have a certain perspective on the object and the context of the conversation while knowing that the interlocutor has his/her view as well. These assumptions about the view of the partner of communication will be called the **presumed perspective of the other**. The logically iterative continuation "A knows that B knows that A knows" normally is not significant for planning and realising verbalisations. Reflexivity is usually limited to one level; it is part of one's own self-awareness and one's integration into the social environment.

Mutual understanding becomes more difficult if the partners have different perspectives and when the experts' perspectives are based on their different training and roles within a team. Then a speaker cannot easily infer his/her partner's perspective from his own specialised knowledge and from his own experiences. In interdisciplinary communication, knowledge about one's communication partner's perspective can no longer be simply inferred by recourse to one's own professional knowledge. This is present in interdisciplinary, interprofessional teams, as typical for hospitals. Although doctors and nurses have quite different training and different social status within the hospital, they must communicate very intensely because of the very fact that they have different tasks. So, there is an obvious tension between consensus and difference of individual perspectives in interprofessional/interdisciplinary cooperation. It is easy to think of examples where interdisciplinary work teams fail to fruitfully exploit this delicate balance.

2 Methods, Subjects and Research Questions of the Study

For our study, we have chosen experts within hospitals because medical work consists of a large amount of verbal communication: of talk between medical doctors, between medical doctors and nurses and between the different groups of experts, the patients and their relatives. As Atkinson (1995: 93; our italics) in his very interesting study on the social construction of medicine has described it: "Such talk is not simply *about* the work of the hospital. It *is* medical work." We report on a study with medical doctors and nurses from paediatric oncology units. These two professions have been trained differently and they have different tasks to fulfil. On the one hand, it can be expected that they have developed different perspectives. On the other hand, the problems they have to solve are closely interwoven: severely ill children have to be taken care of, mostly under pressures of time and uncertainty. Very concerned parents have to be informed and to be involved, and their confidence and compliance must be maintained.

We are certainly aware of the fact that doctors and nurses not only differ with respect to their professional perspectives. There are, for example, also marked differences in social status and power. Or, to give another example, in many European states, the majority of doctors still are male while nursing is traditionally a woman's job. It is well known from sociological research that these factors are important constraints for the interprofessional communication between doctors and nurses (Blickensderfer, 1996; Stein, Watts & Howell, 1990). A longitudinal approach could probably reveal in detail how the development of a person's professional perspective is also shaped by the established power relationships and the social status a specific profession enjoys. From a cognitive psychological point of view, however, we do acknowledge the impact of such sociological variables but we do not concentrate on them in our own research. This is justified by the theoretical assumption that the concept of perspective is the crucial mediating cognitive structure between antecedent variables such as those mentioned above and an expert's manifest professional behaviour.

We examined perspective taking and coping with the difference between professional perspectives in three areas: (1) we analysed problem representation, by asking what doctors and nurses know about the other group's task demands; (2) we analysed whether doctors and nurses benefit from the differences between their perspectives, by asking whether they make use of different information when venturing a prognosis; and (3) we inquired how responsibility for mutual understanding is distributed among the professions, by asking about the feeling of being understood when communicating with experts of the other profession.

2.1 Design of the Research Interview

To investigate mutual understanding between doctors and nurses, we started with an informal observation on the child oncology unit at the hospital of our university. In this preliminary study, we sought to detect points of intersection between the doctors' and nurses' work where a need for interdisciplinary communication would become evident. Our observations revealed two topics in particular that are obviously of special relevance for interdisciplinary communication between doctors and nurses: "parental consultation" and "venturing of prognoses". Based on these topics, we subsequently designed a research interview. One part of this interview contained questions referring to the task requirements of informing and instructing parents about their child's cancer. The other part was mainly concerned with aspects of the decision-making involved in venturing prognoses, i.e. estimating the probability of cure for a particular patient. The interview included open-ended questions which could be answered orally, closed questions and rating scales which we administered in a paper and pencil format (Baumjohann & Middendorf, 1997).

We created two complementary versions of the interview, one for the doctors and one for the nurses. In both versions, our subjects were asked about features concerning their own work, for example, doctors were asked which information they normally use to venture a prognosis. Nurses were asked, for example, how often they would talk to parents about questions or topics concerning their child's care during chemotherapy. At first, participants were prompted to describe their individual professional perspective on a particular topic. Additionally, however, we also required our subjects to generate hypotheses about their work partners' perspective on the topic in question. For example, doctors were asked if they thought that nurses would hold prognoses independently from the doctors' official ones. To complement these questions, we asked nurses if they thought that doctors would be competent to answer parents' questions about homecare during chemotherapy.

2.2 Overview of the Substudies and Subjects

The whole study comprises two substudies. In the first study, 10 doctors and 10 nurses from the child oncology unit at our university hospital participated in the interview (Baumjohann & Middendorf, 1997). The second study included 12 doctors and 13 nurses from two additional oncology units located at other German university hospitals. With a few exceptions, we used the same interview questions for both substudies. However, in the second substudy, we added another subscale to the interview to measure doctors' and nurses' satisfaction with the interdisciplinary communication between the professional groups. Hence, all results reported below concerning satisfaction with communication are based exclusively on the dataset of the second substudy. For all other analyses, however, we combined the datasets of both substudies to create a larger sample which would increase the reliability of our results.

2.3 Research Questions

1. Problem representation. On the one hand, how accurately can doctors assess the task demands nurses must cope with in their everyday work? On the other hand, what do nurses know about the demands doctors must cope with in their work? To address these questions, we compared doctors' and nurses' mutual assumptions about the frequency of parental questions which their respective work partners were confronted with in everyday work.

2. Using the variety of perspectives. Do nurses and doctors use the differences in their professional perspectives to solve work problems? Here, we used the problem of prognosis to investigate this question. In oncology, such prognostic assessments are especially important.

3. Commitment to establish understanding. How is the responsibility for establishing mutual understanding distributed among doctors and nurses? Here we ask for the satisfaction with interprofessional communication, for the feeling of being understood by the working partner and we compare the frequency of "being asked" and "asking" when the issue of palliative vs. curative treatment comes up. This is an important topic for nurses' as well as for medical doctors' work and the perceived responsibility for initiating discussions about it sheds light on the social distribution of the commitment to foster understanding.

3 Results

3.1 Question 1: Problem Representation

On the one hand, how accurately can doctors assess the task demands that nurses have to cope with in their everyday work? On the other hand, what do nurses know about the demands doctors have to cope with in their daily work?

To investigate this question, it is particularly useful to look at those demands which are similar for both doctors and nurses. Parental questions provide a good example of such a demand. We had doctors and nurses rate on five-point scale how often parents ask them questions about different aspects of the child's cancer. In addition we also asked both groups to assess how often they think that their respective work partners are confronted with such parental questions. Thus, doctors and nurses rated their own situation, i.e. their self-perspective, and they also provided ratings for the situation of their work partners, i.e. the presumed perspective of the respective other. Frequency ratings were collected on four different topics: (a) parental questions about medication (e.g. effects and side-effects of chemotherapy), (b) prognosis for the child's illness, (c) painful diagnostic measures (aspiration of spinal marrow, CT), and (d) nursing practice. Questions of this sort are usually raised by parents during more informal communication situations which follow the first "official" conversation between medical doctors and parents during which the parents were informed about their child's disease. Formally, it is part of the medical doctors' duties to explain the diagnosis as well as to discuss treatments with the parents, but we expect that concerned parents do not follow this formal division of responsibilities.

To check whether the demands of parental questions indeed are similar for both groups we analysed the frequencies of parental questions according to doctors' and nurses' self-assessments.

Separate analyses of variance conducted for all four topics detected a substantial difference between the two groups only for questions concerning nursing practice. On the one hand, doctors are apparently confronted less with questions on nursing practice than nurses are ($F(1,43) = 8.08$, $p < 0.01$). On the other hand, according to their own perspective, nurses have to answer parental questions on medication, prognosis and diagnostic measures as often as doctors (Figure 1).

In the next analysis, we compared doctors' and nurses' assessments of their respective work partner's situation. More concretely, we analysed how accurately a doctor can anticipate how often, on average, nurses are confronted with parental questions (Figure 2). Conversely, we also calculated how accurate or realistic nurses' assump-

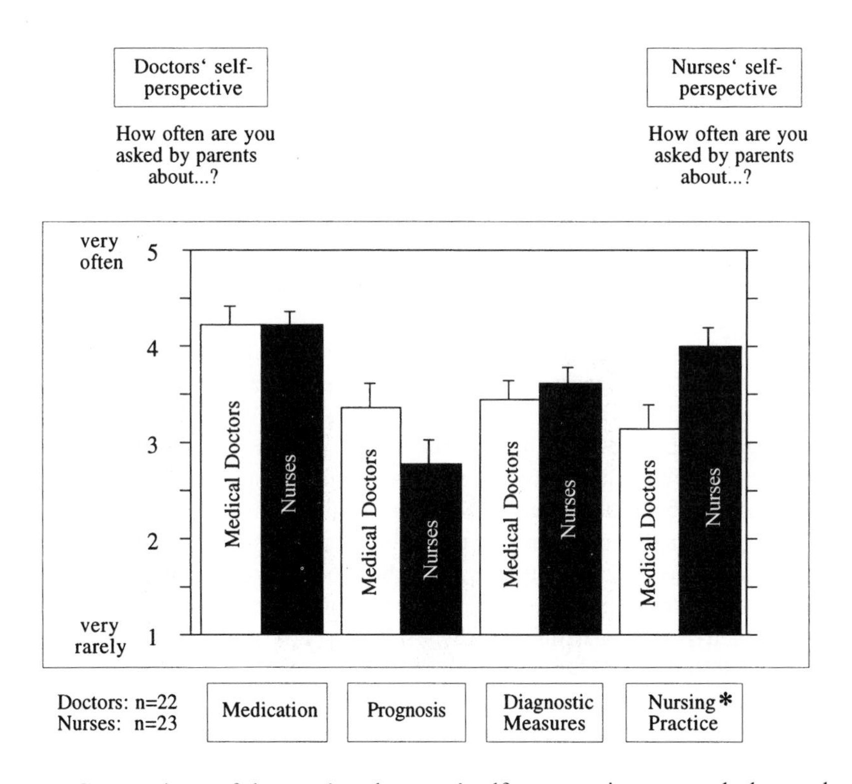

Figure 1: Comparison of doctors' and nurses' self-perspectives on task demands: Self-assessments regarding frequency of parental questions about four topics (mean ratings and SD; * = mean difference significant at 0.01-level)

tions are with respect to the frequency of parental questions doctors have to answer in their everyday work (Figure 3). In other words: we compared each group's perspectives of their work with the presumed perspective of the other group.

No differences were found for medication and diagnostic measures ($p > 0.1$). With regard to the other topics, prognosis and nursing practice, statistically reliable differences emerged. Nurses systematically overrated how often doctors are usually confronted with parental questions on prognosis ($F(1,43) = 7.5$, $p < 0.01$). Whereas, doctors overrate the frequency of nurses being asked by parents about nursing practices and methods ($F(1,42) = 7.1$ $p < 0.01$).

3.1.1 Summary and Discussion of Question 1

It would certainly have been useful, in addition to the ratings, to collect observational data about the objective task demands. In particular, the possibility that doctors' and nurses' assessments are differently anchored cannot be ruled out completely. For example, a nurse's rating of being confronted "often" by questions on prognosis

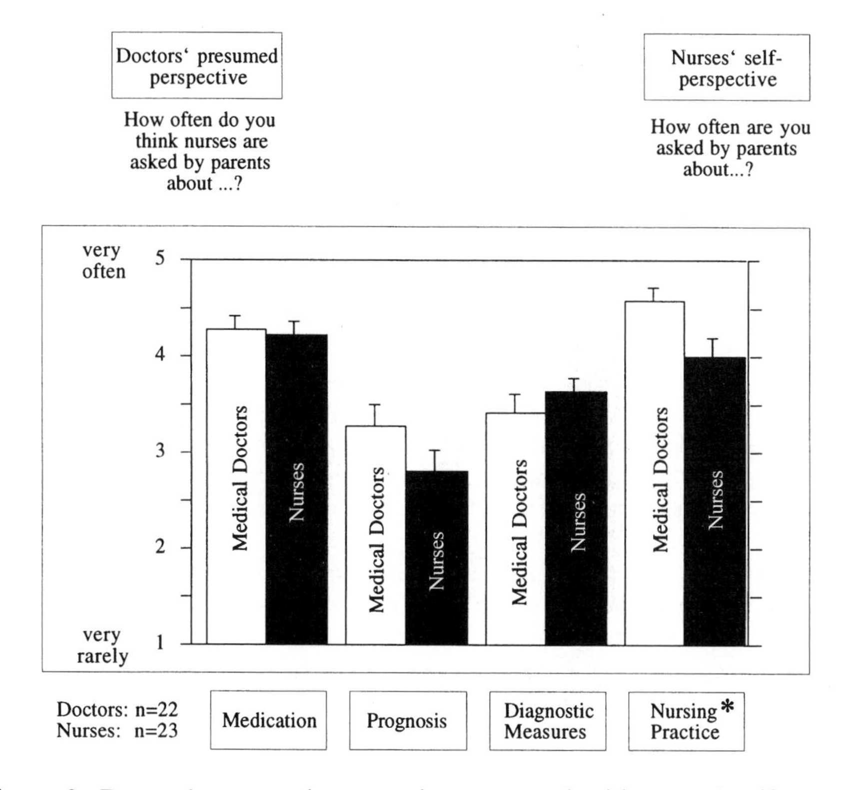

Figure 2: Doctors' presumed perspective compared with nurses' self-perspectives (mean ratings and SD; * = mean difference significant at 0.01-level)

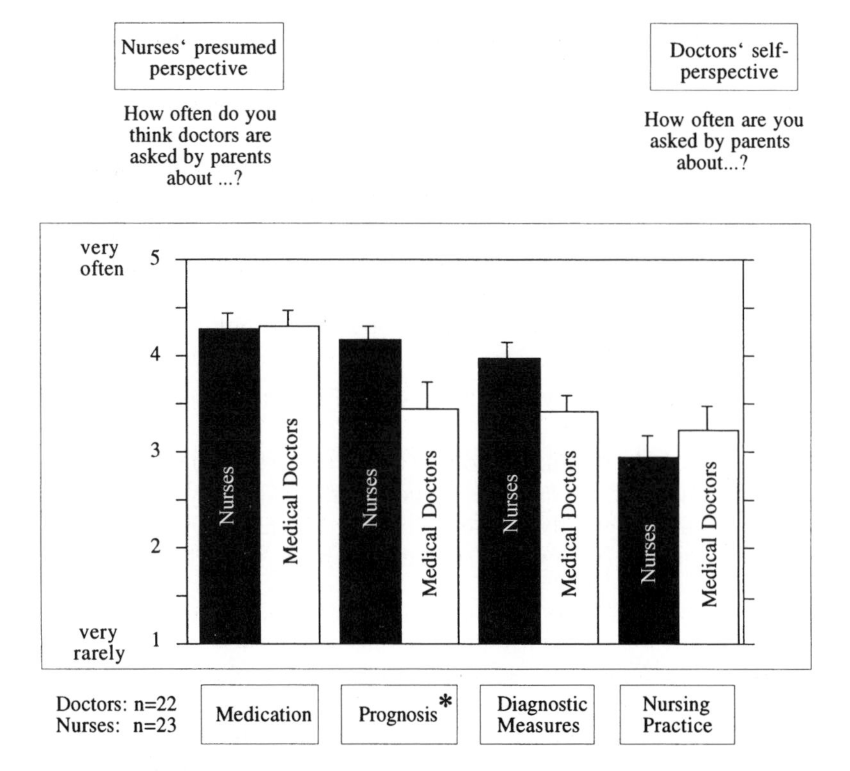

Figure 3: Nurses' presumed perspective compared with doctors' self-perspectives (mean ratings and SD; * = mean difference significant at 0.01-level)

might be based on an objectively lower frequency compared with the real frequency a doctor would presumably consider as "often". However, such different anchors are not really a problem. The crucial point is that *subjectively* the nurse's experience of being confronted is similar to the doctor's although objectively he might have to answer more questions on a particular topic. More generally, from a cognitive point of view it can be argued that on principle a problem representation is never based on objective task demands but always on the subject's *representation* of such demands.

Why do systematic overratings occur for prognosis and nursing practice, and why do they not occur for medication and diagnostic measures? A closer examination of our qualitative interview data reveals that for prognosis and nursing practice both doctors and nurses express clear opinions concerning the question of who is responsible. For both work partners it is no matter of debate that venturing a prognosis is primarily a doctors' task whereas nursing practice is clearly outside the doctors' domain. Hence, systematic overratings occur only with regard to those two topics which are clearly perceived as belonging to the other group's area of responsibility. Obviously, the perception of one's work partners' demands is strongly influenced by the stereotype images of official roles and competencies. When assessing their work partners' situation it is possible that our participants have based their assessments on such considerations of responsibility. For example, when assessing how often a doctor is confronted with parental questions about prognosis, a nurse may have reasoned: "Because venturing prognoses is clearly not my task but definitely a doctor's one they may *consequently* be confronted with such questions rather often". However, a doctor estimating how often nurses are probably asked by parents about nursing methods may have reasoned "*Because* nursing practice is outside *my* responsibility, nurses may be asked such questions more often than I am".

However, such a clear assignment of responsibilities does not exist with regard to diagnostic measures and medication. Nurses are clearly not responsible for decisions of whether a particular medication or diagnostic measure is indicated but very often they know a lot about, for example, the application of medicine or how a diagnostic measure is carried out. These competencies experienced nurses usually possess are also acknowledged by the doctors, and medical doctors rely on them in order to deal with parental questions. Consequently, the responsibilities for parental questions on medication and diagnostic measures are shared much more by the work partners than this is the case for the other topics.

With respect to parental questions, we can record that doctors' and nurses' perceptions of task demands are in fact more similar than is expected by the work partners. In particular, regarding topics such as medication, prognosis and diagnostic measures, nurses apparently feel they are confronted with parental questions as often as doctors. At the same time, however, both groups seem to systematically overestimate the practical relevance of an official allocation of responsibilities and competencies.

3.2 Question 2: Using the Variety of Perspectives

Do nurses and doctors use the differences of their professional perspectives to solve work problems?

To venture a prognosis, i.e. making predictions about the course of a child's cancer, or about the probability of cure, is usually part of a doctor's responsibility. Prognosis is certainly a very problematic and delicate affair. The prognosis ventured for a particular patient will, for example, determine the treatments to be administered; this clearly influences many decisions that may have far-reaching consequences for the patient. From this, it can be asserted that prognosis is an important issue not only for the doctors, but also for nurses.

Venturing a prognosis generally implies making a decision about the probability of survival or cure for an individual patient. Theoretically, in order to generate such a probabilistic statement, different sorts of information must become integrated. In the first place, several probabilities have to be considered, such as, for example, the standard probability of cure for a specific disease (e.g. 80% for general lymphatic leukaemia), the true positive rate of diagnostic measures, and the probability of a specific medical treatment being effective (e.g. effectiveness of chemotherapy). In addition to these objective group-based data, an individual probability has to be derived from factors concerning the individual patient's clinical state and general condition (e.g. minor diseases, his/her psychophysical resilience, and psychosocial factors). Based on this individual estimate the standard probability of cure usually has to be adjusted. Given an individual case, however, it is often difficult to collect sufficient and reliable data about the patient's general condition. Especially with respect to the assessment of a patient's general condition it can justly be assumed that nurses' perspectives on their patients may be helpful for doctors' prognostic decision-making.

Are there any competencies typical for nurses that the work partners consider relevant for venturing prognoses? Is the doctors' representational system flexible enough to allow for information which has been collected by nurses with their specific professional perspective? To address these questions, we first asked doctors whether they thought that nurses would usually develop prognoses at least partly independent from the doctors' official ones. In the case of an affirmative answer doctors were then requested to guess the nurse's presumable prognosis. Secondly, we asked doctors which sources of information they thought nurses would probably draw upon when generating a prognostic assessment. We also requested nurses to guess the doctor's presumable prognosis and to speculate which sources of information doctors would normally use to venture a prognosis. The latter questions are derived from the theoretical assumption that problem representations of different experts are at least partly dependent on the specific kinds of information available under a certain professional or disciplinary perspective.

Questions were administered in an open-ended fashion. Participants' answers were transcribed and subjected to a qualitative content analysis (Mayring, 1995). All answers were categorised and interpreted by two independent raters. Cases of diverging interpretations were decided by a third rater who was also familiar with the interviews. The following statements are selected from the interview transcripts. They are typical examples for doctors' and nurses' hypotheses regarding the question of which sources of information the respective work partner presumably uses to form prognostic evaluations.

Nurse (Interview 5, p. 6, lines 168–175):

> "Hmm. I think a nurse is normally in close contact with the patient and works closely with the individual patient. On the other hand, doctors see all the patients and their judgement is rather based on a patient's record and on the results than on his real clinical state. Because of our close contact with the patient it is easier for us to find out when he is really not alright. You can be sure that his trouble is real though medically there is no positive result. And you know exactly whether it is true what the patient says. In contrast, for the doctors it is often more difficult to decide whether the children really tell the truth or whether they are awkward and just pretending".

Doctor (Interview 4, p. 10, lines 442–446):

> "We probably regard a patient's general condition, or his clinical state, as rather secondary compared to the statistics and empirical results, well, compared to the scientific and medical point of view in general. In this respect, a nurse probably has a different point of view from which she is better in assessing a patient's general condition because she is in close contact with the patient".

First, these examples clearly show the differences between the doctors' and the nurses' professional perspectives. In particular, doctors and nurses differ with respect to the sources of information they draw upon to generate prognostic assessments. Secondly, these differences concerning the *availability* and *usage* of information are perfectly mutually reflected and acknowledged: doctors base a prognostic judgement primarily on objective information, i.e., statistical data and scientific facts. In contrast, their close contact with the patient often enables nurses to obtain information which otherwise would remain inaccessible for the doctors. In the example cited above it is, in the first place, the nurse who can determine how the patient's utterance should be understood: should his complaints be taken seriously or is he just pretending?

The fact that both nurses and doctors show good perspective taking with respect to the *availability of information* ("input-level") leads to the question of whether doctors and nurses are equally aware of their respective work partner's perspective when the prognostic *judgement* itself is concerned ("output-level"). To answer this question we had our participants imagine a particular concrete case of one of their patients and formulate their own prognosis as well as the prognosis their respective work partner would probably venture. Participants' answers were coded by two independent raters according to three categories:

A. No idea about the perspective of the other.
B. Details given, but no difference between one's own perspective and the presumed perspective of the other.
C. Details given, and a difference made between one's own perspective and the presumed perspective of the other.

The raters agreed on 80% of the cases. As before, diverging classifications were decided by a third trained rater who was also familiar with the materials (Figure 4).

Out of all 21 doctors (one missing data) only nine were able to produce an assumption about a prognosis the nurse would probably hold for the imagined case (Categories B and C). Furthermore, six of these nine doctors mentioned prognoses for the nurses that were identical to their own (Category B), that is, only three doctors took into consideration a potential distinction between their own and the nurse's presumed prognosis. In contrast, however, all nurses ($N = 20$, three missing data) were able to produce a prognosis for the patient they had selected as a case of reference. At the same time, nurses also had no problems in developing a hypothesis about the doctor's prognosis. While for a majority of nurses (13) the prognosis mentioned for the doctor was identical to their own (Category B), there were yet seven

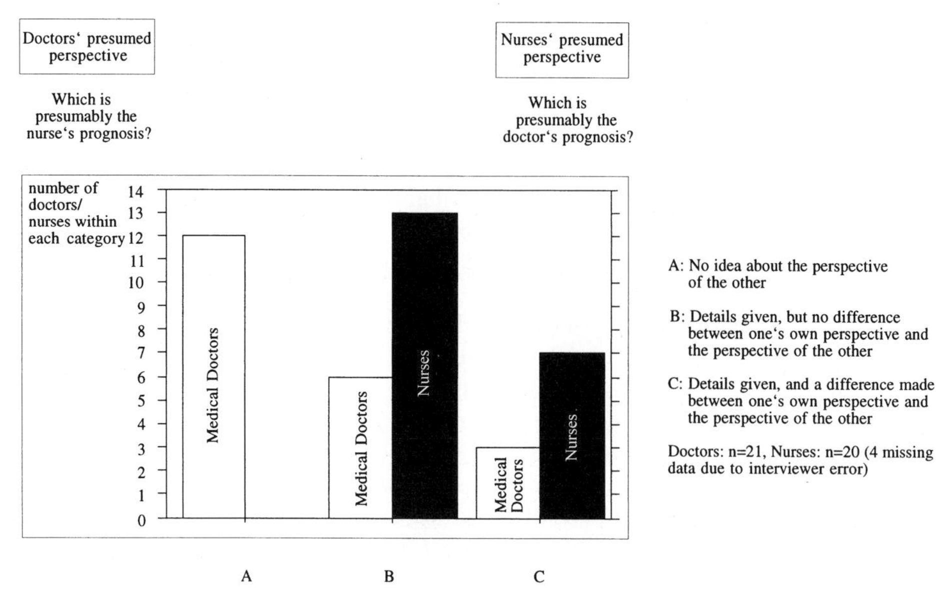

Figure 4: Doctors' and nurses' ability/willingness to differentiate between their own and the presumed perspectives of the other (absolute frequencies of cases given)

nurses whose own prognosis differed from the doctor's presumed prognosis (Category C). For example, some nurses assumed that the doctor's prognosis would be more optimistic than their own. Other nurses described their own prognosis as a subjective "feeling", while the doctor's prognosis was imagined as a probabilistic statement (e.g. "80% chance of cure").

3.2.1 Summary and Discussion of Question 2

It is indeed surprising that the majority of doctors who participated in the study apparently had no idea how nurses would probably think about the prognosis for the particular case that they had in mind. Although doctors and nurses show pretty good perspective taking with respect to the sources of information potentially relevant for prognostic judgements ("input-level"), this recognition of differences in the professional perspectives is obviously not paralleled on the level of prognostic judgements ("output-level").

Nevertheless, answering a more general item on this issue, the majority of doctors would be very appreciative if nurses developed their own opinion about prognoses. In particular, doctors expect valuable contributions from nurses especially with regard to the patients' general condition, his/her psychophysical resilience and his or her coping behaviour. In other words: the medical doctors would appreciate nurses' ideas on prognosis but they generally did not acknowledge that holding a prognosis is already an integral part of nurses' perspective and professional development. Our analysis suggests that doctors lack the representational flexibility needed to acknowledge nurses' partly alternative conceptualisations of prognostic problems.

3.3 Question 3: Commitment to Establish Understanding

How is the responsibility for establishing mutual understanding socially distributed among doctors and nurses?

Prognostic decision-making as discussed above can be regarded as a good example of MR that are socially distributed among different cognitive agents. So far, our results show that nurses' specific professional competencies are not sufficiently integrated into prognostic decision-making. It is possible that doctors' representational system simply lacks the flexibility needed to benefit from nurses' experiential knowledge about prognostic problems. Another reason may be, however, that communication between the professional groups is not effective enough. This latter possibility leads to the question of how the responsibility for establishing mutual understanding is socially distributed among the professional groups. We explored this issue in three steps. First, using a rating scale, we examined how satisfied doctors and nurses are generally with the interprofessional communication. Secondly, we had our participants rate the degree of mutual understanding usually established in conversations with interlocutors of the other profession. Thirdly, we investigated whether there are differences between doctors and nurses with respect to a personal commitment to achieve mutual understanding in interprofessional communication. All analyses to be reported in this section are based on the data collected in substudy 2 ($N=25$).

3.3.1 Satisfaction with Interprofessional Communication

First we looked at our participants' general ideas on what makes communication satisfactory. The interview contained 15 items representing variables of potential relevance for successful communication. Such variables were, for example, the interlocutor's professional knowledge, or acknowledgement of a co-worker's area of responsibility. There were no substantial differences between doctors' and nurses' relevance ratings indicating that both groups closely agree on the criteria on which successful communication generally depends (MANOVA: $F(15,9) = 1.16, p > 0.1$). This result was an important prerequisite for our next stages of analysis. We now have a basis to presuppose that doctors and nurses will apply similar criteria when asked to assess the quality of interdisciplinary communication at their work place. This became the next issue we asked about.

From our analysis of the responses, in contrast to the previous result, we found significant differences between both groups, in particular when those items are analysed which require our participants to rate how happy they are with the communication on their unit. On average, nurses are markedly less optimistic compared with

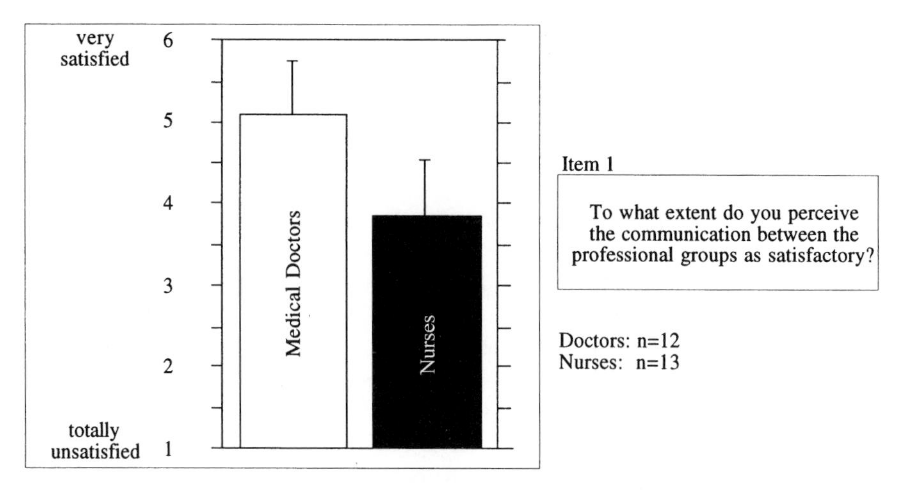

Figure 5: Doctors' and nurses' satisfaction with communication (mean ratings and SD)

doctors who rate their communication with nurses as quite satisfactory (cf. Figure 5, Item 1: $F(1,23) = 20.7, p < 0.01$).

3.3.2 Feeling of Being Understood by the Working Partner

Successful communication requires a certain degree of mutual knowledge, i.e. a common ground, to be established. As a part of this process, the communication partners routinely assess one's interlocutor's understanding of what the speaker intends to convey. Hence, if nurses perceive the interprofessional communication as less satisfactory than doctors they should equally indicate a lower degree of mutual understanding usually achieved in conversations with doctors. To investigate this question, some of our items refer to how well the person felt they were understood by the respective working partner. Participants rated on a six-point scale how often, after a conversation with a working partner of the other profession, mutual understanding has not been established sufficiently. Compared with medical doctors, nurses indicate a marginally significant higher frequency of having such unsatisfactory conversations (cf. Figure 6, Item 2: $F(1,23) = 3.8, p < 0.06$). At the same time, nurses feel more often than doctors that in a conversation with a co-worker of the other profession, he or she misses the point one wants to make, that is, the doctor misconstrues the nurse's communicative intention (cf. Figure 6, Item 3: $F(1,23) = 4.4, p < 0.05$). These differences also prove to be reliable in the multivariate approach where Items 2 and 3 were treated as dependent measures and 'profession' as the independent measure (MANOVA: $F(7,17) = 4.2, p < 0.01$). In addition, when nurses' hypotheses about the correctness of doctors' understanding are compared with doctors' self-perspective, i.e. how well they have understood what the nurse meant, the same result is obtained as before: nurses are significantly less optimistic about doctors understanding their

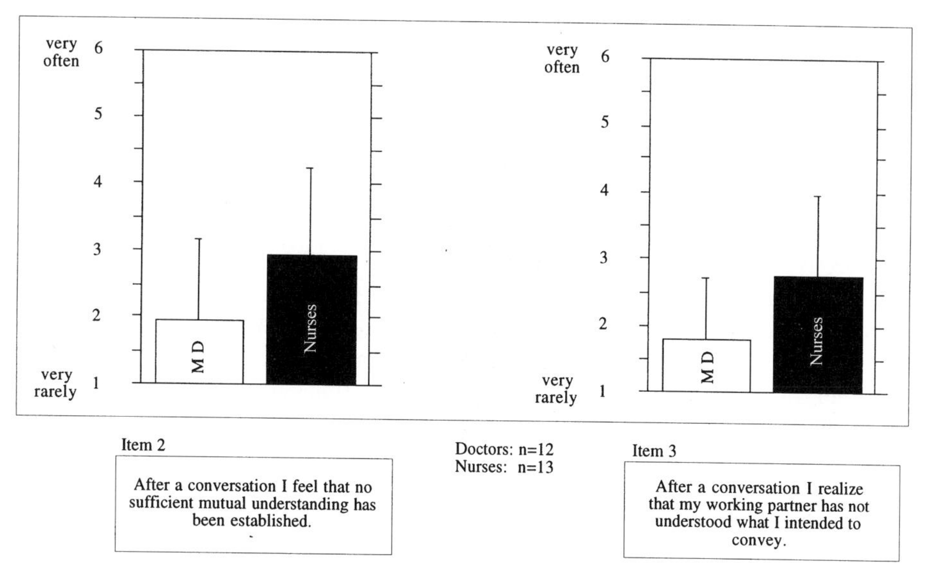

Figure 6: Doctors' and nurses' ratings of established mutual understanding and feeling of being understood by the working partner (mean ratings and SD)

intention than doctors are with respect to how well they have understood the nurse ($t(1,23) = -2.4$, $p < 0.05$).

3.3.3 Feeling of Being Responsible for Mutual Understanding

The crucial precondition for interprofessional communication being successful is certainly a personal commitment to actively contributing to mutual understanding. Are both working partners equally committed to establishing mutual understanding? Or, are there any differences between doctors and nurses with respect to the perception of such a responsibility? Since nurses are less satisfied than doctors with the degree of mutual understanding usually established in interprofessional conversations it would be plausible to assume that they also perceive a stronger commitment to achieve such understanding compared with doctors. This hypothesis is paralleled by the more general assumption that, compared with doctors, being responsible for communication going smoothly might be a normal part of nurses' professional identity.

We explored this question using again the issue of prognostic decisions. In oncology, there are also very severe cases, where a decision about the general aim of the medical treatment has to be made. Sometimes, a cancer proceeds such that therapeutic success becomes more and more unlikely while the patient suffers heavily not only from the disease itself but also from the side-effects of chemotherapy. In this situation, it can be reasonable to change medical treatment from cure to palliation. In the latter case, the central aim of treatment is to ease the patient's pain while it is accepted that there exists no way to rescue him/her.

To investigate doctors' and nurses' commitment to achieve mutual understanding in situations where the question of palliative treatment is at stake, we used a specific rating procedure. First, we had our participants rate how often they are asked by the working partner of the other profession about this question. Secondly, doctors and nurses also rated how often they themselves usually initiate a conversation with their respective working partner on this topic. Separate mean comparisons of these two items were computed for both the doctors and the nurses. The rationale of this analysis is as follows: we assume that a person feels committed to establishing mutual understanding if he/she indicates a higher readiness to talk compared with his/her rated frequency of being asked about the issue in question. Such a difference between the frequency of "asking" and "being asked" was found for the nurses ($t(1,11) = -4.0$, $p < 0.01$) but not for the doctors ($t(1,11) = 0.76$, $p > 0.1$) (Figure 7).

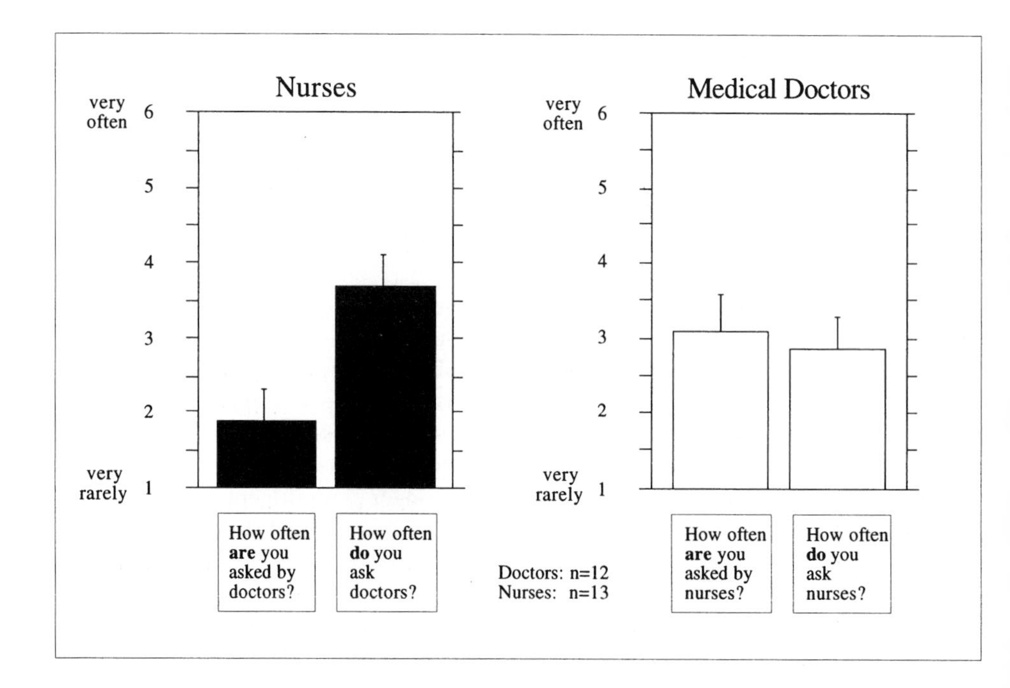

Figure 7: Commitment to establish mutual understanding. Rated frequencies of "asking" and "being asked" separately compared for doctors and nurses (mean ratings and SD)

3.3.4 Summary and Discussion of Question 3

Compared with doctors, nurses are on average less satisfied with the interprofessional communication. Similarly, nurses are significantly less optimistic about the degree of mutual understanding usually established in conversations with a doctor. Obviously nurses show a greater awareness of problems concerning the interdisciplinary communication between the two professions. In line with this awareness of communicative problems, nurses seem to perceive a commitment to achieve mutual understanding between the professions, especially when difficult prognostic decisions are concerned. For the doctors no such commitment could be found. Together, these results show that the responsibility for establishing mutual understanding is unevenly distributed among doctors and nurses.

4 General Discussion

In this study, MR distributed among different cognitive agents have been studied in three different areas. First, in problem representation by asking what doctors and nurses know about the other group's task demands. Secondly, by exploring the use of different perspectives resulting from differences between doctors' and nurses' tasks, and thirdly, by enquiring how the responsibility for establishing mutual understanding is distributed among both professions.

When experts have to cooperate, the multiplicity and the flexibility of their knowledge should be extended. It should include one's own as well as the working partner's task demands. How accurate are working partners' perceptions of the demands others are confronted with? With respect to the domain of parental consultation this question can be answered as follows. Both doctors' and nurses' assessments of their respective work partner's task demands turned out to be only partly correct. Apparently, both groups systematically overestimate the practical relevance of an official allocation of responsibilities. At the same time, the task demands, at least when parental questions are concerned, are more similar than is expected by the working partners. Hence, such a factual change of work demands seems not to be sufficiently acknowledged by one's work partners. In other words, there may in reality be more practical interprofessional similarity (interdisciplinarity) than the professionals are aware of.

Still, these conclusions should be treated with caution because they are based on rating data. Without observational data about the real task demands available, it cannot be guaranteed that nurses and doctors used the same referential system when assessing frequencies of parental questions. So, it is quite possible that, for example, a nurse referred implicitly to a different frequency when she rated how *often* she is usually asked questions about a specific topic than a doctor who provided a similar judgement. It is no doubt of theoretical as well as of practical relevance to fully explore this possibility in further studies.

The knowledge which is necessary for successful cooperation of experts is not simply an extension of the expert's knowledge base. On the one hand, cooperation clearly requires additional knowledge about the partner's perspective. On the other hand, cooperation between professionals providing knowledge from different perspectives may relieve the individual expert of the burden of complexity. However, this holds true only if the experts are open-minded and flexible enough to integrate the contributions from their colleagues. With respect to the contributions of nurses to prognosis, we found only partly such openness. When asked to think about a concrete patient and to remember their own prognosis for this case, only a small proportion of the medical doctors participating in our study was able and willing to imagine a prognosis possibly held by the nurses.

Which are the cognitive prerequisites for doctors to effectively benefit from nurses' experiential knowledge about patients when venturing a prognosis? At first, a certain awareness about the other profession's access to different information is necessary. This prerequisite is clearly given by our sample of medical doctors. Doctors are aware of the fact that nurses often have privileged access to information about the patient's general condition because of their close contact with the patient. However, these specific professional experiences of nurses with their patients apparently have no significant impact on doctors' prognostic decision-making. At least they did not result in ideas about possible alternative views which may have been developed by nurses.

A further prerequisite may be a certain conceptual openness which allows for the integration of different types or formats of representations. The medical doctors' format of their observations about patients is constrained by the features of scientific facts and data, and is mediated mostly by instruments. The nurses' format of their observations is constrained by the features of immediate, personal experience, including a more episodic structure and qualitative judgements. The transition between both formats is of course fluid and learning to integrate both types of information is part of the development of expertise (cf. Boshuizen & Van de Wiel's discussion of "knowledge encapsulation", chapter 12). Nevertheless, there are differences in content as well as in format of the processed information which are structured by the specific professional perspectives of medical doctors and nurses. It can be assumed from research on everyday reasoning that openness for different types of information, in other words, for the multiplicity of representations, depends also on the epistemological attitude which is implicitly held (Kitchener, 1983; Kuhn, 1991). By epistemological attitude we refer to the basic assumptions concerning the data normally considered as reliable and relevant, the types of arguments which are accepted without further justification and the reasons taken for granted. It ensures the coherence of diagnostic interpretations and of treatment decisions (see also Alpay, Giboin & Dieng (chapter 9) on the importance of "coherence" in interdisciplinary teams).

Working on a hospital unit usually implies that one is working under time pressure and that there is always a large amount of complex information to deal with. The flow of information is crucial for effective work but time pressure and workload may sometimes impair successful communication. Certain organisational routines (e.g. ward rounds) ensure the basic exchange of information but they cannot guarantee mutual understanding. Therefore, we asked if there is a distribution of the *responsibility for mutual understanding* among medical doctors and nurses.

The nurses in our sample were less contented with the communication with medical doctors than their work partners were. Also, they showed a greater awareness of problems concerning the interdisciplinary communication between the two professions. In particular, our data suggest that the responsibility for effective interprofessional communication is rather a part of nurses' institutional roles than of doctors'. There are many possible explanations for this imbalance. But it is beyond the scope of this chapter to discuss the impact of institutional roles and hierarchies on the distribution of responsibility for mutual understanding between experts. Nonetheless, with some caution, we may conclude that nurses' stronger commitment to mutual understanding cannot be regarded merely as a result of gender differences because, traditionally, paediatrics is not dominated by men compared with most other medical subdisciplines. Correspondingly, we had more women (12) than men (10) among the doctors in our sample, while at the same time all but one of the nurses were female.

One possible cause that should be mentioned is related to the importance of MR for expert cooperation. Any representation of professional knowledge requires a specialist language, and learning these languages is part of becoming a professional within a certain domain. Within hospitals, the specialist language of the medical doctors is clearly the standard language. The relation between this standard (specialist) language and the languages of the other professions in a hospital may be similar to the relation between standard languages and dialects. Hence, if this analogy holds true this would imply that the burden of maintaining mutual understanding rests mainly on those who do not speak the standard language.

The language and dialect analogy also sheds light on the issue of learning. Medical doctors' specialist language is not only a medium of communication but at the same time it embodies a conceptual system. Hence, the specialist language is an expression of the theories which have been developed within medicine. As a conceptual system, it offers a medium to subsume new observations, in other words, to learn by assimilating new experiences. Medicine is an experience-based discipline the knowledge of which is largely based on the discipline's own practice (i.e. on experiences with thousands of patients). For this reason, medical concepts are epistemologically open for the assimilation of personal experiences. They offer possibilities to reconceptualise personal observations and immediate experiences. This process of encapsulating new "data" into the theoretical knowledge takes a long time; it is part of learning to work and to think as a medical doctor (Boshuizen & Van de Wiel, chapter 12).

Nurses, however, do not acquire their **own** "specialist language" to discuss topics beyond nursing, e.g. the patient's general condition. Such a separate specialist language would allow them to express their observations and assumptions about patients in their own "dialect". As far as theories about diseases, symptoms and treatments are taught in nursing schools, this knowledge is again put into a conceptual framework that is soaked by the epistemology of medicine. In other words, the conceptual systems which have been learned by medical doctors and nurses are differently suited to feed representations of personal experiences into the process of interprofessional communication.

The aforementioned explanation, is, of course, only a hypothesis, but it marks a possible direction of further research. It would be fruitful to enquire into the impact of specialist languages on the distribution of multiple representations in expert teams which are composed of different professions.

In the first part of the chapter, we started with the question of how the concept of MR changes if it is used for the analysis of knowledge held by cooperating experts. In order to avoid conceptual confusion and to avoid further widening of the, already very broad, concept of MR, we suggested the notion of "perspectives" to refer to knowledge which is distributed among different persons and shaped by different (professional) approaches. Taken all together, the following aspects may be interesting for further consideration. Knowledge about the task demands of the other, representations of the presumed perspectives of the other, a certain epistemological orientation which allows for the processing of different types of information and knowledge about the possibility of different languages used by differently trained professionals. These features may be helpful to preserve the cognitive heritage of classic research on expertise, on the one hand, and to relieve the expert from his loneliness on the other.

Acknowledgements

The data reported in this chapter were collected by Monika Baumjohann, Hildegard Hospital and Sandra Middendorf as partial fulfilment of the requirements for the *Diplom* degree in Psychology at the University of Muenster. All data were completely reanalysed in preparation for this chapter. Special thanks to Riklef Rambow for many fruitful discussions and valuable comments on earlier drafts of the chapter and to Henny Boshuizen and Maarten van Someren for their interest in our work. We would also like to thank Günter Seib for improving our English and Anne Runde for technical assistance.

11

One Person, Multiple Representations: An Analysis of a Simple, Realistic Multiple Representation Learning Task

Hermina J. M. (Tabachneck-)Schijf and Herbert A. Simon

1 Introduction

In the introductory chapter of this part of the book, the term "representation" was defined as "a set of data, plus a set of operators to transform those data". In this chapter, learning is examined in the context of two such representations, namely the verbal representation and an abstract visual representation, graphs.

One of the previous chapters (Alpay, Giboin & Dieng, chapter 9) exemplified how *multiple* agents can deal with coordinating multiple representations in order to accomplish a cooperative task. These multiple agents, all experts in a different field, had to teach a single agent to collect needed data for them from one event, from which each accomplished a different goal, after which they collaborated to coordinate all the conclusions in a final report that explained the event and described possible interventions to prevent it from happening again. In another chapter in this part of the book (Boshuizen & Van de Wiel, chapter 12), studies are described where a *single* agent needs to explain a single event, pulling the needed data from multiple sets, each describing events like this at a different grain size. Knowledge within each dataset was acquired separately from the other, forming separate representations of the event. The different levels of explanation, and thus the different representations of the event, needed to be coordinated, intermingled as it were, in order to learn to explain the event.

The topic of the current chapter is similar to that of Boshuizen and Van de Wiel: we examine the behaviour of a single agent, who is learning to combine different representations in order to learn to explain an event. In the experiment described in this chapter, students were given a single (verbal) or two (verbal and graphs) representations of learning material in a carefully designed environment. The purpose of the experiment was to discover to what extent the students could learn to reason qualitatively with the verbal representation, whether adding the graphical representation would help them learn to reason qualitatively, and to what extent the students would coordinate, would intermingle the two representations during the reasoning.

This experiment is built on our results from a previous experiment. Since the design and results of the earlier experiment are relevant for the design and interpretation of the current experiment, we will describe it briefly below. Summarising it here, the outcome was that novices, on questions where information from both representations was needed for a correct answer, namely questions requiring qualitative reasoning, by and large only reasoned within a single representation. Sometimes they would reason in two representations, one after the other. They only attempted to coordinate information from two different representations if they were confronted with different outcomes from the reasoning in the two representations. The coordination consisted of redoing the reasoning in the two representations, one after the other. Only twice (of 132 answers) did subjects really make an attempt to tie information, and reasoning, in the two representations together at a smaller grain size. It was an obviously laborious undertaking.

After this (to us very surprising) rather bad performance in the questions requiring qualitative reasoning, we wondered whether novice subjects could be made to produce coherent qualitative reasoning using multiple representations. Therefore, in the current experiment, the opportunity to coordinate representations was much increased (in fact, as much as we could, we forced the subjects to coordinate representations during the tutorial) and to make sure subjects understood what was required, the type of reasoning we expected was modelled. The experiment in this chapter describes the results of this attempt. First, however, we provide the description of the earlier experiment.

1.1 The Earlier Experiment: Novices' Behaviour with Coordinating Two Representations

In this experiment, novices in economics read a tutorial explaining basic economics principles of supply and demand in words, then were presented with problems, such as finding an equilibrium, stating the consequences of shifting a supply schedule, or explaining how equilibrium is maintained. The tutorial was carefully designed to be as informative as possible, and to minimise the number of references to pictorial elements. Data for the problems were presented in informationally equivalent forms (line graphs, algebraic equations or tables), in a between-subjects design. Problems presented with the data were the same for all groups. Answering some of the problems required only numerical computations on information displayed in the data; other problems required qualitative explanations that pulled on material in both the data and the tutorial.

The purpose of the earlier experiment was three-fold. First, to establish that subjects were indeed mentally working with the same data representation they were given (i.e. that subjects who received one representation were not mentally translating them to another.) Secondly, if that could be established, to assess whether and how subjects' problem solving was affected by the different data representations. Lastly, to see how well subjects coordinated the information in the verbal tutorial with the information in the data, which were in tabular, equation or graphical formats; such coordination was implicitly required to correctly answer some of the questions.

By tracing the subjects' thought processes with talk-aloud protocols (Ericsson & Simon, 1993), it was determined that they were indeed working mentally with the representation we had given them (Tabachneck, 1992; Tabachneck & Simon, 1992). The terms and calculations subjects used could not be made to fit to another of these three representations and still fit their assigned representation perfectly.

Subjects' problem solving was indeed affected by the different representations. First, as mentioned, subjects used different terms and calculations. Actually, their protocols just looked completely different when they were using the data to answer questions. Secondly, significant statistically on numerical questions only, subjects using line graphs did better than subjects who used equations or tables. Although at first we were inclined to ascribe that to better understanding with the graphical representation, some further sleuthing (brought on by writing a computer model describing the data, plus some data from the subjects using equations and tables that we at first could not explain) uncovered that the reason for this was that the subjects often used rather superficial pictorial reasoning on these questions, relying on a simple visual strategy to answer them (i.e. find and act on those parts of the display that changed from the before- to the after-situation). With the line graphs we used this amounted to reading off numbers at intersections of supply and demand lines, and the strategy met with success. With equations and tables, where changes in the display required computation after reading off the numbers, the strategy produced incorrect answers. The same superficial strategy just happened to net the correct answer from the graph — it was not due to better understanding at all. Rather, the lack of coordinating representations resulted in wrong answers in two representations, and not in the third, because it happened to present the answer in a visually computable manner.

The third issue we wanted to examine is whether, and how, subjects coordinated the verbal tutorial with the data. To correctly answer questions requiring verbal explanations, subjects should have linked the tutorial, the problem statements and the data displays to already stored LTM memories of terms like "sale" or "price". An analysis of the verbal protocols showed that subjects in all three conditions provided rather unsatisfactory answers to the explanation questions. Even though all information was readily available, these novice subjects barely coordinated the verbal information in the tutorial with the information in the datasets when both sources were needed to answer the questions properly. Instead, their answers overwhelmingly used only one representation at a time (Tabachneck, 1992). For these questions, subjects either used material from the verbal representation, or described a particular strategy they had followed on the numerical questions which was based on the data representations, without providing a rationale for the strategy in terms of the underlying economics. For instance, one subject's answer to the question "Why does the price go up or down?": "Well, I looked at the equations and they differed by 1 (referring to the intercept values, directly comparable in these very simple linear equations) so I figured the price went up by a dollar".

In some answers, however, both the verbal and data representations were used. Most of this use evinced a fairly simple type of coordination; for example, subjects would make a connection between a label and a line, or a legend and a symbol — elements that were usually in the same data screen. On some occasions, subjects produced a slightly higher level of coordination, comparing outcomes of different types of reasoning. This occurred when verbal reasoning led to the conclusion that a variable would increase, but subsequent reasoning from the data inferred that the variable decreased. Such contradictions produced repetitions of both types of reasoning and very rarely (2 out of 132 answers), an attempt to really interweave the different reasoning modes. Even then it was not a matter of expressing the one representation in terms of the other, but going through the two types of reasoning in closer detail and comparing outcomes of subcalculations. This proved to be a very effortful endeavour.

In summary, the subjects worked with the representations they had been given; their problem solving differed accordingly and they did not succeed in coordinating the representations very well. Moreover, they seemed to lack the understanding that such coordination was needed in order to answer the questions correctly. Although they could freely do so, they referred back to the verbal introduction only rarely, thus allowing themselves very little opportunity to coordinate the representations. Lack of coordination with the verbal information led to a lack of understanding of what the elements in the visual display really meant. Hence, any inferences made from the data displays were shallow, and often wrong. Simple perception *could* have yielded the answers, but *only* to someone who had learned to notice the relevant features of the visual display, possessed the appropriate inference operators, and could translate back and forth between the operators and their real-world economic interpretations, i.e. their verbal-introduction counterparts (see also Tabachneck & Simon, 1996). The novices, who lacked this knowledge, could not obtain the answers.

Economics experts, on the other hand, readily coordinate the information in this way (for a part parallel, part serial computer model describing our hypothesis of an economics expert's memorial structure of these principles, please read the postscript to this chapter, and Tabachneck, Leonardo & Simon, 1997).

The design of the current experiment will be discussed next, as well as the materials the subjects were given to learn from. Results of the experiments and a discussion follow.

2 Experiment Design

There were two conditions in this experiment. In the control (C) condition, subjects received only a verbal tutorial. In the experimental (E) condition, the tutorial was enriched with graphs.

2.1 Subjects

The subjects were all male undergraduate students at Carnegie Mellon University in Pittsburgh, PA, USA. Only males responded to our advertisement, but since there were no relevant gender-related differences found in the first experiment, this was not deemed a problem. Subjects were paid $10 for the 2-hour session, were all native English speakers, and had received no economics instruction beyond a survey or other basic type of course in high school. None of the subjects had participated in the first experiment.

In total, 12 subjects participated in the study, seven who were audiotaped only, and five who were also videotaped. Three of the audiotaped subjects had to be left out of the analysis, since they did not follow the instructions and failed to read some crucial material; another of the audiotaped subjects mistakenly skipped problem 3. At that time a small correction to the experimental material was made so that subjects were forced to follow instructions. The last five subjects were videotaped in addition to being audiotaped. In the end, four subjects were in each condition (C and E). Four of the subjects were maths/computer science majors, two were in mechanical engineering, one was a physics major and one majored in industrial management. Six of the subjects were at the end of their freshmen year, one was a sophomore, and one a junior (respectively years 1, 2 and 3 of their study).

2.2 Materials

As this time subjects had to be able to do problems only with the verbal tutorial, we redesigned it to be sure that the teaching materials were sound. The contents were reduced to logical "building blocks", and the reasoning tested for logical rigor with mathematical software (*Mathematica*, software from Wolfram Inc.). After adjusting the building blocks, each was constructed back into normal text, buttressed with examples, and became one tutorial section. To further improve understandability, a summary paragraph was added to each section. Thus, each verbal section consisted of a summary, a main explanatory part and an example part.

For the pictorial material, a graph was designed for each of the example sections in such a way that it graphically captured all the data and reasoning of that example section. To promote coordination of the two representations, references between the text and the graph were added to the E subjects' materials. Because of a desire to keep the text part exactly the same for both groups of subjects, these references were situated in the graph rather than in the text. Where dynamic changes occurred (surplus, shortage, schedule shifts), more than one graph was shown to depict the various stages. The graphs thus depicted the same information as the verbal material. Subjects did not *have* to use the graphs in their reasoning; but as the expert found a graphical representation of the material most helpful (this was corroborated by the presence of graphs on this material in nearly every introductory economics textbook we examined), we hypothesised that having the additional representation would be helpful. (Other papers in this book review research that agrees with this point of view. Also, see section 6, the postscript on CaMeRa.)

2.3 Procedure

The experiment consisted of five parts:

1. *Pre-test.* In the first 10 min (approximately), subjects filled out a questionnaire assessing their knowledge of economics. This questionnaire was adapted from a questionnaire used by Glaser and his colleagues in their research with Smithtown, a computer-based tutor designed to research discovery-based learning in the domain of economics (see Shute, Glaser & Raghavan, 1989, for a review of the Smithtown research). The full test consisted of a pre- and a post-test, both containing 15

questions with a three-part multiple-choice answer. Subjects also rated their confidence in the answer on a five-point Likert scale. The post-test questions were of the same deep structure but a different surface structure than the pre-test questions. There was no time limit for taking these tests; subjects took about 10 min to complete each half of the test set. The pre- and post-test texts are not included here.

2. The second part of the experiment consisted of instruction in giving a verbal protocol (see Ericsson & Simon, 1993). Also, subjects were asked about their economics knowledge, and their SAT scores.
3. *Main task*. In the third part of the experiment the subjects completed the main task. The main task is described below.
4. *Post-test*. Subjects then filled out the post-test questionnaire.
5. *Draw-a-graph task*. This was followed by a "surprise" task. All subjects were given access to the C version of the E tutorial, and were asked to try to graphically represent the reasoning reflected in the example problem solution. The purpose of this task was twofold, namely: (a) whether subjects who had studied the graphs could reproduce them, and (b) whether subjects who had not seen the graph could translate the reasoning from a verbal into a graphical format. The prediction was that subjects who had seen the graphs would be able to reproduce it reasonably well, but that subjects who had not seen the graphs would have a very difficult time doing so. Since ability to memorise was not a factor of interest, subjects had full access to the text (but not the graphs) of the example problem solution and all the cards preceding it during this task. Thus, they did not have to (re-)construct the graphs from reasoning stored in memory.

2.4 The Main Task

The main task was programmed in *Hypertalk* on a Macintosh IIx computer (software and hardware by Apple computers). The program consisted of a *stack of cards* (a series of linked screens) in the *Hypercard* application. Subjects could control some aspects of the flow of access to the various cards, but not others. They could at any time during the problem solving go back to most of the cards in the stack that they had already viewed, one at a time. The task material consisted of 15 cards:

1–5: an explanation of demand, supply, equilibrium, surplus and shortage;
6: a summary of the surplus and shortage reasoning*;
7: an explanation of shifts in general*;
8: an explanation of the demand positive shift;
9: *problem 1*. The effects of the demand positive shift (the answer to this problem was fully worked out as an example of the reasoning required on the other, quite similar problems. This was also the problem Ss had access to in the draw-a-graph task.);
10: an explanation of the demand negative shift;
11: *problem 2*. The effects of the demand negative shift* (on this problem, the answer to problem 1 was accessible as a reference);

12: an explanation of the supply positive shift;
13: *problem 3*. The effects of the supply positive shift*;
14: an explanation of the supply negative shift;
15: *problem 4*. The effects of the supply negative shift*.

All but the cards marked with an asterisk were accompanied by examples and in the E condition, accompanied by graphs as well. The main text of each section first appeared, with the summary in the left hand margin (problems 2, 3 and 4 did not have a summary section). The example text belonging to each section was accessed by clicking on a button marked "LOOK AT EXAMPLE". The example text replaced the main text, but subjects could switch between them at will.

The graphs were contained in a separate *Hypercard* stack, appearing below the main stack on an Apple Portrait monitor, which is whole-page size. The main stack took up the top half of the screen; the graphical stack the bottom half. Generally, a graph appeared when a subject would access the example. In some cases (explanation of surplus, shortage and the positive demand shift, as well as the effects of the positive shift) there was more than one graph per card; in those cases, there were extra buttons distributed throughout the text marked "UPDATE GRAPH". Subjects were instructed to first read through the text above each "UPDATE GRAPH" button before clicking on it. These instructions were followed appropriately. Excerpts of the text and graphs used in the experiment can be found in Appendix 1.

For the most part, subjects could take as much time as they wished on all the tasks. Only one exception to that rule was made: it was obvious that the subjects in the E condition, looking at graphs in addition to the text and example would take longer to read the tutorial. We did not want an eventual superior performance on the E condition to be attributable to more time-on-task, and therefore ran some pilot subjects in the E condition, noted the average time to work through the example, including the graph, and set the minimum time subjects would have to take to look over the examples in the no-graph condition to that average time. Subjects were instructed (and prompted where necessary) to keep thinking aloud about the example and the problems when they were through reading until the minimum time was up. This minimum time was only enforced the first time C subjects read the information. Subjects in both conditions were asked to read aloud as well as think aloud throughout the experiment.

2.4.1 Performance Evaluation Measures

The first problem was worked out for the subjects and part of the tutorial. The question was: *reason through the effects of the positive shift in the demand schedule*. Schematically, this is how we hoped subjects would learn to reason. The headers to the main parts are emboldened. This schema provided 21 criteria on which to rate the subjects' performance:

Problem 1, effects of a positive shift in demand:

1. **Start with the system in equilibrium**
2. **Unchanged schedule point of view:** (1) price same (old equilibrium price),
3. (2) supply same (old equilibrium quantity)
4. (3) demand up (old equilibrium quantity + change)
5. therefore supply quantity < demand quantity: shortage.
6. **Shortage reasoning**: if shortage then (1) price up
7. (2) supply quantity up
8. (3) demand quantity down.
9. therefore price should be higher than the original equilibrium price,
10. and quantity should be between old equilibrium quantity and old equilibrium quantity + change
11. **Changed schedule point of view**: (1) price up (price at supply amount below),
12. (2) supply up (old equilibrium quantity + change),
13. (3) demand the same (old equilibrium quantity)
14. therefore supply quantity > demand quantity: surplus
15. **Surplus reasoning**: if surplus then (1) price down
16. (2) supply quantity down
17. (3) demand quantity up.
18. therefore price should be lower than price of supply quantity at old equilibrium quantity + change,
19. and quantity should be between old equilibrium quantity and old equilibrium quantity + change
20. **Conclusion:** price should be above price in (9) and below price in (18),
21. and quantity is between old equilibrium quantity and old equilibrium quantity + change

A similar set of 21 criteria was set up for the other three problems. To score the subjects' answers, we awarded 1 point to reasoning corresponding to each of the numbered lines (21 in total). If the subject only expressed half of a criterion, e.g. as a conclusion, "price is above a certain price", but not "price is below another price", half a point was given. If a subject only stated a correct conclusion without any other explanation, they received 2 out of the possible 21 points (i.e. credit for sentences 20 and 21).

2.4.2 Expectations for Results

We expected that the E group would coordinate the two representations, and thus we expected to find the use of graphical representational terms in their performance. If they were not using the graphical representation in their problem solutions, their performance would look nearly the same as the C group.

It was difficult to form a hypothesis on whether the C group or the E group would perform better. One type of research would predict that the E group would be expected to do better. Here, researchers seem to agree that encoding material in several representations promotes understanding, whether by providing redundancy and "webs of referential meaning" (e.g. Kaput, 1989), or by providing a more optimal reasoning medium for different aspects of problems (Larkin & Simon, 1987). Experts, with their knowledge encoded in several representations, appear to agree.

On the other hand, in our previous experiment we had witnessed the difficulties subjects encountered in coordinating representations, and felt that perhaps the effort of coordination would nullify any benefit, at least for these novice subjects in the short time allowed in this experiment. The C group, focusing on one representation, might thus perform better.

2.5 The Draw-a-Graph Task

In the draw-a-graph task, a subject received a page with some instructions and a graph-blank consisting of a box with tick marks and cross-hatching. There were no labels on the axes or the tick marks. Subjects were again asked to think out loud. The text on the page asked subjects to draw a graphical representation of the example problem solution (subjects had access to the C example problem solution and all the cards before that), and gave some hints as to what should be on the graph:

1. label the axes
2. sketch a demand line — use a solid line, put a D at each end
3. sketch a supply line — use a solid line, put an S at each end
4. sketch the positive demand shift, use a dashed line, put a D2 at each end
5. show the effects of the shift on the equilibrium price on the graph; don't forget surplus and shortage reasoning.

2.5.1 Performance Evaluation Measures

The evaluation scheme was designed accordingly. As you can see, point 5 "blossomed" into line 6 through 21, but subject had ample support for those steps in the example problem solution.

1. label x-axis (price or quantity)
2. label y-axis (quantity or price)
3. sketch correct demand line (D)
4. sketch correct supply line (S)
5. sketch correct shifted demand line (D2)
6. indicate old equilibrium
7. indicate new equilibrium
8. sketch or indicate line at $4.-

9. indicate shortage
10. indicate price going up
11. indicate supply quantity going up
12. indicate demand quantity going down
13. indicate where supply and demand will meet
14. sketch or indicate line at $6.-
15. indicate surplus
16. indicate price going down
17. indicate supply quantity going down
18. indicate demand quantity going up
19. indicate where supply and demand will meet
20. indicate price conclusion
21. indicate quantity conclusion.

2.5.2 Expectations for Results

The C subjects had never seen the graphical representation, so it was expected that they would not do well on the task, and would be quite slow at it. Given their mathematical background however, and their knowledge of graphs in general, we did expect them to be able to do some of the task.

We expected the E subjects to do much better and be much faster on this task than the C subjects.

3 Results and Discussion, Main Experiment

3.1 Background and Test Data

A family alpha-level of 0.05 was set; for exploratory purposes, we held to an alpha-level of 0.1. Table 1 shows the subject background information and the results of their performance on the economics pre- and post-test.

Subjects' SAT scores varied from 640 to 760 for the maths portion and from 470 to 720 for the verbal portion. There were no significant differences between the C and the E subjects. The scores on the economics pre-tests varied from a low of 7 to a high of 14, with a mean of 10.9 (out of a possible 15). Unfortunately, the subjects in the C conditions were closer to ceiling performance on the pre-test than the subjects in the E condition (mean C, 12.3; mean E 9.5; one-way ANOVA, $F=5.11$, $p=0.064$). Post-test scores varied from 12 to 14, mean of 13 — no differences between the groups. The gain between the tests was significant at the exploratory level (paired samples t-test, $n=8$, t-value -2.27, DF 7, alpha 0.06) and quite consistent: only one subject lost one point from pre- to post-test, all the others gained. The average post-test scores for the groups were very similar, 12.75 vs. 13.25 for C and experimental, respectively.

Subjects' confidence in their answers also increased. This gain was significant (paired samples t-test, $n=8$, t-value 8.92, DF 7, alpha < 0.01). On a scale from 1

Table 1: Main task background data and economics pre- and post-test results

S#	Cond.	SATs math	SATs verb	Major/year		Econ in high sch.	Econ test score pre	Econ test score post	Econ test score gain	Econ test confid. pre	Econ test confid. post	Econ test confid. gain
13	C	740	570	maths	jr	1 yr srv	11	13	+2	2.5	1.3	+1.2
14	C	760	720	ind.mg	fr	1 sem	14	14	0	2.0	1.0	+1.0
19	C	690	470	mech E	so	1 sem	13	12	−1	3.1	2.1	+1.0
20	C	640	630	maths	fr	none	11	12	+1	2.5	1.9	+0.6
Avg	**C**	**707**	**597**				**12.3**	**12.8**	**+0.5**	**2.5**	**1.6**	**+0.9**
15	E	750	700	physics	fr	none	11	13	+2	2.6	2.3	+0.3
18	E	*730*	*580*	maths	fr	1 sem	7	14	+7	2.7	1.9	+0.8
21	E	640	560	maths	fr	1 sem	9	14	+5	2.1	1.3	+0.8
22	E	730	620	mech E	fr	none	11	12	+1	2.6	1.7	+0.9
Avg	**E**	**712**	**615**				**9.5**	**13.3**	**+4.2**	**2.5**	**1.8**	**+0.7**

Cond: C — Control, verbal only; E — Experimental, verbal plus graphs.
SAT scores in italics are self-reported and could not be verified.
Majors: maths includes computer science undergrads; ind.mg: industrial management; mech E mechanical engineering.
Econ in highsch: economics courses taken in highschool; srv, part of survey course; sem, semester.
Econ test: economics principles test, pre, pre-test; post, post-test; gain, post-test less pre-test.
Note: control subjects are closer to ceiling in pre-test. Test is multiple choice, 3 answers/question.
Econ test confid.: ratings of subjects' confidence in the correctness of their answer; scale 1 (very confident) to 5 (pure guessing — could not eliminate any answers).

(very confident) to 5 (guessing) scores went up from an average of 2.5 to an average of 1.7. Subjects were more confident on nearly every item of the tests. There were no significant differences between the C group and the E group.

In summary, subjects did appear to have learned something. The next sections will be more specific as to what exactly they learned, and how successful we had been in inducing the E subjects to coordinate the verbal and graphical materials.

3.2 Computer Recorded Data

Performance on problem 1, worked out for the subjects as an example, was not scored. The answers to problems 2, 3 and 4 were scored.

The computer-recorded data includes the total time taken for the main experiment, which was broken up into the time to read the tutorial screens (i.e. the demand, supply, surplus, shortage, surplus + shortage summary and the shifts cards), vs. the time to do the problem-solving part of the experiment. This data, along with a summary of the data from the verbal protocol analysis, which we will discuss in detail below, is in Table 2. The problem-solving part consists of the cards supplying information on positive and negative shifts in the demand and supply, the card showing the example problem solution (problem 1 was done for the subjects), and problems 2, 3 and 4.

Table 2 shows that there was little difference between the graph and the no-graph group in *total time taken, time taken to solve the problems*, or *time taken to read the tutorial*. Only the *time taken to solve the problems* was subjected to statistical analysis, since C subjects took an average of 23.1 min, and E subjects 32.4 min. A one-way ANOVA showed no significant difference ($F(1,6) = 1.23$; alpha $= 0.31$), due to the low n and high variability in the time taken (from 15.1 to 46.8 min).

There were no significant differences in *criteria met* between the two groups ($F(1,6) = 0.34$; alpha $= 0.58$). The pattern of the *criteria met results* will be described later, since these data were gathered in the verbal protocol analysis.

An overview of the statistical data both for the main and the draw-a-graph task are given in Table 3. The results for the draw-a-graph data will be discussed in the draw-a-graph section.

In conclusion, from the statistical analyses we see little difference in the two groups' reasoning. These analyses, however, cannot tell the story of whether subjects managed to coordinate the verbal and graphical materials, and in what ways other than outcome measures C subjects' reasoning differed from the E group's reasoning. For this we turn to the results of the protocol analyses.

Table 2: Main task, results overview

		Time taken in minutes			**Prot. an. Results**
Subject #	**Cond.**	**Tutorial**	**P1–4**	**Start-Finish**	**# of met criteria**
13(1)	C	13.6	15.1	28.7	7
14	C	11.9	21.6	33.5	41
19	C	18.2	34	52.2	19
20	C	15.7	29.8	45.5	35.5
Avg.		**14.9**	**25.1**	**40.0**	**25.6**
15(2)	E	20.8	27.8	47.5	17.5
18	E	15.2	30.5	45.7	19.5
21	E	17.1	46.8	63.9	26
22	E	9.1	24.3	33.5	21
Avg.		**15.6**	**32.4**	**47.7**	**21**

(1) Missed two examples, supply positive shift and negative shift.
(2) Missed one example, supply negative shift.
Cond: C — Control, verbal only; E — Experimental, verbal plus graphs.
# of met criteria: total number of criteria met.

Table 3: One-way ANOVAS, main experiment and draw-a-graph task

Dependent variables	**Independent variable = condition**						
	Verbal only		**Verb. + graph**				
***n*=4 per group**	**mean**	**SD**	**mean**	**SD**	**DF**	***F*-ratio**	**Sign. *F***
Main exp:							
time Probl. 2,3,4	25.2	8.4	32.4	10.0	1, 6	1.23	0.31
# criteria met	25.6	15.5	21	3.6	1, 6	0.34	0.58
Draw-a-graph:							
Time	14.3	0.6	8.3	5.8	1, 6	4.23	0.09#
# criteria met	6.8	5.9	13.8	6.0	1, 6	2.92	0.14#

approach significance.

3.3 Verbal Protocol Analysis

As steps 20 and 21, the final conclusions, constitute what would normally be considered the outcome measure, we will look at them first.

3.3.1 Final Conclusions

An overview of the final conclusions given by subjects (criteria 20 and 21) is shown in Table 4. Correct conclusions are emboldened.

Subjects gave a variety of answers. Some subjects gave ranges as well as exact interpolated answers, others gave one or the other answer. The range-answer was targeted. An exact interpolated answer by itself was not counted as correct. Some subjects did not get both of the boundaries for the new equilibrium price or quantity, but many did. On problem 2, only two subjects got the answer entirely correct, one C and one E. On problem 3, four subjects were completely successful, three C and one E; another E was nearly successful. On problem 4, four subjects (three C and one E) solved the problem correctly.

3.3.2 Other Criteria

An overview of the results of the criterion analysis for problems 2, 3 and 4, is given in Table 5.

Total scores ranged from a low of 7 to a high of 41. There was no significant difference in *number of criteria met* between the conditions. Subjects did fairly well on the first half of this task. The second half was not as well reasoned through. Most of the subjects had a good handle on the consequences of a shortage or a surplus, and could reason about when a shortage or surplus would occur. The problem was in establishing the second half of the reasoning. Many subjects just could not reproduce

Table 4: Main task, final conclusions

S#	Cond.	CS	Problem 2	CS	Problem 3	CS	Problem 4
13	C	4	$3,400 lbs*	1	$3,600 lbs*	2	$5,400 lbs*
14	C	11	**2 < P < 4;** 200 < Q < 400	14	**2 < P < 4;** **500 < Q < 700**	5	**4 < P < 6;** **300 < Q < 500 –**
		5		5			$4,400 lbs
19	C	5	no conclusions	7	**2 < P < 4;** **500 < Q < 700**	7	**4 < P < 6;** **300 < Q < 500**
20	C	14	**2 < P < 4;** **300 < Q < 500 –** $3,400 lbs	12 5	**2 < P < 4;** **500 < Q < 700**	9	**4 < P < 6;** **300 < Q < 500**
15	E	6	**2 < P < 4;** 200 < Q < 300	4.5	**2 < P <** 3; **500 < Q < 700**	2	**4 < P < 6;** **300 < Q < 500**
18	E	7	**P < 4**; Q?	4	no conclusions	7	no conclusions
21	E	7.5	**2 < P < 4;** 100 < **Q < 500**	8	$2-; **500 < Q** < 900	13	$6,500 lbs
22	E	7	**2 < P < 4;** **300 < Q < 500**	9	**2 < P < 4;** **500 < Q < 700**	5	P = 6, Q > 500

C — Control, verbal only; E — Experimental, verbal + graph; S# — subject number; CS — number of correct reasoning statements (see Table 5).
Correct parts of conclusions in bold.
*Correct interpolated answer, but a range answer was required.

the part that dealt with establishing where the supplier would have to peg the price to fully meet the changed demand (in the changed demand problem), or, conversely, what quantity would be demanded if the supplier insisted on pricing the entire change in the supply schedule through to the consumer (in the changed supply problems).

Below is a brief description of the subjects' behaviours while solving problems 2, 3 and 4. The C subjects never used any graphical terms. Of the E subjects, interestingly, two neither drew graphs to help their reasoning nor used graphical terms, while the other two did draw graphs. Of the latter two, one reasoned almost exclusively in the graphical representation. The other, however, actively attempted to coordinate the two representations.

3.3.3 Control Subjects

Subject 13, verbal protocol analysis. This subject is the worst of the set in meeting criteria: seven in total, out of 63, for all the problems together. He does have all the correct answers, however, but in an interpolated format, in spite of remembering (at a point where the example solution is available to him) that the example problem solution contained a range. He does not check the example solution. Instead, he says that he has no idea how we obtained the range. S13 appears to use the given

Table 5: Main task, criterion match results

Verbal only

subject number		13			14			19			20		
problem number		2	3	4	2	3	4	2	3	4	2	3	4
begin system in equilibrium	1				C	C	C	C	C	C	C	C	C
unch. sched.: price same	2	C			C	C	C		C	C	C	C	C
unch. sched.: qty. unch. sched. same	3				C	C	C			C		C	
unch. sched.: qty. chgd, sched. up/dn	4		C		C	C	C	C	C	C	C		
result 2,3,4: surpl. or short. (s/s)	5	C			C	C	C	C			C	C	C
s/s: price up/dn	6	C			C	C	C			C	C	C	C
s/s: supply quantity up/down	7	C			C	C	C				C	C	
s/s: demand quantity up/down	8							C			C	C	C
res. 6,7,8: price lowr/highr t.eq.	9				C	C		C			C	C	
res. 6,7,8: qty betw. eq & eq +/- ch.	10				/	/						/	
ch. sched.; price up/down	11			C	C	C	C		C		C	C	C
ch. sched.: qty. chgd sched. same	12			C	C	C	C		C		C	C	
ch. sched.: qty unch. sched. up/dn.	13					C	C						C
res. 11,12,13: surpl. Or short.	14				X		C						
s/s: price up/dn	15						C						
s/s: supply quantity up/down	16						C						
s/s: demand quantity up/down	17										C		
res. 15–17: price lowr/highr t.eq.	18					C					C		
res. 15–17: qty betw. eq & eq +/- ch.	19												
concl.: price betw. a and b	20				C	C	C		C	C	C	C	C
concl.: qty. betw. c and d	21				X	C	C		C	C	C	C	C
		P2: 4 P3: 1 P4: 2 all: 7			P2: 11.5 P3: 14.5 P4: 15 all: 41			P2: 5 P3: 5 P4: 5 all: 19			P2: 14 P3: 12.5 P4: 9 all: 35.5		

Verbal plus graph

subject number		15			18			21			22		
problem number		2	3	4	2	3	4	2	3	4	2	3	4
begin system in equilibrium	1	*C*	C	C	C	C	C	C	C			C	
unch. sched.: price same	2							C		C	C	C	C
unch. sched.: qty. unch. sched. same	3		C		C		C	C		C		C	
unch. sched.: qty. chgd, sched. up/dn	4		C		C	C	C	C	C	C	C	C	C
result 2,3,4: surpl. or short. (s/s)	5	C			C	C	C	C	C	C			C
s/s: price up/dn	6	C	C		C		C	C	C	C	C	C	C
s/s: supply quantity up/down	7	C				C	C	C		C	C	C	C
s/s: demand quantity up/down	8					X	C			C	C		
res. 6,7,8: price lowr/highr t. eq.	9	C			C	X						C	
res. 6,7,8: qty betw. eq & eq +/- ch.	10					X							
ch. sched.: price up/down	11				C	X				C			
ch. sched.: qty. chgd sched. same	12							C		C			
ch. sched.: qty unch. sched. up/dn	13				C					C			
res. 11,12,13: surpl. Or short.	14	*C*						XC	C	C			
s/s: price up/dn	15		C						C				
s/s: supply quantity up/down	16												
s/s: demand quantity up/down	17												
res. 15–17: price lowr/highr t. eq.	18												
res. 15–17; qty betw. eq & eq +/- ch.	19												
concl.: price betw. a and b	20	C	/	C	/			C	CX	CX	C	C	CX
concl.: qty. betw. c and d	21	C	C	C				/	/X	CX	C	C	X
		P2: 8 P3: 6.5 P4: 3 all: 17.5			P2: 8.5 P3: 4 P4: 7 all: 19.5			P2: 9.5 P3: 6.5 P4: 11 all: 26			P2: 7 P3: 9 P4: 5 all: 21		

C — correct; / — half correct; X — error that was not immediately corrected. Empty cells: omissions.
Normal script: info from the verbal protocols; italics: info from the drawn figures, not verbalised.

data as tables and simply interpolates the matching numbers without bothering with much reasoning.

Subject 14, verbal protocol analysis. S14 is one of the best subjects. He meets 43 of the 63 criteria, 8.5 out of 10 on the first part of the reasoning (*from the point of view of the unchanged schedule...*) on P2 and P3, and 7 of 10 on P4. On the second half (*from the point of view of the changed schedule...*) he improves with each problem, and meets 3 criteria on P2, 6 on P3 and 8 on P4. Most of the criteria he does not meet are the intermediate conclusions at the end of the first and the second part of the reasoning. He does, however, get 5 of the 6 final conclusion statements correct.

Subject 19, verbal protocol analysis. This subject only satisfied 19 of the 63 criteria in the set of problems. There are almost no criteria met in the second part of the reasoning; in spite of that, he gets the correct conclusions in the last two problems by looking at the data in the problem cards.

Subject 20, verbal protocol analysis. S20 is pretty solid in the first part of the solution, and patchy in the second part. He amasses 35.5 points out of the total of 63, just better than half. S20 also voices all the correct conclusions.

3.3.4 Experimental Subjects

Subject 15: uses mostly graphical reasoning. S15 does not meet many of the criteria, just going by his verbal protocol (14.5 of 63). He mentions nothing from the second part of the reasoning, and there is scant verbalisation of the first part. However, much of the reasoning is done from the graphical representations that he spends considerable time drawing; the reasoning drawn on the graph has been credited as well (3 more criteria). The following gives a closer look at how this subject coordinates the graphical and verbal representations.

P2: *High emphasis on graphical representation.* In this problem, S15 uses very little verbal qualitative reasoning from the tutorial; he reasons mostly from the graphical representation. S15 first spends some time trying to draw the correct graph, using the data provided, then checks with the demand positive shift example graph and picks up information on the axis labels and line directions of supply and demand. He then draws some more, while checking the information in both the main card and the example card of the example problem solution. Concludes that the results here should be the opposite in the example solution: "Where before, you had a higher price and a higher quantity, logically, you should have the lower price and lower quantity when you have the negative shift", so he writes in "unchanged supply, shortage" and "changed demand, surplus". Then S15 draws a few numbers next to the axes and states the full correct conclusions.

P3: *High emphasis on graphical representation, followed by an attempt to cross-check in the verbal representation.* Again S15 reasons mostly with the graphical representation and in graphical terms. After he is done drawing, he tries to check his answer with a brief attempt at verbal reasoning. Figures out the correct placement of the shifted supply curve in the *negative shift in supply* card; first draws it on the wrong side. Looking at the problem 2 card, he labels the axes and draws in the price and quantity amounts (rather sloppily), checking on the example in the *positive shift in demand* card. Tries to read the price and quantity boundaries off the graph, gets the price wrong, probably because of sloppy axis markings.

P4: *High emphasis on graphical representation; an attempt to coordinate with the verbal tutorial is not successful.* Again, has problems drawing the shifted supply line; hesitates as to what side it should go on. Gets it on the correct side. Comparing the two intersections, he concludes that the price will be greater, but the amount lower. Says " . . . doesn't make sense, because — supply had a negative shift". Checks on *negative shift in supply* card (no graph available in there), is somewhat confused on the explanation, returns and decides to just look at the graph for an answer. Cites correct conclusions "so looking at — quantitatively looking at — perhaps from four to six dollars . . . and — 300 to 500 lbs"; correct because the labelling was less sloppy on this graph.

Subject 18: uses exclusively verbal reasoning. S18 does not draw graphs while answering the problems; he just jots down some numbers. He was fairly successful at the first part of the reasoning (17 criteria met) but only met 2 criteria in the second part of the reasoning, and only has 1/2 point on all of the conclusions. At times, the reasoning here looks like "term salad", throwing around terms without much of an idea of what they really mean. Although S13 looked worse than S18 on paper, he at least got the correct conclusions, albeit in a non-wanted format, and whatever reasoning he did was logical. There are no graphical terms used in the reasoning, and S 18's reasoning looks much like the C subjects'.

Subject 21: explicitly attempts to coordinate the verbal and graphical representations. S21 passes 18 criteria in the first part of the reasoning, and 8 in the second part, for a total of 26 of the 63. He gets 5.5 of the 6 conclusions correct. This subject spends a lot of time attempting to draw a correct graphical representation of the problems. His protocol reveals that *meaningfully* translating the verbal problem statement into a graphical representation is an error-prone process, as well as a slow one. He takes nearly 47 min, much more than subjects using only a single representation (all between 15 and 34 min, mean about 26 min).

P2: *Interactive reasoning, actively attempting to connect the verbal reasoning to the graphical lines and areas.* On the second problem, S21 begins by drawing a graph. He is not very precise with the line-up of the quantity tick marks and the position of the lines on the graph, and he ends up giving the wrong price range. S21 reviews the example problem solution thoroughly, and gets most of the criteria in the first half of the reasoning, some in the second. Works very interactively with the graph, going back and forth to the example problem, and connecting bits of verbal reasoning to graph areas.

P3: *More reliant on the graph, successfully incorporating bits of verbal reasoning.* Begins with drawing the graph again. Has some problems with the quantity direction

— increasing or decreasing along the x-axis because he is confused about what line to assign to supply or demand (cannot look at the example any more). Again, is not careful with the lineup of the quantity tick marks and the demand/supply lines. Draws a close-up of the area bounded by (resp., $ and lbs) (4,500), (2,500) (2,700) and (4,700), with an X from corner to corner. This is a correct graph, but S21 gets confused and wants to add in the 900 lb mark mentioned on the changed supply, ending up with an X with endpoints (4,500), (2,500), (2,900) and (4,700). Thus, he incorporated the 900 mark into the demand, not the supply. He then changes that to (4,500), (2,700), (2,900) and (4, 700) — a straight vertical line with a negatively slanted line through it. It is unclear what is done with these close-up graphs, except that the conclusion is that the price should go from $4.- to $2.-. Subject now concentrates back on the whole graph. He has connected the triangular area above the new equilibrium price/quantity to "surplus", and the triangular area below to "shortage". S21 now reasons almost entirely from the graph, incorporating bits of verbal reasoning, and not without success, but it makes him very dependent on how well he drew it. Conclusion is "decrease price from $4 to $2, and — change their — supply will have to tone down from — they'll have to find a better — from 500 to 900 lbs, with the optimum equilibrium price somewhere around 350 and 600 lbs".

P4: *More reliant on the graph, successfully incorporating bits of verbal reasoning.* Starts drawing the graph, getting much faster. Puts in demand and old supply line, and tick marks on the axes at $2, $4, and $6 and 300, 500 and 700 lbs. Draws in the surplus/shortage triangles, this time properly lined up. The reasoning sounds very nice, both parts of the reasoning, it sounds like he understands this. Bits of verbal reasoning are interlaced, and he appears to understand the verbal (i.e. economics) meaning of the graph lines and areas. However, it also becomes clear that he has not picked up the fact that the intersection between the demand and the new supply is the new equilibrium. He says: "and somewhere in here — I think — I hope — so — the new equilibrium price should be — approximately there — or it could be there" while putting two marks on the demand line, one below the new equilibrium, one just above. And the conclusion: "effects are when you have a shortage, and the conclusion has been stated that you need to — increase — oh wait. The demand. You must increase the demand. So, the only way to do that is to increase the price from $4 to $6 dollars and increase the supply from — 300 to 500 — to eliminate the shortage. Yup." It is interesting that despite the elaborate graphs and having so many pieces of the reasoning, S21 never quite gets the correct conclusions. It just seems to get too complicated: exactly what we had expected with the double-representation load.

Subject 22: uses exclusively verbal reasoning. In contrast to S15 and S21, but like S18, S22 does not use any paper and does not draw any graphs. He does frequently review the tutorial text and the problem example. He meets 21 of the 63 criteria, is fairly solid on the first half of the reasoning, but very weak on the second half. Gets 5 of 6 conclusions correct.

In summary, only one of the four E subjects really attempted to coordinate the two representations, trying hard to connect the verbal pieces of reasoning to the graphical lines, intersections and areas. There were clear effects of this attempt: the subject took much longer than the other subjects, and in the confusion of trying to deal with two representations, he gets none of the conclusions quantitatively correct (although qualitatively, and graph-wise, he apparently knows what he is talking about). Two experimental subjects decided to not use the graphical representation in their replies; their solutions indeed look much like the control subjects'. This is to have consequences for their behaviour on the draw-a-graph task, described below.

4 Results and Discussion, Draw-a-Graph Task

The line-up on the evaluation measure makes it look fairly easy and straightforward to draw such a graph — just follow the problem example. It proved not to be easy. The problem example only mentions a few data points: the old equilibrium at (4,500), the new demand at (4,700), and, on the next page, the new supply at (6,700). This gives two points only for the supply line (if you access the second part of the example — half the subjects did not), but only one point for the demand and the new demand. Subjects who had not seen such graphs before would have to either remember or look up the price/quantity relationships of supply and demand in order to get the slope of the lines correct. Even assignment of labels to the axes was decidedly non-trivial. Subjects vacillated between demand/supply, which led to the problem of what to do with the shifted demand, and price/quantity, in which case they wondered about the correctness of putting supply and demand quantity on the same axis. Correctly indicating the surplus–shortage reasoning was beyond three of the four C subjects, and surprisingly also beyond two of the four E subjects: those subjects who had not drawn graphs during the problem solving phase. We believe that both the subjects and we have gained new respect for people who create new representations.

4.1 Computer Recorded Data

An overview of the time data and number of criteria met while drawing the graph are given in Table 6. In this section, we will look at the time data. The criteria analysis will be discussed in the protocol analysis section.

As expected, subjects in the C condition were slower (14.3 vs. 8.3 min, average). This difference did not quite reach statistical significance — a one-way ANOVA showed $F=4.2$, DF 1,6, and an alpha of 0.09. Each C subject's time was around 14–15 min, with very little variability. The E subjects who had used graphs in their problem solutions, S21 and S15, took respectively 4.4 and 7.1 min. Of the E subjects who had not used graphs while solving the problems, S18 took nearly 17 min, and S22 took 4.8 min. Subject 18 was a slow talker, however, and the length of his protocol in words is not much different from that of some of the others in the E group. The verbal protocol analysis will provide some insight into these numbers and into the subjects' processes.

Table 6: Draw-a-graph task results overview

S#	Cond.	Time	# of met criteria
13	C	13.8	0
14	C	13.9	14
19	C	15.0	8
20	C	14.4	5
Avg.		14.3	6.8
15	**E**	**7.1**	**19**
18	E	16.8	11
21	E	4.4	18
22	E	4.8	7
Avg.		**8.3**	**13.8**

Cond: C — verbal only; E — verbal plus graphs.
# of met criteria: total number of criteria met. See Table 7 for more detail.
 # reviews: number of times a subject reviewed any previously read card from one of the problem cards.

4.2 Verbal Protocol Analysis

The results of the criteria analysis are given in Table 7.

Table 7 contains the number of criteria met by each of the subjects. Quite noticeable is S13's total lack of points. This is because S13's approach was very different, and although it contained some elements of truth, the criteria could not be matched to it. Also noticeable is the scarcity of coverage on the bottom half of the table — perhaps not surprising given that subjects did not do well on this in the main task either.

The subjects who had seen graphs before appeared to do better than the subjects who had not, although the difference did not quite reach statistical significance (13.8 vs. 6.8 points, on the average, but much variability — one-way ANOVA: $F = 2.92$, DF 1,6, alpha = 0.14). S18 and S22 (the subjects who had seen graphs but had not used them in solving P2–4) look much like the subjects from the C condition. Apparently even in the learning phase these subjects had not paid much attention to the graphs.

To illustrate the problems the C subjects had, we will go into the protocols of S13, S14 and S19 a bit more deeply (S20 is much like S19). Basically, once they assigned the correct axis labels, and remembered the price:quantity relationships of supply and demand, they could draw the supply and demand lines. Only one (S14) managed to portray a bit of the surplus and shortage reasoning. S14 was the subject performing best in the main task.

Table 7: Draw-a-graph task criteria match results

Criteria for draw-a-graph	**Control**				**Experimental**			
Subj. #	13	14	19	20	15	18	21	22
1. label *x*-axis (price or quantity)		C	C	C	C	XC	C	C
2. label *y*-axis (quantity or price)		C	C	C	C	C	C	XC
3. sketch correct demand line (D)		C	CXC	C	C	C	C	XC
4. sketch correct supply line (S)		C	XC	X	C	C	C	C
5. sketch correct shifted dem. ln. (D2)		C	XC	C	C	C	C	C
6. indicate old equilibrium		C	C	C	C	C	C	C
7. indicate new equilibrium		C	C		C	C	X	C
8. sketch or indicate line at $4.-		C	C				C	
9. indicate shortage		XC		X	C	/	C	
10. indicate price going up		C			C	C	C	
11. indicate supply quantity going up		C			C		C	
12. ind. demand quantity going down					C	C	C	
13. ind. where sup. & dem. will meet					C	C	C	
14. sketch or indicate line at $6.-		C					C	
15. indicate surplus		C		X	C		C	
16. indicate price doing down		C			C	/	C	
17. ind. supply quantity going down					C		C	
18. indicate demand quantity going up					C		C	
19. ind. where sup. and dem. will meet					C		C	
20. indicate price conclusion					C			
21. indicate quantity conclusion					C			
totals	0	14	8	5	19	11	18	7

C — correct; X — incorrect; / — half correct; XC incorrect, then correct; CXC correct, then incorrect, then correct.

4.2.1 Control Subjects

Subject 13. No points. S13 illustrates perfectly how difficult it can be to translate one type of representation into another. He checks the *supply*, *shifts*, *surplus* and *shortage* cards in his quest for the correct graphical representation. He has a very hard time with the graph, beginning with points 1 and 2, assigning labels to the axes. The axes, in turn, become (in *x*, *y* order): (a) supply and demand, (b) price and quantity and (c) back to supply and demand. Each of these hypotheses became a whole section.

In section (a), S13 first discovers that there is no axis to assign the changed demand to. He does not like that, and tries some more of the text out on the graph. Checks out the information in the *shifts* card. Returning, he plots two points on the graph, ($4.-, 700 lbs and $4, 500 lbs), using the *x*-axis for the price and *y*-axis for the quantity.

(b) He then re-assigns the *x*-axis to price, but keeps demand as label on the *y*-axis. Supply, he says "will be a hidden variable, it will just be used". Now, he plots the old and new demand lines, using the *x*-axis for price and the *y*-axis for quantity. The lines are not parallel; rather, they cross at the origin and fan out from there, D through (4, 500), D2 through (4, 700). On checking his lines, however, he finds that at $4.-, the quantity goes from 500 to 700 lbs, but at 500 lbs, the price goes from $4 to $2. Now, he puts D2 below D and tries again. This time, at $4.-, quantity goes down, and at 500 lbs, price goes up. In the example, at $4.-, the quantity goes up, and at 500 lbs, the price goes up.

(c) He abandons this axis marking and returns to supply/*x*-axis and demand/*y*-axis. Price disappears entirely. Both axes are labelled with quantity amounts from 100 to 1000 on the *x*-axis, and 100 to 800 on the *y*-axis. He now draws a line from (100,100) to (800,800) and labels it D. The old D line, drawn from (100,100) to (500,600) is relabelled D2. He then checks this idea. D turns out to be a line on which all equilibria have to lie, namely where supply and demand quantities are equal. The D2 line lies in the shortage area, where supply is smaller than demand, and is relabelled "shortage". A surplus line goes on the other side of D, where supply quantities are larger than demand quantities. The final picture looks like this:

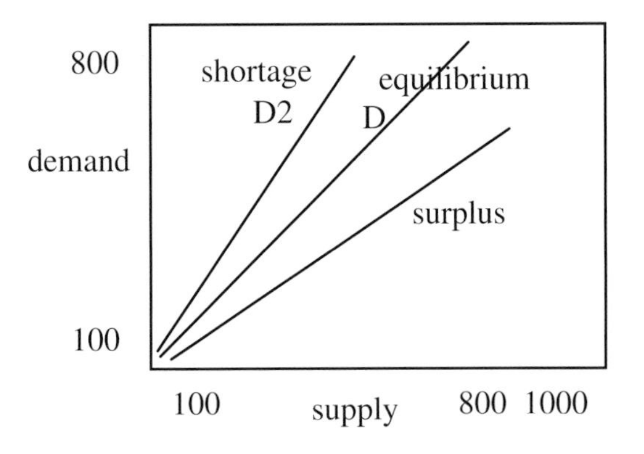

Awarding points to this original approach was difficult. Although parts of this representation could be translated into the wanted representation, really, nothing is correct.

Subject 14. The best subject among the C group, and also the best subject overall in the main task, does surprisingly well (14 points). S14 ends up with a pretty close graph, although it takes him a lot longer than the E subjects who had worked with the graphs in their problem solutions. It takes several attempts to get the axes labelled correctly. He cannot portray the surplus–shortage reasoning, however.

Subject 19. Eight points, criteria 1 through 8. Begins by labelling the *y*-axis with "lbs" and the *x*-axis with "$", and identifying the old equilibrium with a dot and axis markings. He is not sure whether those are the correct assignments, and hesitates a while. Tries to draw the demand line, but there is a lack of points (one point does not define a line). Draws in the other known points at (4,700) and (6,700). Remembers the definitions of demand re inverse quantity/price relation, and draws a line with the correct slope. Then gets confused and relabels that line supply, and draws the demand and changed demand with a supply slope. Catches the error and straightens it out. Identifies new equilibrium, but cannot map surplus/shortage reasoning.

4.2.2 Experimental Subjects

To illustrate the E subjects' drawing, S15's and S22's protocols are summarised below. S21's is much like S15's (both used graphs in their problem solving), and S18's is much like S22's (both did not use graphs). S18 and S22 looked more like the C subjects than the other E subjects, with hesitations and inability to portray some or all of the shortage/surplus reasoning.

Subject 15. S15 drew graphs in the main part of the experiment, so he already has some experience. S15 satisfies every criterion except drawing the lines at $4.- and $6.- (19 points); arguably, because there *are* dots drawn at the intersections of where those lines would be and the changed demand and supply lines. Labels the *y*-axis "price", *x*-axis "quantity", draws in the old equilibrium point, the demand line, supply line and changed demand line with no problems. Uses arrow system to show the surplus and the shortage with their effects. Correct final conclusions indicated. During the drawing, S15 looks at the information in the *positive shift in demand* card four times. He also looks at the *shifts* card twice, and at the *surplus + shortage summary* once, and clicks around in the *example problem solution* quite a bit.

Subject 22. Seven points, criteria 1 through 7. Did not draw graphs while solving the main experiment's problems. Remains within the *example problem solution* card. Labels *y*-axis "supply" and *x*-axis "price", the opposite of the graphs. Draws a positively sloped line, calls it demand, and soon relabels it correctly as supply. Draws the demand line, and relabels the *y*-axis "amount of product". Draws new demand line, and identifies the old and new equilibrium points. Attempts to draw the surplus and shortage but does not get very far.

In summary, as expected, the C group took longer time to draw the graph than the E group. The latter group had at least picked up the general format of the graph, with three crossing lines and one line (the changed schedule) parallel to its unchanged counterpart, and the axis labels. The first group had to figure out the correct slope of the lines, since they were only given one point per line, and also figure out that the changed demand line was parallel to the unchanged line. They also had to figure out which labels to assign to the axes. Whether a subject started with price and quantity or with demand and supply on the axes seems to be a "head or tails" issue. Neither start came with complete confidence in the accuracy of the assignments. Even the E subjects were somewhat confused as to the labels: both S21 and S22 initially labelled the *x*-axis "supply" rather than "quantity".

5 Summary and Discussion

The issues were, (1) can subjects follow and reproduce well-constructed qualitative reasoning? (2) will adding a graphical representation help subjects or not? and (3) are subjects able to coordinate the two representations? We will briefly summarise our findings below.

(1) Did the subjects learn to reason qualitatively about these problems? Yes and no. Yes, because most of the subjects in both conditions understood and replicated the first half of the reasoning, comparing the changed schedule with the unchanged schedule at the old equilibrium price. Five of the eight did a rather good job, and the pieces of the reasoning that the other three voiced were also at least appropriately used. Subjects did well on the surplus/shortage reasoning, often verbalising, and sometimes implying all three consequences. We were especially surprised with how well the first half of the reasoning about the changed demand transferred to reasoning about changed supply. Even though the basic reasoning schema remains the same, nearly all the slot values change.

But, also, no, subjects certainly did *not* learn the whole reasoning. Performance on the second half of the reasoning was patchy at best. Only one subject, S14, met the majority of the criteria, and only on the fourth problem. Subjects did not do as well with the two-viewpoint approach as they did with the surplus/shortage consequences. They were generally aware that there *were* two viewpoints, and most of the subjects realised that there were two boundaries. Many managed to get the boundaries correct (i.e. the final conclusions) even if the second half of the reasoning was patchy or lacking altogether, generally reasoning from the data provided in each problem statement instead of using verbal qualitative reasoning. How did subjects find the second part? We hypothesise that first, they were aware that the answers should be in a range format, so they knew to look for the other boundaries. The other boundaries *could* be deduced from the data in the problem statements; also, they could be found from a correctly drawn graph if one remembered, and drew correctly, how the reasoning was expressed on the graph. Thus, in both the C and in the E formats the conclusions could be gathered by means other than qualitative, verbal reasoning. Still, these are very encouraging results considering that subjects only spent an average of 15 min learning material that may take weeks in an economics class.

(2) Did having graphs available during the tutorial help? It is not clear that the graphs *helped* performance — more likely, they hindered performance; but what is far more clear is that they *altered* subjects' behaviour. Two of the E subjects (S15 and S21) made heavy use of the graphical representation to help them solve the problems. Both these subjects' reasoning in the problems solving was closely tied to the graphical representation: they made errors that were based on faulty graph drawing. S15 only used the verbal representation until he had his graph drawn; after this he reasoned almost exclusively with the graphical representation. S21, however, really tried to do the verbal reasoning using the graph as basis. Unfortunately, he was especially prone to sloppy drawing. Reasoning off the graph therefore, netted him the wrong answers. The graph-drawing experience appeared to have helped these subjects in the draw-a-graph task, since they drew almost perfect graphs, although the qualitative

verbal reasoning was not very good at times. The other two subjects in the E condition were not as attuned to the graphs. They did not draw graphs during the problem solving, and did their reasoning almost all verbally, even though they referred back often to the graphs when they had them available. Still, on the draw-a-graph task their graphs were drawn more quickly and easily than the C subjects'; however, they failed to represent surplus and shortage adequately, as did three of the subjects in the C group. The E subjects not drawing graphs made no references to graphs in their verbal protocols. Their reasoning appeared to be based on the tabular and verbal data.

Thus, it does not appear that the graph helped subjects do the qualitative reasoning better. Perhaps even to the contrary. The behaviour during the problem solving, however, was structured very differently for those E subjects who drew graphs. These subjects spent large amounts of time translating the problem into a graphical format (and made errors in the process), and then used the graph in their reasoning (translating again), sometimes with less than good consequences; subjects in the C condition only had the verbal reasoning to concentrate on.

(3) How well were subjects able to coordinate the two representations? Two of the E subjects based their problem solving on them; the other two did not. The subjects who had spent time coordinating the verbal and graphical representations as a result of their problem-solving behaviour clearly had an easier time drawing the graph, with better results, than those who had kept their reasoning on the problems in a verbal mode. In fact, those subjects were not all that much better (although faster on some components) at drawing the graph than those subjects who had never seen the graphs in the experiment.

It takes effort to translate the verbal problem statements into graphs; this was obvious from the verbal protocols. Apparently this effort is needed to coordinate two representations; the connections between representations are not automatically or easily made. This conclusion has been reached by many authors in this book, e.g. Boshuizen and Van de Wiel (chapter 12) and Ainsworth, Bibby and Wood (chapter 7). Experts who use several representations interchangeably, would do well to realise in their teaching that making such connections should be considered material to be learned, as much so as "the meat" of a topic.

6 Postscript: CaMeRa

The work just described inspired us to construct a computer simulation to help clarify the use of multiple representations in problem solving, focusing on their role in visual reasoning (described in Tabachneck-Schijf, Leonardo, & Simon[1], 1997). Data in the experimental literature, our own work (Tabachneck, 1992; Tabachneck & Simon 1992, 1996) and concurrent verbal protocols were used to guide construction of a

[1]A preliminary version of CaMeRa was discussed in a poster presented at the 1994 Cognitive Science Conference (Tabachneck, Leonardo & Simon, 1994).

linked production system and parallel network, CaMeRa (*C*omputation with *M*ultiple *R*epresentations), that employs a "Mind's Eye" representation for pictorial information, consisting of a bitmap and associated node-link structures. Propositional list structures are used to represent verbal information and reasoning. Small individual pieces from the different representations are linked on a sequential and temporary basis to form a reasoning and inferencing chain, using visually encoded information recalled to the Mind's Eye from long-term memory and from cues recognised on an external display. CaMeRa, like the expert, uses the diagrammatic and verbal representations to complement one another, thus exploiting the unique advantages of each.

The CaMeRa model was designed to capture as accurately as possible current, experiment-based hypotheses about how human perception and reasoning employ the several sensory modalities, perceptual processes and memory structures to accomplish their tasks. Here, to provide a link between the experimental work and the modelling work, we will borrow some material from the CaMeRa paper, namely that which briefly recaps the principal experimentally supported features of the processes and structures that we have considered, and for each, indicates briefly how it has been incorporated in the model. The description of the implementation and other details of the model are in Tabachneck-Schijf, Leonardo and Simon (1997).

Modalities are separate and different.

Empirical evidence. There is much support in the literature for separate and different modalities. Different types of short-term memories (Atwood, 1971; Baddeley, 1986; Brooks, 1967, 1968) such as a "visual–spatial scratchpad" and an "auditory loop" appear to have different capacity limits (e.g. Baddeley, 1986; Zhang & Simon, 1985). Evidence for different processes used in verbal, as compared with visual, representations can also be found in research on individual differences in abilities to process verbal, pictorial and spatial information (e.g. Just & Carpenter, 1985; Paivio, 1971; Riding & Calvey, 1981), as can demonstrations of the computational advantages of having such different representations (Larkin & Simon, 1987). Finally, Kosslyn offers strong evidence for both separation and difference in verbal and visual perception and cognition, separation increasingly supported by neurophysiological evidence (Kosslyn, 1980, 1994; Kosslyn & Koenig, 1992). Zhang and Norman (1994) also found differences in the way their subjects worked with visual and spatial information.

The model. Each mode of representation in the CaMeRa model has its own data structures and operators. Thus, if a pictorial stimulus is perceived, it is encoded and stored in a pictorial data structure and interpreted with pictorial rules. Similarly, if a sentence is read, it is stored in a verbal-semantic representation. If a thought is recalled from memory, it will be processed in verbal form in verbal STM, and in pictorial form in pictorial STM (assuming the information has both a pictorial and verbal LTM representation). Pictorial operators cannot modify verbal data structures and vice versa, although each may, through associative links, retrieve information from the other. The modalities interact like two layers of memory drifting alongside each another, tied together temporarily by links in working memory, and disconnected again after the information is no longer needed.

Mental images resemble visual stimuli closely.

Empirical evidence. In studying the representations people use, we must distinguish between external and internal information, between information in the environment and information in the brain. To understand a drawing of an "A" above a "B", the external drawing must be transformed into an internal representation, the mental picture contained in the Mind's Eye (Kosslyn, 1980).

Much research supports the idea that information encoded into STM by perception is represented in basically the same form as the identical information transferred from LTM to STM, so that a mental image produced by looking at an external drawing is represented in STM with the same type of pictorial structures as a mental image generated from the memory of the drawing. We call the hypothesis based on this evidence the Mind's Eye hypothesis.

Evidence for the Mind's Eye hypothesis is of three kinds: behavioural, neurophysiological and computational (see also Glasgow, 1993; Tabachneck & Simon, 1996). We call attention especially to the research of Kosslyn and his colleagues (1980, 1992, 1994), who have demonstrated through numerous experiments and simulations in all three areas that images evoked by perception and memory produce the same configuration in the Mind's Eye, and use the same procedures. Computational accounts have also been given by Baylor (1971) and behaviourally by Finke and Shepard (1986).

The model. At the working memory level in CaMeRa, the medium, the data structures and the rules that operate on them are the same both for images derived from perception and images drawn from memory. We have not yet implemented mechanisms that would cause differences in resolution between images of these two kinds, but these will appear as we implement limitations on working memory.

All inferences are made on working memory contents.

Empirical evidence. In a sense, this is true by definition. That is to say, working memory is sometimes defined as the system that supplies the inputs for processing, functions with which to do the processing, and a storage area to receive the output of processing. The empirical content of the claim derives from consistent evidence that memories involved in current processing have different parameters for acquisition rates, access times and durability than memories used for more permanent storage. For example, the large body of evidence for the structures hypothesised in EPAM (Elementary Perceiver and Memorizer) points to short-term memory acquisition times of a few hundred ms, but long-term memory times of about 8 s per chunk;

STM retention times of about 2 s (without rehearsal), but indefinite LTM retention times (Richman, Staszewski & Simon, 1995).

The model. In CaMeRa, all the inferencing is done in working memory (the auditory and visual short-term memories). In order for the information in data to be used, it must be copied to the working memory. When learning processes are added to the present model, all changes in memory structures will also be made initially in the working memory. Whether these changes are permanent or not will depend on the type of reasoning that was done with the data. The LTM record of data used merely for computation would not be altered, but if inferencing showed that data were erroneous or incomplete, corrections would be learned and introduced into LTM.

Experts can use and integrate multiple representations.

Empirical evidence. An immediately noticeable quality of expert problem solving is the ease and frequency with which an expert uses multiple representations. An example of this, from data we have gathered, is the behaviour of an economics expert[2] who was asked to explain to a student some principles of supply and demand, of equilibrium and of the effects of a shift in the supply schedule. While engaging in this explanation, the expert was constructing a graph on a blackboard to illustrate the various principles, interspersing his drawing with verbal explanations. A part of the protocol the expert generated is reproduced in Appendix 1, along with our analysis of how the pictorial and verbal parts of this protocol are interconnected, and how the CaMeRa model accounts for the expert's reasoning. The corresponding reasoning using exclusively verbal means without a diagram is described formally in Appendix 2. It can be seen to be substantially more complex than the diagrammatic reasoning.

When the expert was asked to give a similar explanation without either using or referring to pictorial elements (in fact, sitting on his hands to prevent gesturing), he was unable to do so in three consecutive trials. He made visual references within 3 or 4 sentences,[3] and, when interrupted, reported he had been constructing a mental diagram and reading information from it.

[2]Relying on the precedent of Ebbinghaus, who used himself as subject, the third author of this paper served as the expert in economics in these experiments. On both occasions of giving a protocol, he was not warned that one would be elicited, hence could make no advance preparation. The expert protocols were obtained after the basic structure of CaMeRa had been fixed, whereas the protocols of novice behaviour were gathered before modelling work began. All the protocols, like the other evidence cited in this paper, should be interpreted as guides to the construction of the model and tests of its sufficiency to produce the basic phenomena that had been observed, rather than as conclusive tests of its accuracy or generality.

[3]After the difficulty he experienced in the experiment, the expert, with considerable additional work later in the day, was able to put down on paper an explanation completely void of visual referents. He found this not to be an easy task, and he reports that he found himself unable to avoid using his own mental imagery to guide his thought while carrying it out.

The model. The way in which this behaviour is modelled by CaMeRa is explained in considerable detail below. In general terms, the verbal and visual information of the modelled expert are linked together associatively by referents to one another. When visual data that carries verbal information (e.g. labels on variables) is accessed by recognition or otherwise, the verbal information that it might refer to is activated next. The visual and verbal "layers" of memory are closely tied together when the information is activated in working memory.

Novices have difficulty using and integrating representations.

Empirical evidence. In economics, equations, tables and graphs are widely used to enhance, enrich and illustrate verbal explanations. The experiment reported in this paper, and the earlier experiment, reported in detail elsewhere (Tabachneck, 1992), illustrate novices' difficulties in interpreting verbal and pictorial representations and achieving integration of the two.

The model. Although the model does not currently emulate a novice, and cannot yet learn, it was built with the expert/novice differences in mind. CaMeRa can easily be degraded to become a novice model by eliminating all the referents between visual and verbal material.

Appendix 1: The Text of the Improved Tutorial

1. Demand

Summary Text

Demand. The amount of a product people will buy at a given price. Consumers will buy more of a product as the price decreases: the quantity varies indirectly with the price. Individual demand schedule: set of prices and the amount of a product demanded at each price for one buyer.

Market demand schedule: sum of all individual demand schedules.

Main Text

Demand. The demand, which is the amount of a product a consumer will buy, is related to its price. In general, at a high price consumers will buy a smaller amount than at a low price; at a low price they will buy a larger amount than at a high price.

An individual demand schedule states the amount of product demanded by a consumer at each price. On this schedule, price will decrease as quantity increases: price and quantity are inversely related. Individual demand schedules are highly variable; you can imagine that demand schedules of someone who earns $200,000 a year will be quite different from those of someone earning $30,000 a year.

Generally, we combine the demand schedules for a certain product for all buyers by adding the quantities they wish to buy at each price. This combined schedule is called a market demand schedule.

Example Text

Example, demand schedule. If a product were priced at $5 a unit, a person might buy one unit. Conversely, if that same product were priced at $3 a unit, that person might buy two units, and someone else, who could not afford the product at $5, might now buy one unit. Similarly, the entire population of buyers may buy 5000 units if it were priced at $3; 4000 units at $4; 3000 units at $5; and so on.

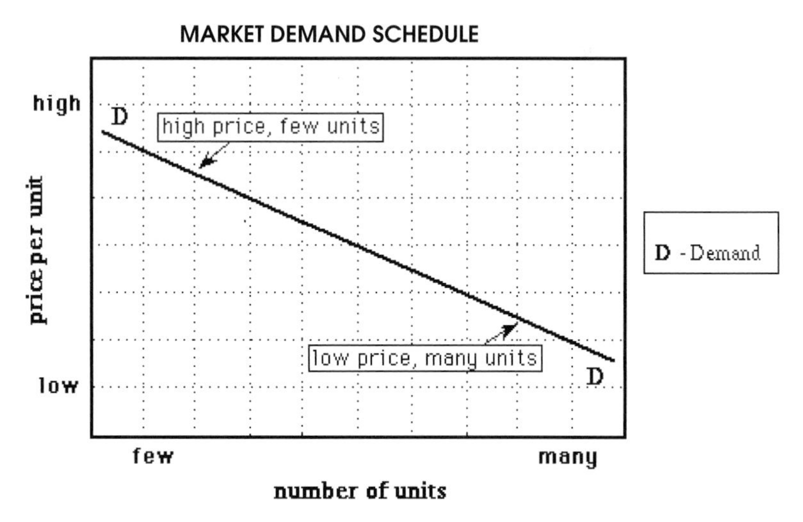

Notes: *The demand description is straightforward. On the graph, we indicated the correct labelling of each end of the line — some of the subjects in the first experiment were not able to do this.*

2. Supply (not included here)

3. Equilibrium (not included here)

4. Surplus

Summary Text

Surplus. The amount of a product supplied at a certain price is larger than the amount of a product demanded. To eliminate surplus, lower price: suppliers will make less of the product and consumers will buy more of the product. Equilibrium price: price where quantity supplied = quantity demanded.

Main Text

Surplus and its effect on the equilibrium. Note: you will be expected to be able to reason through what happens to the demand and supply in case of a shortage or a surplus.

Sometimes a situation occurs in the market where the quantity supplied is larger than the quantity demanded. This is called a surplus. When there is a surplus, the suppliers are left with extra goods that the consumers will not buy.

Suppliers will want to eliminate the surplus, since an unsold inventory cuts profits; hence, they will offer to reduce the price. As the price decreases, producers will make less of the product for market (since the supply schedule states that supply varies directly with price) but buyers will purchase more (according to the demand schedule), thereby reducing the surplus. Thus, when there is a surplus, the price will gradually go down, thus producers will supply less, and consumers demand more, until the surplus is eliminated and the supply again equals demand.

The price at which the quantity supplied is equal to the quantity demanded is called the equilibrium price.

Example Text

Note: before clicking the "UPDATE GRAPH" button, read the text that precedes it.

Example, surplus:

For instance, take the market in hay. The current market price per bale is $10. The market is in equilibrium, and hay-farmers in North Dakota are supplying 10,000 bales of hay at $10 a bale, which are all sold to cattle-ranchers in North Dakota at that price. There are no ranchers who want to buy more, and no farmers who want to sell more at that price.

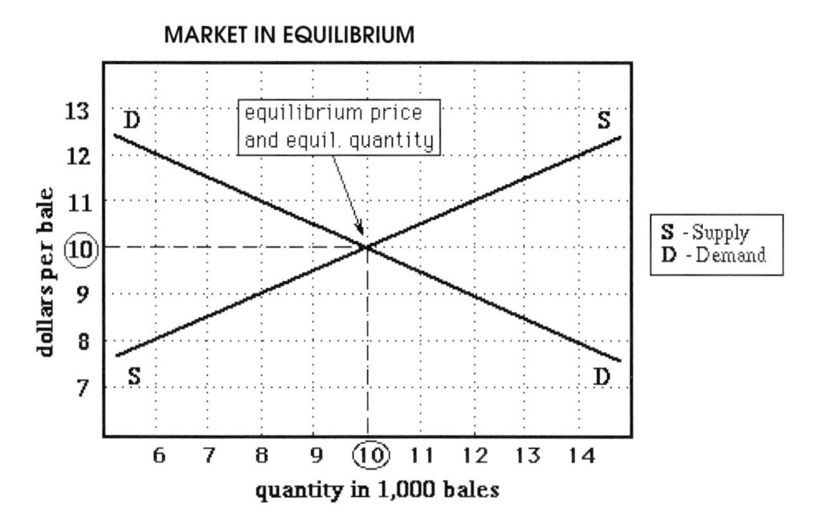

Notes: *The example refers to bales of hay. We start the subject off with a picture of the equilibrium point.*

Let us assume that, somehow, hay bales were mispriced at $12 a bale. At that price, hay-farmers would decide to plant more hay (hiring extra hands, etc.) and end up supplying 14,000 bales (from the market supply schedule: at $12 suppliers will supply 14,000 bales). But, at $12 a bale cattle-ranchers will only buy 6000 bales (from the demand schedule). 14,000 – 6000 = +8000: there will be a large surplus of hay.

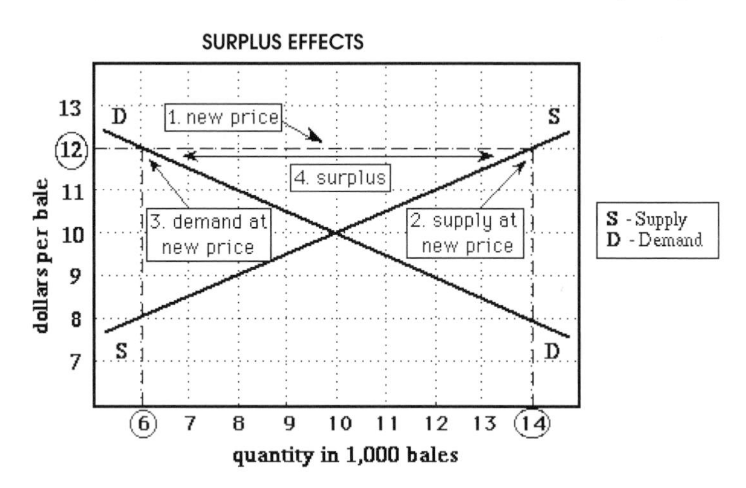

Notes: The new price (the cause of the surplus) is drawn in as a dashed line. All the points are labelled in order of occurrence in the example text. It is the intent to establish the surplus and shortage reasoning as a "chunk" in the subject's memory, so that it can be used effortlessly when the shift reasoning requires it. First, it is noted that a surplus is undesirable for suppliers, who will lower the price to try to eliminate it . . .

Hay-farmers do not want to be stuck with all that hay in inventory, so the next year, they might lower the price, say to $11 a bale, and consequently decrease their plantings. Now, cattle-ranchers will buy 8000 bales, and hay-farmers would supply 12,000 bales. There is still a surplus.

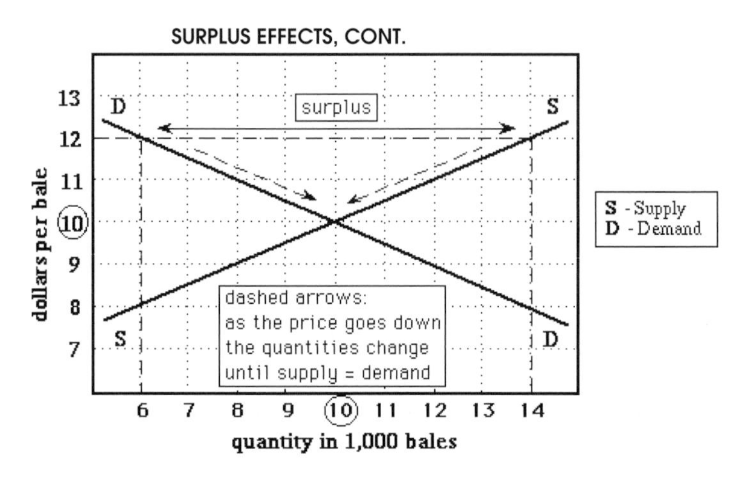

Notes: *. . . and then, both the demand and the supply work together as a now localised feedback system to force the price back to the equilibrium . . . Note the simultaneous dual propagation of the supply* and *the demand toward the equilibrium point. This is one of the crucial points in the reasoning. Such dual propagation prevents losing track of the equilibrium point.*

In order to sell all their bales, the hay-farmers will have to lower their price to $10 a bale, the price at which the cattle-ranchers will buy 10,000 bales of hay, and hay-farmers will limit supply to 10,000 bales of hay.

5. Shortage (not included here)

6. Shifts in Market Schedules and Their Effects

Summary Text

Shifts. A change in quantity of a market schedule that affects all prices on the schedule. A schedule shift will always disturb the equilibrium. Positive shifts: higher quantity at each price. Negative shifts: lower quantity at each price.

Main Text

The entire market schedule changes when a large number of consumers or suppliers are affected by certain economic changes. Such changes are called shifts. Any shift in a market schedule will necessarily disturb the equilibrium. When one of the schedules, say demand, shifts, the other schedule, here supply, remains the same. Changes that affect consumers' buying behaviour are shifts in the market demand schedule, and changes that affect suppliers' manufacturing behaviour are changes in the market supply schedule.

If the change affects consumers, at every given price all consumers together would wish to buy more (or less, respectively) than they bought previously at the same price. If the change affects producers, then at every given price all producers together would want to produce more (or less, respectively) than they produced before at the same price. We will assume that the basic relationship between price and quantity that characterises supply or demand (direct for supply, indirect for demand), still holds in any shifted schedule.

As you are reading the text in the next sections, note that positive shifts mean a higher quantity at each price, and negative shifts mean a lower quantity at each price for both supply and demand schedules. Also note that the new equilibrium price always falls between the old equilibrium price and the price at which the old equilibrium quantity would be demanded, or, respectively, supplied under the changed schedule. Similarly, the new equilibrium quantity always falls between the old equilibrium quantity and the quantity that would be demanded, or, respectively, supplied by the changed schedule after the schedule shift at the old equilibrium price.

Note: *(no example included here)*

*7. **Demand Schedule Positive Shift***

Note: *All subjects were given an example (Demand Schedule Positive Shift) of how they were to reason. The problems were closely analogous to the example problem. While doing the second problem on their own, subjects could still access the example problem. During the third and fourth problem, the example problem was no longer available.*

Summary Text

Reasoning example available on first problem, not on second or third. Your reasoning should be as detailed as ours. Demand, positive shift: quantity demanded at each price increases.

Main Text

Information for Example Problem, Problem 1.

Notes: *We will illustrate what happens when the market demand schedule shifts positively, and show you the effects on the equilibrium price and quantity, with an example. You will then tell us what the effects will be when the market demand shifts negatively (you will be able to look at our answer for the positive demand shift), and tell us about the effects for negative and positive shifts of the market supply schedule without being able to see the previous answers. Thus, make sure that you understand the reasoning on the first example; the second part is a bit difficult. We expect that your reasoning through the effects of the other schedule shifts will be as detailed as our reasoning through the positive shift in the market demand schedule.*

Market demand schedule — positive shift:

An economic change has affected the demand schedule, and shifted it positively. When the demand schedule has shifted positively, the quantity demanded at each price increases. This means that some buyers want to obtain more units at any given price; also, there are more people willing to buy the product at a given price.

Example Text

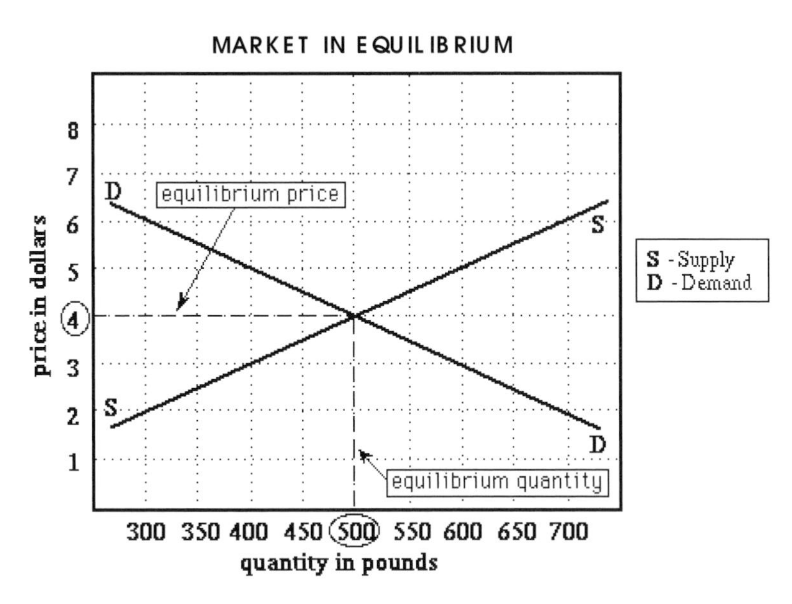

Notes: *Once more, start with the old equilibrium price.*

The government has decided to abolish the income tax, and thus everyone has more money available. You decide that, instead of 2 units of a product a week, you can afford to buy 3 units at $4 each; alternatively, if the price were to go up to $5 a unit, you could now still afford those 2 units, where before, you would have only bought 1 unit at $5. Reasoning along similar lines, you might now buy 5 instead of 4 units at $2, 4 instead of 3 units at $3, or 1 instead of none at $6.

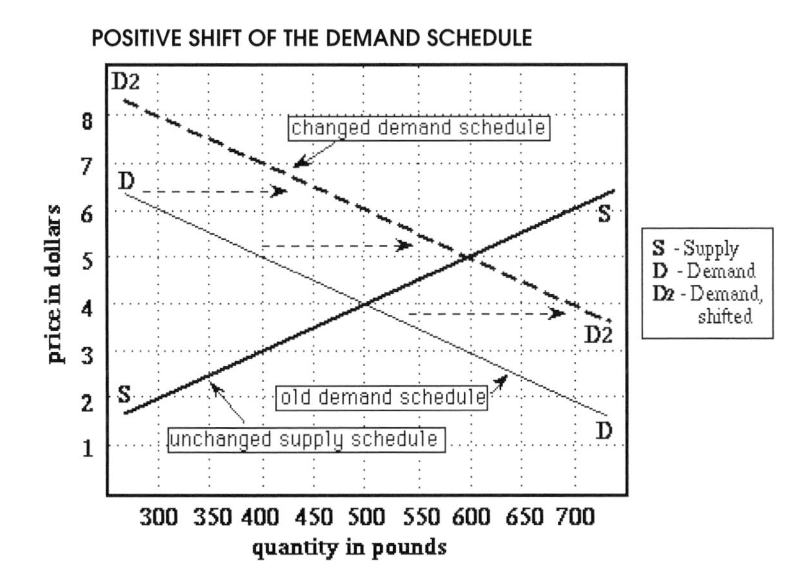

Notes: *The arrowed lines between the old and the new demand line are drawn horizontally to show the shift in quantity demanded at each price. Subjects have a tendency to "see" a shift occurring along a line perpendicular to the two demand lines. The lines are again redundantly marked (with signs and in the legend) to ease coordination of information within the graph.*

Many people think the same way as you do and if you combine all their individual demand schedules, consumers are able to buy more units of the product at any given price. Thus, the market demand schedule has shifted positively.

8. Example Problem — Problem 1: Effects of Demand, Positive Shift

Summary Text

Begin with market in equilibrium; shifts disturb the equilibrium.

At the old equilibrium price, assess effects on market if the unchanged schedule were in effect — surplus or shortage? You describe what happens in case of a shortage or surplus.

At the old equilibrium price, assess effects on market if the changed schedule were in effect — surplus or shortage? You describe what happens in case of a shortage or surplus.

Conclude where the new equilibrium price and quantity will fall.

Main Text

Before the shift, the market was in equilibrium. To understand how much the equilibrium quantity and price change will be, and in what direction, we must consider two points of view: what would happen to supply and demand if the unchanged supply schedule persisted at the old equilibrium price, or if the changed demand schedule were in effect at the old equilibrium price.

From the point of view of the unchanged supply schedule. At the old equilibrium price suppliers would still be producing the same amount of the product, but the demand schedule has shifted positively, so quantity demanded has increased. Thus, quantity supplied minus quantity demanded is negative: if the market price would remain at the original equilibrium price there would be a shortage, with all the consequences thereof.

From the point of view of the changed demand schedule. At the old equilibrium price, the quantity demanded is up. But to supply that larger quantity, the price has to go up as well. At that new, higher price, the new quantity demanded (on the changed schedule) would then begin to decrease. Now, quantity supplied minus quantity demanded is positive; hence, if the market price would go up as needed to supply the new higher demand, there would be a surplus, with all the consequences thereof.

Conclusion. Thus, when the market demand schedule shifts positively, the new equilibrium price is higher than the old equilibrium price but lower than the price would be if the change in demand were completely implemented. The new equilibrium quantity is more than the old equilibrium quantity but less than the market quantity would be if the demand change were completely implemented.

Example Text

There is nothing mysterious about the numbers in this example: all data are taken from the original demand and supply schedules, and from the changed demand schedule, which we will not enumerate here.

Let us consider $4 sirloin steak as the product. Remember that everyone is now buying more sirloin steak because they have more income since they do not have to pay federal income tax any more. For the sake of simplicity, we assume a small population of a few thousand people.

Before the repeal of the income tax, consumers were buying 500 lbs, and farmers were supplying 500 lbs of sirloin steak per week at $4/lb.

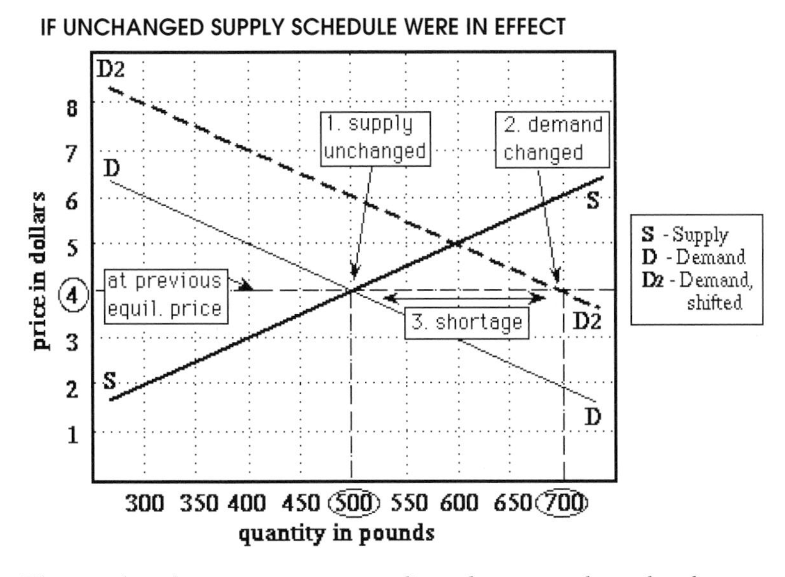

Notes: *The graph is becoming more complicated — note that what happens during a shortage, is not again illustrated. The graph would become too confusing. The shortage information is available to the subject at any time during the experiment. Note how the argument keeps track of the old equilibrium price as a point to relate to. There are two points of view, each establishing one boundary. This one establishes that the new price is higher than the old equilibrium price.*

Let us see what happens if the unchanged schedule, supply, stayed in effect. At the old equilibrium price, $4/lb, after the shift, the supply schedule is unchanged so producers will still supply 500 lbs, but the demand schedule has changed and at that price, consumers now want 700 lbs per week. Quantity supplied minus quantity demanded (500 lbs − 700 lbs = −200 lbs) is negative, a shortage.

Before reading on, please reason through what happens when there is a shortage. Click the "SHORTAGE" button, below, for a review, if needed.

Note: the following text became available after clicking on a button "solution" which the subjects were encouraged to click on after reasoning the problem through for themselves.

*[If you reasoned correctly, you will have found that at the new equilibrium the price of a pound of sirloin steak will be higher than $4, and that the equilibrium quantity will be somewhere between 500 and 700 lbs.]

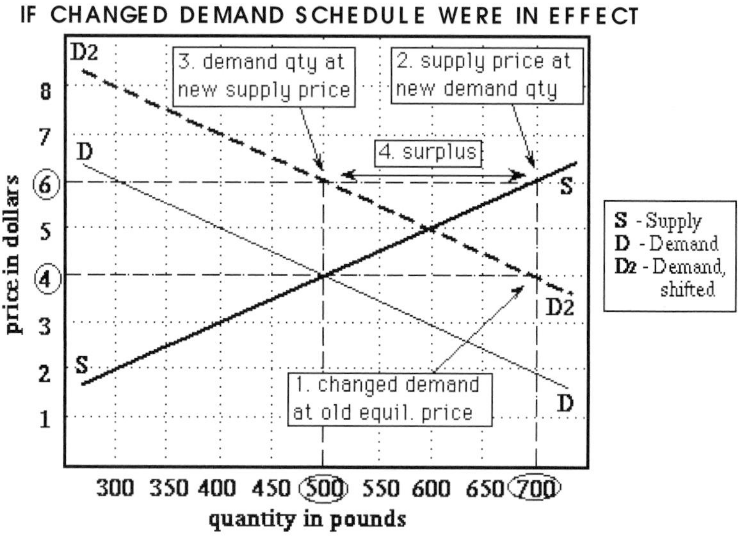

IF CHANGED DEMAND SCHEDULE WERE IN EFFECT

D2
3. demand qty at new supply price
2. supply price at new demand qty
4. surplus
1. changed demand at old equil. price
S - Supply
D - Demand
D2 - Demand, shifted
price in dollars
quantity in pounds
300 350 400 450 500 550 600 650 700

Notes: *Having found the lower bound of the range the price will be in, we now look for the upper bound, so it can be shown that the price will not increase by the entire amount of the demand shift.*

Now, let us see what happens if the changed schedule, demand, went into effect. We know that at the old equilibrium price, consumers now demand 700 lbs of sirloin steak. Suppose that farmers would try to supply that entire 700 lbs. We know that they cannot do that at $4/lb. In order for the farmers to supply 700 lbs of sirloin steak, they would have to price it at $6/lb. But at $6/lb people only want 500 lbs (data from the changed demand schedule): at that price, there would be a surplus (700 lbs–500 lbs = +200 lbs), positive, a surplus.

As before, right now, out loud, please reason through what happens to the price, supply quantity and demand quantity when there is a surplus. Click the "SURPLUS" button, below, for a review, if needed.

*[If you reasoned correctly, you will have found that, at the new equilibrium, the price will be below $6/lb and the equilibrium quantity will be between 700 and 500 lbs.]

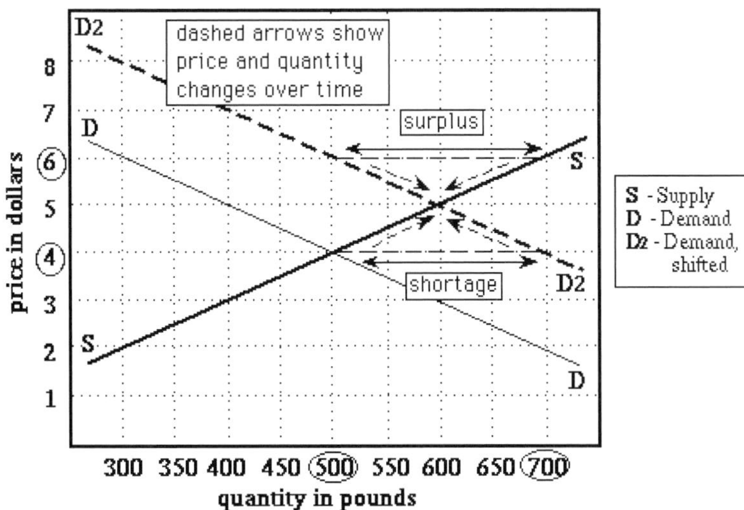

Now we put the two strands of reasoning together: the conclusion is that the new equilibrium price must fall between $4 and $6, and the new equilibrium quantity between 500 and 700 lbs.

Note: Only the problem statements of problems 2 and 3 have been included, not problem 4 or the cards explaining the particular type of shift.

Problem 2: Negative Shift in Demand

Main Text

Problem 2: Demand — effects of a negative shift. This time, consumers have less income available and they demand a lower quantity of goods at any given price. Again, consider the market in sirloin steak. Before the negative shift in demand, it was in equilibrium at $4/lb, with consumers buying 500 lbs and producers supplying 500 lbs of sirloin steak.

Some figures you may (or may not) need from the unchanged supply, unchanged demand and changed demand schedules:

UNCHANGED SUPPLY: At $6 producers supply 700 lbs
At $4 producers supply 500 lbs
At $2 producers supply 300 lbs

UNCHANGED DEMAND: At $6 consumers demand 300 lbs
At $4 consumers demand 500 lbs
At $2 consumers demand 700 lbs
CHANGED DEMAND: At $6 consumers demand 100 lbs
At $4 consumers demand 300 lbs
At $2 consumers demand 500 lbs

Please reason through the effects of a negative shift in the demand schedule. Be as detailed as we were when we reasoned through the effects of the positive shift in the demand schedule. To review the text and example for the example problem, click the "EFFECTS-SHIFT" button, below.

Problem 3: Positive Shift in Supply

Main Text

Problem 3: supply: effects of a positive shift. The quantity supplied at each price has increased, so we say that the market supply schedule has shifted positively. Back to sirloin steak: there is a new hay-combine on the market that cuts the cost of producing hay in half. The hay farmers pass on the savings to the cattle ranchers. Before the positive shift in supply, the market was in equilibrium at $4/lb, with consumers buying 500 lbs and the ranchers supplying 500 lbs of sirloin steak.

Some figures you may (or may not) need from the unchanged supply, unchanged demand and changed supply schedules:

UNCHANGED SUPPLY: At $6 producers supply 700 lbs
At $4 producers supply 500 lbs
At $2 producers supply 300 lbs
UNCHANGED DEMAND: At $6 consumers demand 300 lbs
At $4 consumers demand 500 lbs
At $2 consumers demand 700 lbs
CHANGED SUPPLY: At $6 producers supply 900 lbs
At $4 producers supply 700 lbs
At $2 producers supply 500 lbs

Please reason through the effects of a positive shift in the supply schedule. Be as detailed as we were when we reasoned through the effects of the positive shift in the demand schedule.

12

Using Multiple Representations in Medicine: How Students Struggle with them

Henny P. A. Boshuizen and Margaretha W. J. van de Wiel

Medical expertise development requires the acquisition and integration of large amounts of theoretical and experiential knowledge. This process takes many, many years. It had been estimated that it takes at least 21,800 hours to become an experienced general practitioner in the Netherlands (Boshuizen, 1989). Most other medical specialities will require more time since their training period is longer. Even general practice will take more nowadays since the training period has been extended by one year.

Boshuizen and Schmidt (1992) have hypothesised that in the course of this process several knowledge-restructuring processes take place. One process is the integration of knowledge of different subject matters. On the one hand, in many medical curricula different subject matters are taught separately, e.g. anatomy, physiology, epidemiology, pharmacology, internal medicine, neurology, etc. On the other hand, the problems physicians are confronted with in practice require that physicians apply this knowledge in an integrated way. Even the simplest problem of a painful throat requires this integrated knowledge application. The physician should know how a normal throat looks like and which structures can be observed, what normal causes of this kind of pain are in different age groups, how the diagnosis is furthermore affected by the epidemiological situation, how the different causes of pain in the throat can be treated, what the side-effects of this treatment can be, etc.

Earlier we wrote the following about expertise development:

> ". . . students . . . also have to learn to use that knowledge in tasks relevant for their future profession. In medicine, they must learn to take a patient's history, do a physical examination and order lab tests in order to make a diagnosis. These responsibilities require a base of well-organised knowledge, as well as cognitive, perceptual and psychomotor skills. Physicians must also be able to decide on patient management and treatment. Moreover, physicians need the knowledge and skills to communicate to the patient their assessment of the situation and the management strategy to be taken, because justification and explanation, adapted to the level of comprehension of the patient, play an important role in patient compliance. Typically, the knowledge and skills needed for these tasks are not learned through direct teaching, nor are they learned from books, but through demonstrations, practical training, and while the stu-

> dents are immersed in medical practice during their clerkships or internships (Patrick, 1992). An important aspect of at least part of this learning is that it occurs in an implicit way and results in tacit knowledge that is hard to verbalise and externalise (Reber, 1989). Jones, Calson and Calson (1986) have shown that acquisition of formal knowledge through clinical experience was very limited.
>
> Not only procedural learning, but also perceptual learning, is important in medicine. A great deal of medical perceptual learning is formalised. Students explicitly learn to differentiate and recognise heart sounds, patterns on ECGs, the structures they feel during physical examination, etc. However, the less obvious, more intricate stimulus configurations, especially those involving more than one sensory system, must be often picked up from the environment. Examples are the patient with diabetes type 1 with poor metabolic control, who is smelling of acetone, or the hypothyroid patient with her or his deep grating voice, slow speech and yellowish skin, or the "secret drinker" with liver problems who has a specific smell that is often masked by perfumes, mouth rinse or after-shave and who has a specific appearance characterised by a yellow shade of the whites of the eyes. Linked to conceptual and procedural knowledge, this perceptual learning plays an important role in the development of diagnostic and patient management proficiency." (Boshuizen et al., 1995)

Although described from a different perspective and with a vocabulary belonging to a different theoretical domain, this quote pinpoints some of the problems of coordination of representations: the difficulty of verbalising and externalising the knowledge and skills that have been learned through a different mode, the importance of perceptual learning in this domain and again the difficulty of linking that perceptual knowledge to the verbal mode and representations. It also indicates that much of this coordination by the student goes unsupervised. Apart from the obvious examples such as learning to read ECG traces or recognising heart sounds, the coordination between the different representations, for example, the sounds heard, the mental model of the anatomic structure "under the skin", the mental simulation of the fault that might have occurred causing a disease in that mental model must be done by the student without much assistance by an expert. Yet, this whole enterprise might be the key to success in learning medicine.

The examples given here indicate that the integration of representations is not an overnight process. It starts when students first study their textbooks, but it proceeds for years: when they have to integrate prior knowledge with specialist handbook knowledge and articles, and when they are dealing with patients. This dealing with patients adds an extra dimension to the learning process. It has to combine text-based learning and perceptual information processing. Having to apply prior knowledge on patient findings requires the application of several kinds of knowledge, aimed at building a coherent explanation of the patient findings. When available knowledge is not in a matching format, this can be a difficult enterprise.

Research on the development of medical expertise (Boshuizen & Schmidt, 1992; Schmidt & Boshuizen, 1993a,b) has suggested that this process of knowledge coordination and integration requires the application of basic science knowledge in clinical problem solving. In this theory that focuses on cognitive change, it is conjectured that the small-scale basic science concepts which students learn first and that explain clinical phenomena, are "encapsulated" under larger scale or higher level clinical concepts, thus being the bridge between these two knowledge bases. The underlying assumption is that this process takes place during clinical reasoning, that is, during diagnosing a case.

Boshuizen and Schmidt (1992) hypothesised that the application of knowledge in real cases triggers this reorganisation process. By the end of the first stage of knowledge acquisition, students have a knowledge network that allows them to make direct lines of reasoning between different concepts within that network. The more often these direct lines are activated the more these concepts cluster together and the more students become able to make direct links between the first and last concept in such a line of reasoning, skipping intermediate concepts. These intermediate concepts often happened to be the detailed biomedical concepts. In the first publications about this knowledge restructuring process, it was termed "knowledge compilation" (Schmidt, Norman & Boshuizen, 1990), a term emphasising the short-cuts in the lines of reasoning (Anderson, 1987). In later studies it was found that step-skipping was not the only thing happening. Not only were detailed concepts deleted from the lines of reasoning, new, overarching concepts emerged. Many of these concepts have (semi-)clinical names, such as micro-embolism, aorta-insufficiency, forward failure, portal hypertension or extra-hepatic icterus. These concepts link detailed knowledge about pathological structures or processes to clinical manifestations, providing a powerful reasoning tool. Seen superficially, the use of such concepts leads to step skipping in clinical reasoning, however, since these concepts coordinate a wealth of both detailed biomedical and clinical knowledge, they also provide depth and coherence to clinical reasoning. Hence in later articles the term "knowledge compilation" was replaced by "knowledge encapsulation". This term better denotes the overarching aspect of the concept (e.g. Boshuizen & Schmidt, 1992; Schmidt & Boshuizen, 1993a). Later Van de Wiel (1997) found that higher expertise levels were associated with "better filled" encapsulations giving further proof of this new view.

The process underlying knowledge encapsulation had not been described until recently (Van de Wiel, 1997). In discussions in the research group two ideas were scrutinised. One idea was that encapsulating concepts would be connected through inhibitory links to the underlying biomedical knowledge base. The advantage of this view is that it easily explains why experts use hardly any biomedical knowledge in their clinical reasoning. A disadvantage of this view is, however, that it cannot explain the flexibility with which experts can shift between reasoning in "encapsulated mode" and reasoning or explanations in "extended mode". Hence the other and older idea of simple Hebbian strengthening of links became a favourite. In this view links between concepts that are applied together in the diagnosis, explanation or treatment of a patient are strengthened, leading to clusters of concepts that are closely linked; the encapsulating concepts that summarize the knowledge in these clusters are, in fact, no more and no less than concepts that have acquired a central status in such a cluster, since they are directly and strongly linked to sets of clinical features, other encapsulating concepts and causes of disease. The latter theory can explain that experts do not apply biomedical knowledge in routine clinical reasoning, but can activate this knowledge when necessary. In this explanation it is assumed that routine cases activate a path of encapsulating concepts between signs and symptoms and diagnosis. In more difficult cases such a path cannot be set up instantaneously, hence spreading activation activates the underlying more detailed concepts leading to a reasoning process in which biomedical concepts play a role.

In the context of this book the assumption of simple Hebbian strengthening seems a little bit naive, since the knowledge involved in the knowledge encapsulation process can have many different kinds of representations that first need to be coordinated.[1] Many of the examples of encapsulations given above can illustrate this, but let us take "portal hypertension" to investigate the problems we are facing. This concept refers to an elevated blood pressure in the portal venous system, the diseases and pathophysiological mechanisms causing this elevated pressure, and the resulting signs and symptoms to be found in patients. A process of cellular changes in liver tissue, known as liver cirrhosis, is the most common cause of portal hypertension, leading to obstruction of the blood flow through the liver and the congestion of blood before the liver in the portal venous system. The high blood pressure in the portal vein leads to leakage of fluid to the abdomen and increased blood flow through the collateral veins connected with the portal vein. Dilated collateral veins, which may incidentally bleed, can be detected in the oesophagus, the abdominal wall and in the rectum (piles). To understand these processes students must have biomedical knowledge about infection processes, cell destruction and tissue healing, and the properties of the newly formed tissue. They must also have knowledge about the blood circulation around and through the liver, and they must be able to visualise how this blood circulation will be affected in case of an obstruction at some place. To perform a successful mental simulation of this altered circulation, they must also apply knowledge about the way the body generally reacts to permanent shortage of blood at one place and oversupply at another site. The knowledge involved here has multiple representations. There is visual and spatial knowledge about the anatomy of the liver and the blood vessels. Students also have mental models of the blood circulation that can be "run" in standard mode and by

[1]In our defence we can say that theories of associative networks that assume the strengthening of links between nodes in the network (e.g. Anderson, 1993; Kintsch, 1988) are usually confined to the conceptual level of representation, including concepts and propositions. Anderson's theory also includes temporal strings, or spatial relations, referring to different kinds of representations. These theories thus focus on the verbal representations and pass over the fact that multiple representations of a concept may exist and need to be coordinated in the learning process. For instance, the work of Koedinger and Anderson (1990) addresses the development of expertise in solving geometric problems. Their article contains several figures of different kinds of triangles to illustrate differences between novices and experts, but no details are given on the integration of visual and verbal representations, although they are incorporated in a type of schema developed as a result of continued experience with solving this kind of problem.

propagating changes through them. There is also verbal and visual knowledge of the signs and symptoms of a patient with liver cirrhosis (the process leading to portal hypertension), etc. In particular, the use of mental models of anatomical structures, linked to verbal representations is well documented, although not explicitly stated in terms of the coordination of multiple representations (Boshuizen & Schmidt, 1992).

In other chapters in this book (e.g. Savelsbergh, De Jong & Ferguson-Hessler, chapter 13; Boshuizen & (Tabachneck-)Schijf, chapter 8) we can see that the integration of multiple representations is not a self-evident process. Different representations require different inferencing mechanisms and different representations can only "interact" with each other by translating outcomes of one process into a format that can be read by the other inferencing process. Simple Hebbian strengthening of links between concepts that are active at the same time seems impossible in cases of representations of different formats, unless they have been translated into a common format.

Not only do the inference processes seem to be different, the 'part' of working memory where these different inferencing processes take place seems to be different as well. Baddeley (1997) has shown that working memory is not a unitary structure that can simultaneously keep 7 ± 2 (Miller, 1956) items in storage or can process only a limited amount of information per unit time. This view needs to be adjusted in several ways. One is that working memory can be broken down into at least three component parts: two slave systems — the visual sketch pad for the processing of visual information and representations (e.g., mental imagery) and the articulatory loop for auditory/verbal information and representations — and a control mechanism, the central executive. One might conjecture that visual information is processed in the visual sketch pad and verbal information in the articulatory loop, and that co-ordination takes place somehow in or by the central executive. However, research in this area does not substantiate these conclusions to this extent. It is obvious that different parts and processes have to be able to communicate, but how this is done is not clear. In other tasks it is vital that there is no interaction; this is, for instance, the case in the combined action of telephoning and driving. This is further evidence that merely simultaneous activation of concepts is no guarantee that links between these concepts will be made or strengthened.

Another related problem is that managing and coordinating different representations at the same time seems to require extra cognitive capacity as compared with the use of one representation only (Sweller & Chandler, 1991). The latter problem suggests that it may not even be very likely that students who do not completely master the different knowledge bases relevant for a specific problem activate the different representations, because that would result in heavier demands on their cognitive capacities, and hence stick to reasoning in one representation mode.

These two problems, the requirements that must be fulfilled before integration of two different representations can take place and the problem that students might not even use different representations in the first place, have led to the questions we want to explore in this chapter, i.e. what triggers the use of multiple representations in clinical reasoning and how are these multiple representations coordinated? In order to investigate these questions we "followed" a medical student who dealt with two clinical cases relating to one underlying medical problem, polycystic ovaries syndrome (PCOS), and compared that with two medical domain specialists.

1 Multiple Representations in the Domain Involved

Polycystic ovaries syndrome is a disorder in the regulation of follicle development. The description in a relevant textbook (Vander, Sherman & Luciano, 1990) strongly suggests that many different kinds of representations play a role here. Evidently, the relevant normal biomedical knowledge is characterised by multiple representations: the macro and micro anatomic knowledge involved (i.e. of the ovaries and the follicles that are produced there) has both a verbal and a spatial nature; knowledge about growth and development of these structures has a temporal, a spatial and a verbal aspect, while knowledge about regulation of these processes involves knowledge of dynamic interactions between continuous variables (the amount of hormones produced) and feedback systems that regulate the interactions between the production of these hormones and that trigger the development of structures in the ovaries and the uterus. The latter processes are often described verbally and by means of abstract graphical representations.

Mental imagery research by Kosslyn (1980) and Shepard and Metzler (1971) suggests that people can make visual mental representations of such descriptions and inspect and mentally manipulate these representations. Mani and Johnson-Laird (1982) showed that the more determinate a description was the more likely it was that readers would form a mental model of that and make visuo-spatial inferences from it. This skill to run mental models of variables is, however, limited. Mental rotation tasks involving one variable can be performed easily (Shepard & Metzler, 1971). However, simulation studies with economics students show that situations involving more than two independent (numerical continuous) variables are very hard to deal with mentally (Mandl, Gruber & Renkl, 1993). Even advanced students make errors with it in a simulated factory environment. Therefore, it is not unreasonable to assume that a domain characterised by the time-dependent process of follicle growth, affected by six hormones that mutually interact and that are produced by five organs or structures is difficult for students to learn, although both themselves and their teachers indicate that it is not too difficult (Dawson-Saunders et al., 1990).

The clinical knowledge is rather straightforward. It involves the following signs and symptoms: menstruation is irregular or even absent, often associated with overweight, facial hairgrowth and acne, enlarged ovaries and a disturbed balance between the hormones LH and FSH. The cause of this condition is disputed. Some assume that the cause is located in the pituitary gland where the LH and FSH are produced, others assume that the key to these problems must be found in the impaired conversion of androsteendiol into oestradiol in the ovaries, leading to a situation in which follicle maturation is disturbed.

In a well-organised knowledge base the following three levels of knowledge: clinical, pathophysiological and normal biomedical, must be coordinated and the same must be done with the multiple representations. Known limitations of the human information processing system (e.g. the inability to run more than a small number of variables simultaneously) suggest that not all knowledge bases are integrated. We expect that the integration of biomedical, pathophysiological and clinical knowledge starts from the clinical problems met in real practice: in this case the epidemiology dictates constraints on the kind of pathologies that are possible and hence on the number of theoretically possible combinations.

2 Method

The subject was a fourth year medical student. The study took place at the end of the academic year, completing the preclinical phase of the undergraduate training. During these 4 years the student has repeatedly studied the biomedical, the pathophysiological and even the clinical knowledge associated with PCOS. This student was presented with two different cases of polycystic ovaries syndrome. Her task was to diagnose the cases while thinking aloud and to explain each case afterwards. The text of the cases can be found in Appendix 1.

Think-aloud protocols were transcribed and segmented into propositions. The same was done with the explanations. Analyses focused particularly on the kind of representations applied (verbal, pictorial), the kind of knowledge applied (basic biomedical and applied clinical) and the coherence of the representation formed, both within one mode as across modes (text, pictorial). The aim was to identify the interface used (if any) where these coordinations took place.

Two domain experts, both registered gynaecologists, also participated in the study. Their think-aloud protocols and explanations were used as reference.

3 Results and Discussion

3.1 The Student's Efforts

The use of multiple representations in diagnostic reasoning will first be illustrated by the think-aloud protocol of the fourth-year student generated while proceeding through the first case of PCOS. The protocol is presented in the following format: in the left column the text of the clinical case is represented while in the second column the response of the student is provided. In the third column this verbal response is coded in terms of the kind of knowledge applied: D refers to data exploration and interpretation; C refers to the application of clinical knowledge; B refers to the application of biomedical knowledge; Rep refers to repetitions of information given earlier; PA refers to plans for action (further investigation, treatment, etc.); M refers to metacognitive remarks. Diagnostic hypotheses that have been put forward by the student are printed bold. The protocol is interrupted by comments of the authors.

case text	verbal response	code
History:		
The patient is a 23-year-old woman.	She is in a reproductive period of life, if I am not mistaken, eh.	D
She works as a physiotherapist at a boarding-school for physically handicapped children. She has no long-established relationship, but has a different female friend every 2–3 months.	So she has a homosexual nature.	D
She doesn't want any children. Now she turns to you because her menstruation has stayed away for 4 months.	At first you wouldn't think of a **pregnancy**, because it is mentioned that she has a female friend,	C
	but then you should ask her if she is pregnant by any chance.	PA
Her menarche occurred at 13 years of age and ever since she has irregular cycles, (28–38/5–9).	So the menarche is in fact normal,	D
	the cycle is irregular, 28–38 days,	D
	so that's not extreme,	D
	it usually is about 28 days as well,	B
	5–9 days during which she bleeds	D
It has already happened a few times that her menstruation stayed away for 2–4 months.	She has been menstruating for 10 years now	D
	and during this period her menstruation stayed away for 2–4 months, now and then.	Rep
	But it is only after 6 months that you can speak of secondary amenorrhea,	C
	so that is not yet the case.	C

So far the student has more or less built up an image of the kind of patient and the kind of problem she is confronted with using primarily clinical knowledge. There are no indications of the use of other than verbal representations. She has made a clinical evaluation in the sense that according to the 'rules' this is not yet a case of secondary amenorrhea.

case text	verbal response	code
But she has finally taken some steps now also because her menstruation mostly starts very unexpectedly the past year (cycle: 35–56 days, / 2–9 days),	that's very long indeed	D
	unexpectedly,	Rep
	so there is probably **something wrong with the hormone regulation**	B
	she couldn't handle it, in the sense that she produces little progesterone.	B

At this point she starts an evaluation in pathophysiological and biomedical sense, generating hypotheses about the fault. Remarkable is that the interpretations and explanations given are rather superficial. Several bits and pieces of verbal knowledge seem to be activated instead of a few coherent systems that might be represented as mental models.

She doesn't use contraceptives because she only has sex with women.		
She does not have galactorrhea,	what's that again, galactorrhea?	M
	Amenorrhea, galactorrhea, I can't recall it so quickly.	M
She doesn't suffer from hot flushes.	Hot flushes are seen with a disturbed hormone regulation	C
	especially after the menopause,	C
	but that can be caused by several disturbances, I think.	B
Her weight has been too high for many years, 79 kg at a height of 1.70 m.	um, um	
	About 60 kilo is normal at 1.70,	B
	so it's still on the low side,	C
	then you would think of, eh, of the **syndrome of the polycyclic ovaries**,	C
	because there never was a regular cycle	C
	and because she doesn't feel the cycle coming	C
	and is overweight.	C

Based on the findings mentioned, the diagnostic hypothesis of PCO is generated, although the name is a little bit corrupted.

Physical examination:

slightly increased hairiness of the face,	so this is hirsutism	C
slightly increased hair growth on the upper lip and pronounced side-whiskers.		
She has slightly androgenic pubic hair growth and increased hair growth on the upper legs.		
Mamma: normal in size and consistency on both sides.	Why do you pay extra attention to that? Eh, perhaps something can be wrong with it.	D
Vaginal examination: normal uterus in mobile anteversion, anteflexion.		
Adnexa slightly enlarged on both sides (valued at 4 up to 5 cm in diameter)	About 3 is normal, I think.	B

The results of the physical examination are merely interpreted, and not yet related to the diagnostic hypothesis of PCO. Remarkably, she recognises the clinical finding of hirsutism already at the first sign. At this point the lab findings are given and we see an analysis of these findings in several steps: first the biomedical knowledge is activated and the functions of the different hormones are remembered, bit by bit.

Laboratory:

TSH normal;	TSH means the function of the thyroid gland,	B
	the direction of the function of the thyroid gland, from the pituitary to the thyroid gland,	B
Prolactin is normal;	prolactin has something to do with the breast,	B
FSH is low/normal;	FSH is low/normal.	Rep
	FSH is responsible for the follicle maturation	B
	during the first part of the cycle	B
	and it is low/normal.	Rep
LH	is responsible for the ovulation	B
and has increased.		

But halfway she finds support for the diagnostic hypothesis of PCO, and links the outcome of a laboratory test of free testosterone with clinical findings of increased hair growth described in the physical examination.

and the LH / FSH ratio increased;	so that points to those **polycyclic ovaries**	C
free oestradiol is slightly increased;		
free testosterone increased;	that's again.........., then I return to the hair growth, which has slightly increased;	C
androstenedione is slightly increased	as well;	
T4 is normal;		
pregnosticon negative.	I don't know what pregnosticon means	M

After all hormones have been dealt with, the student makes a second, far more detailed round on the lab results, in which she elaborates on the hormone interaction. First she tries to explain the pathophysiological process right away, but soon she realises that there is a problem. Although she mentions the feedback of oestradiol on LH there are no indications that she runs a model to find out where or why things go wrong.

	TSH is normal	Rep
	T4 is normal,	Rep
	so you don't have to think of the thyroid gland,	C
	FSH is low/normal	Rep
	and LH is increased	Rep
	LH/FSH is increased,	Rep
	that is to say that the follicular maturation doesn't get under way very well,	B
	and there is a high LH because of that feedback of oestradiol	B
	on LH normally is, eh, let's see, normally is negative	B

Having recognised that her pathophysiological knowledge is insufficient, she shifts to her knowledge about normal functioning and uses a pictorial representation (Figure 1) and a verbal one to figure out the normal interaction at the organ and hormone level.

Pictorial response

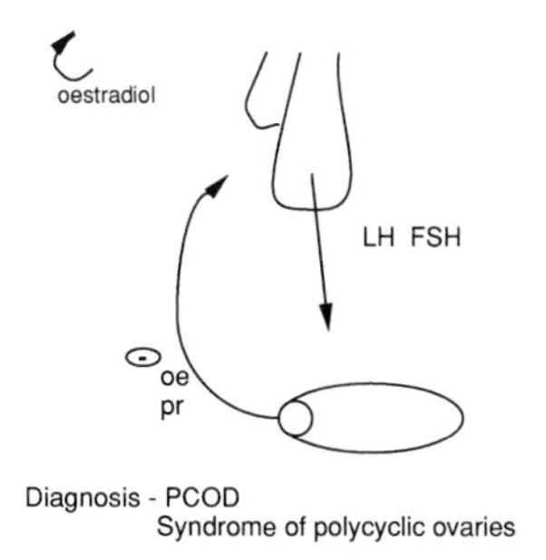

Figure 1: Pictorial representation constructed by student 4.4 of the normal functioning in thinking aloud

that is to say if you.....,if there is LH	B
and FSH from the pituitary gland	B
which go to the ovary,	B
then follicular maturation will arise under the influence of FSH	B
and that maturing follicle produces oestrogens	B
and less progesterone,	B
which have a negative feedback on the pituitary gland,	B
however, there is a LH peak before the ovulation,	B
then there is an ovulation,	B
and then the follicle becomes a corpus luteum	B
and produces progesterone especially.	B

It is not until this stage that she tries to map a mental model of the female anatomy and physiology on the patient's signs and symptoms. Looking at the pictorial representation of the reproductive system she has drawn, she concludes that there must be a disturbance in the hormonal feedback mechanism. However, she is not sure what role the fat tissue and testosterone play in it.

What function has this pictorial representation? Given the fact that she has tried before but did not succeed very well, it can be assumed that the function of this combination of pictorial and verbal representations was to find a stronger basis for the pathophysiological reasoning. Yet the pictorial representation does not bring her very far. The reader might see that the picture only covers the first part (until "pituitary gland") of the protocol section given above. The picture does not, and cannot, reveal the normal time-dependent interactions between the several hormone levels and the follicle. She deals with this part only verbally, starting from "however, there is . . .". Exactly at this point the PCO problems occur: no cyclical changes, no LH peak, hence no ovulation. This aspect is not elaborated further.

However, the student seems more certain at this point and applies this knowledge to the present patient in the following part, to see what aspect of the whole process may have caused this disturbance, and remembers that fat tissue plays a role. This aspect is slotted in, without going into much detail, probably because she forgot about that. She also restates the diagnosis, probably deciding that enough energy has been spent on the case.

LH eh FSH is low/normal in this patient	Rep
so follicle maturation is insufficient	B
LH is increased,	B
so the feedback is not effective,	B
let's see, yes, there also was peripheral	
production of oestrogen in the fat tissue,	B
that testosterone also plays a part in it,	B
but I don't remember it exactly,	M
that also interferes with that feedback,	B
anyhow, and that always points to the	
polycyclic ovary syndrome,	C
and the high testosterone also argues for it.	C
So the diagnosis I want to make is the	
polycyclic ovary syndrome.	C

While proceeding through the case the student made several annotations in the text. She underlined important clinical features, added some normal values and wrote down some of her interpretations. These reminders helped her in constructing a representation of the patient's problem.

The role of the pictorial representation in the think-aloud protocol is not clear. She used this representation when she was uncertain about her verbal knowledge. Remarkably, the use of these representations did not activate new or extra knowledge that had not been activated already through verbal representations. However, the protocol does not show any evidence that this disappointed her. Her pictorial representations seem to have had three functions. One was fortifying the verbal representation by drawing and talking at the same time. Another was to aid memory. Finally, it also might have sensitised the student to the part of the mechanism where the fault might take place.

When asked to explain the pathophysiology underlying the case the student basically elaborated the explanation she had been constructing in the final part of the think-aloud session. Again she mainly used her biomedical and pathophysiological knowledge to explain the case, although she first tried to explain the clinical process, but she gave up before she really got started. Instead, she started to draw a picture representing the organs and hormonal feedback mechanisms involved in the reproductive system (Figure 2). She explained that in case of PCO all organs and hormones are present, but that the hormone regulation is disturbed. On the basis of her biomedical knowledge of the normal functioning of the reproductive system she tried to reconstruct what was happening in the patient. While explaining, she was looking at the picture and adding elements to it. The verbal explanation illustrated very well that she moved back and forth between her knowledge about normal and pathophysiological processes.[2]

[2]The explanation protocol has been divided in sections by "**I**"; the sections that refer to normal biomedical knowledge are printed in italics and the sections that refer to pathophysiological knowledge are printed as standard text; in the sections that are underlined the student indicates that she is not sure or does not know.

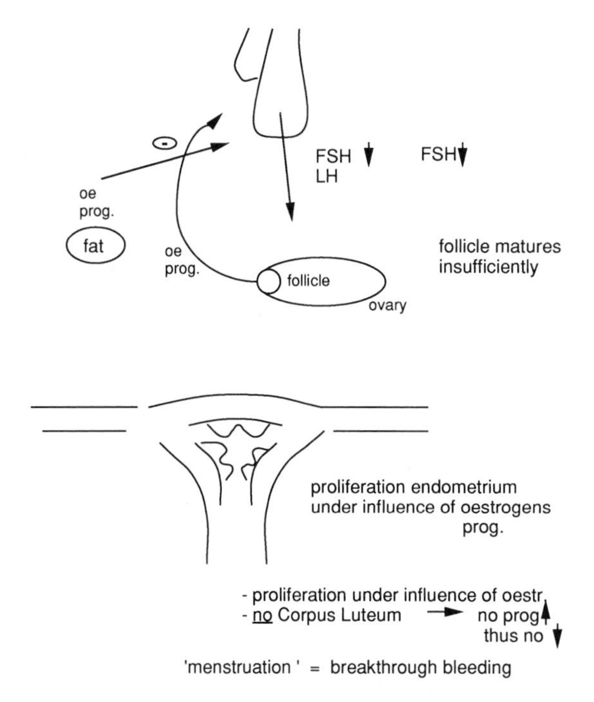

Figure 2: Pictorial representation constructed by student 4.4 in explaining the case

Well, the mechanism of the syndrome of polycyclic ovaries, works such that, eh, there is, eh, follicle—, let's see, there is, in principle there is a good reproductive system, I would say. So pituitary is present, the hypothalamus, the pituitary and ovaries are present, but there is a disturbance in the feedback mechanism, it becomes deregulated, that can be caused by too much fatty tissue, on the one hand, fat plays an important role, and there was something else that plays an important role, eh........ I cannot remember so quickly what also plays a role, **I** *but anyway, it normally is the case that one's pituitary secretes FSH and LH, I have already said that these affect the ovary in which a follicle is going to mature, during the, and that maturing follicle produces oestrogens and progesterone, and later the corpus luteum as well, which inhibit the feedback.* **I** Now, what was the problem with those polycyclic ovaries? that the follicle which produces..., in the fatty tissue oestrogens and androstenedione are also being produced, so those inhibit, those inhibit all the FSH, so there is not enough FSH, and because there is not enough FSH, follicle maturation is insufficient, the follicle matures insufficiently. That's why never an ovulation occurs in the endometrium in the uterus, there is eh, **I** *there normally is proliferation of the endometrium under the influence of oestrogens, so there will be proliferation, and under the influence of progesterone there will be more vascularisation, more glu, glycogen will be stored, so will be prepared for implantation.* **I** So in polycyclic ovaries you have oestrogens that are, that come from the fatty tissue on the one hand, on the other hand, the maturing follicle produces some oestrogen, so there is proliferation under the influence of oestrogens. **I** *Normally there is an ovulation, because the FSH, or no, because the progesterone peak stops all at once. And a menstruation normally is an extraction bleeding because of a decreasing progesterone.* **I** Now, because no ovulation occurred with polycyclic ovary syndrome, there is no corpus luteum, no increase of progesterone, so no fast decrease too, **I** *which normally appears because of the negative feedback on FSH and LH, through which the corpus luteum normally ceases to exit after 12 up to 14 days.* **I** Now this is not the case, eh, so there is no regular cycle, there are bleedings, a sort of breakthrough bleedings, because the endometrium proliferates in such a way, that one day it will be too much and then there will be a menstruation, a sort of pseudomenstruation, so a menstruation between inverted commas, is an, eh, breakthrough bleeding, and that explains why you don't feel it coming, **I** *because normally that increase of progesterone causes tense breasts, and abdominal pain and that sort of things, that you can notice something by the mood,* **I** with a syndrome of the polycyclic ovary there is no progesterone peak and extraction, and that's why you don't feel the menstruation coming and it is very irregular too. Yes?

In this *post hoc* explanation several differences from the think-aloud session are noticeable. The first difference is that several phenomena described in this case are analysed further and more deeply. One of these phenomena is the disturbance in the feedback loop due to problems with follicle maturation and oestrogen production in fat tissue.[3] Missing in the think-aloud protocols, but explained well here, are the changes at the uterus level. Another remarkable finding is that the explanation given is not a direct readout from memory: the student has to think during the explanation; assumptions and linkages have to be made on the spot, and checks on the line of reasoning have to be made; she even has to admit that she does not know some aspects (the role of fat tissue) well enough. The coordination process is still going on. Finally a very striking phenomenon is that she again shifts back and forth between the normal functioning model and the pathophysiology of PCO. Linkages with clinical knowledge are not apparent.

So far her dealing with the first case reveals some initial coordination between several knowledge bases and representations. An important aspect of both the clinical reasoning and the explanation was the scaffolding of one representation by means of another. While drawing and talking at the same time the information implicit in the pictorial representation can be made available to the verbal representation, from which point new elaborations and inferencing, and hence learning, can take place.

3.2 The Second Case

After having dealt with a few cases with different underlying pathology but similar complaints, she is unexpectedly presented with a new PCO case. In the next case we see a similar approach to the first case but with a few remarkable differences. The text of the case and the think-aloud protocol can be found in Appendices 1 and 2. One of the most remarkable differences is that in this protocol no drawings are made, and we do not see an active activation of relevant knowledge in the sense of explicit explanation of individual findings. Apparently, elements in the knowledge bases are activated more easily. Again she first builds up an idea of the kind of patient and problem she is

[3]This is a misconception. Oestrogen is not produced in fat tissue. Instead fat tissue converts androgen, produced by the ovaries into oestrogen. The fact that ovaries produce androgens is found nowhere in the protocol. Maybe the student does not know this.

dealing with. Based on the gross irregularities in the menstrual cycle her hypothesis is PCO. This hypothesis is brought forward at an earlier stage than with the first case. Another difference is that we see that biomedical knowledge is scarcely used and when used, it is not in the sense of backing up the clinical reasoning as was done earlier. This does not seem necessary because that line of reasoning was sufficiently spelled out in the previous case, and she refers to that learning episode implicitly ("again ...").

In her *post hoc* explanation we see that she starts at the point that caused difficulties in the first PCO case: the role of fat tissue in hormone "production". Again she switches back and forth between pathophysiological and normal biomedical knowledge trying to make links between both databases, but now she starts from the pathophysiological part. Again she is able to explain the uterus part of the story, but the hormone disturbances are still problematic. The drawing she makes is very similar to the one made while explaining the first case. While explaining the case she finds out more and more that her knowledge is not sufficient for this task, and although in some respects she does a better job than in the first case, she keeps saying that she does not know. In the end she is more confused than when she started to explain the case. This is illustrated in the following text:

oh, a few moments ago I couldn't give a precise explanation either, **I** eh, well, on the one hand there is too much fatty tissue, and because of that there are too many androgens, oestrogens, of course those androgens do not come from the fatty tissue. Yes, they can be produced in the fatty tissue and in the adrenal cortex, well, with the polycystic ovary syndrome it is especially the fatty tissue, because those people have to lose weight, then they will do better, the ovary, **I** *so normally under the influence of FSH and LH the follicular matures, eh, but those oestrogens and androgens already have a negative feedback, those oestrogens and androgens from the fatty tissue already have a negative feedback on the FSH, which therefore decreases,* **I** which is low / normal in this case and because of that there is insufficient follicular maturation, eh, so there also is a somewhat negative feedback in this case, **I** *oestrogen and progesterone, which normally have a negative feedback on the pituitary.* **I** There is insufficient follicular maturation now, and there is little oestrogen and progesterone from the ovary, but there are those androgens and oestrogens from the fatty tissue which cause the negative feedback. There will not be a LH peak, oh, there is a LH increased..........., **I** How is that again?, I still don't know, **I** *the oestrogen has a ...., normally there is that LH peak and because of that the ovulation,* **I** How was that in the first case? How is that possible? Yes, eh, I had that before, did the LH also increase in that case? I don't even remember. **I** Anyhow, there is not enough oestrogen and progesterone coming from the follicle, because it doesn't mature, and that's why there is that furthering of the mucous membrane, is, eh, occurs under the influence of oestrogen, but there is no progesterone, so that a possible bleeding that occurs is an, oestrogenic breakthrough bleeding, and you don't feel that coming either, well eh, there usually is no ovulation with a PCO-syndrome, yes. **I** *Normally you will find a negative feedback from the oestrogens and progesterone on the pituitary gland for FSH and also for LH,* **I** but LH is increased now, **I** and now I don't remember well anymore how it was, **I** *Normally there also is......., maybe because that eh, but is that feedback negative, but it is positive just before the ovulation,* **I** so it is possible that she might have an ovulation later. **I** Well, I don't know exactly, how it is with LH, but I think it fits into the picture. Well, I am not really clear about that. **I**

The answer to our question as to what triggers coordination of multiple representations is emerging slowly from these protocols and explanations. It seems that merely solving a clinical case does not provide strong triggers. As we have seen in the first case, the student used some biomedical knowledge to scaffold the clinical reasoning process, a finding that is in agreement with earlier research findings in this group of fourth-year students (Boshuizen & Schmidt, 1992). A new finding was that this scaffolding might also help to launch further reasoning on the verbal representation. In the second case the pictorial representation was no longer needed, since clinical reasoning was sufficient for the task at hand, i.e. to diagnose the case. Some implicit coordination might have occurred, but not much when we take the approach used in the *post hoc* explanations into account. This is an important conclusion, since in many formats of medical education it is assumed that through solving cases, knowledge will be integrated, or in the language of the present chapter, representations will be coordinated. This is also an important finding for our own theory of knowledge encapsulation, since in this theory the application of biomedical knowledge in clinical reasoning was a crucial step in the integration process.

In the *post hoc* explanation we see that the recognition that she does not know something starts the process of matching biomedical to pathophysiological knowledge: the switching back and forth between these two knowledge bases in the explanation shows this effect. What we do not see in the explanation is whether there is again some linkage with the clinical knowledge; although she explicitly refers to "the present patient" there is hardly any indication that explicit linkages between individual or groups of case items are matched to pathophysiological processes or systems. There is one exception, in the final part of the second explanation where she explicitly accounts for the fact that this patient does not have premenstrual sensations. A conclusion can be that an impasse in an explanation, witnessed by a metacognitive feeling of "not knowing" triggers efforts after coordination, and that in one such episode only two knowledge bases are coordinated.

This learning mechanism is not unique. SOAR (Laird, Rosenbloom, & Newell, 1986) also assumes that failure is the key to new learning. Yet, educationally and maybe even professionally, there is also a problem inherent to a coordination process that is triggered by the metacognitive assessment that the knowledge is insufficient. This problem is that the knowledge that is going to be coordinated cannot be complete; hence the coordination process cannot be completed successfully at that stage. It must be picked up again at a later moment when more is known, but the student again runs into trouble.

The explanations given by this student show remarkable similarities with the findings in a recent study by Groothuis, Boshuizen and Talmon (1998). In this study in which students were asked to explain the menstrual cycle it was found that their knowledge was sketchy, generated in bits and pieces, and was especially lacking detail regarding the follicle development part of the cycle. Furthermore, and regarding the present study maybe most importantly, the students did not integrate the mechanisms at the cellular level in their explanations, which according to the experts in that study were the key to understanding the mechanisms at hand. Our student in this study did not make connections with that knowledge either.

3.3 Comparison with the Experts

Superficially speaking the expert (G4) protocol (see Appendix 3) is not so different from the second protocol generated by the student. At first a representation is built up of the patient and the problem. Very soon the expert categorises this case as a PCO-like case. There is one issue that remains unclear to him, i.e. why the menstrual problems have increased over the last year. He would like to know if the patient has gained weight recently, because this might have aggravated the problem. This is the same issue as the student was pointing at. The difference is that the student seems to assume that the cause for the problem can be found in a disturbance of hormone "production" in the fat tissue, while the expert only sees it as a condition that may have aggravated the problem. In his explanation we see that in his view the problem is located at the follicle level itself and at the cellular processes taking place. His explanation is characterised by a very fast progressive deepening and back to the clinical level:

1. he started at the ovary and the hormones that are produced or converted there (not in the fat tissue),
2. through to the cellular level where the problem with receptor development is located;
3. back to the follicle level, linking the explanation to the implications this pathophysiological process has for treatment and the problems that can take place then.

The think-aloud protocol of the other expert (G1) is also very short (see Appendix 4). His hypothesis that this patient has PCO syndrome emerges very soon as well. Interestingly, his think-aloud protocol shows the application of some concepts that seem to bridge clinical and pathophysiological knowledge. A key term seems to be "hyperandrogenic amenorrhea" which links the pathophysiological hormonal effects in the ovaries with the amenorrhea. He also uses a combination of clinical and pathophysiological knowledge when explaining the woman's large breasts. In his *post hoc* explanation we see a somewhat different approach than the one taken by his colleague: he emphasises the interaction between the pituitary gland and the ovaries and how this interaction is turned into a vicious circle.

These expert protocols show a few phenomena that are missing in the student's protocols: the coordination between clinical and pathophysiological knowledge, the time-dependent aspects of the hormonal interaction that has gone astray, and the linkage with treatment and problems with that. No references to normal biomedical knowledge are needed to back up the pathophysiological line of reasoning.

Another remarkable difference from the student's protocols is that the experts' protocols do not show any indication that the coordination is fabricated on the spot, although expert G4 says that still a lot is unknown and speculative. However, this is the state-of-the-art of the domain, not of his own knowledge. Coordination of his knowledge bases seems to have taken place earlier. Neither do the experts run into a situation where they have to admit to themselves that they do not know the answer well enough.

4 Discussion: Multiple Representations in Medicine; a Second Look

The aim of the above analyses was to find out what triggers coordination of multiple representations and how that takes place.

Our conclusion made on our way to this point was that first of all representation coordination seemed to be triggered by impasses that the student ran into when applying a specific knowledge base. At that moment she shifted "back" to a more basic knowledge level. For this student this occurred both when applying clinical knowledge and when trying to apply pathophysiological knowledge, hence reverting in both cases to her knowledge about normal biomedical functioning. The fact that she did not recognise some misconceptions, lacked the relevant overarching concepts such as "hyperandrogenic amenorrhea" used by one of the experts and was contented enough with her probably incomplete clinical knowledge prevented her going further with her efforts after coordination.

So far our analysis has also revealed that coordination of representations in medicine is not "merely" the translation and coordination of representations that have been developed to do more or less the same job but with different tools. The latter might be the case in economics or physics, where mathematical and graphical representations or graphical and verbal representations of one topic are used. In medicine the situation is different in so far that not only do different representations have to be coordinated but also different knowledge bases. These knowledge bases have been derived from external bodies of knowledge that have been developed from different sources, from different perspectives and with different purposes. For instance, biomedical knowledge has been largely developed by laboratory scientists aiming at a progressive deepening of the understanding of normal and abnormal phenomena. The clinical knowledge has been largely gathered by generations of physicians who focused primarily on diagnosis and treatment of diseases at different stages of their course. No doubt these knowledge bases are qualitatively different in the sense that they do not tell the same story by different means. Instead, they tell different stories, but one story can be used to better understand the other. However, students themselves have to find out, largely unassisted, where bridges between these knowledge bases can be built and how one story can be used to better understand the other. Books can help them, but apparently students have to construct at least part of the coordinations themselves. This apparently does not come without effort, given the fact that both in the present study and in the one by Groothuis et al. (1998) the link, explicitly made by most physiology books, between the organ and the cellular level was absent in the students' explanations.

Maybe it is one of the difficulties for medical students that their task is not to coordinate complete knowledge bases, but instead to coordinate those parts where keys can be found in their biomedical and pathophysiological knowledge bases that can be used to better explain clinical findings and to underpin choices for treatment. The fact that by the end of their fourth year of training this has not yet occurred suggests that on the one hand, this is more difficult than one might have expected. On the other hand, if it is so important as it appears from the structure and content of the books and from the experts' clinical reasoning and explanations, then other educational means should be used to provoke integration. If we are right that impasses trigger this phenomenon, then educational means should be used that tempt students into situations where they recognise the holes and misconceptions in their knowledge. Since the student in this study did not find fault with her clinical knowledge, tasks other than clinical reasoning tasks should be offered to her. Explanations to fellow students or to themselves seem to do a better job (Van de Wiel, 1997), but more creative formats should be developed.

Another issue that is not raised by the present analysis but that goes back to the introduction to this part of the book (Boshuizen & Schijf, chapter 8) should also be dealt with here. In chapter 8 the issue was brought forward that one of the ways of coordinating multiple representations is to feed these representations into an on-going process in which both kinds of representations have roles and responsibilities. In clinical reasoning this might be the case when dealing with real patients in real situations. Contrary to when dealing with paper cases, like the ones used in the present study, in real clinical situations the physician has to gather the information by watching, hearing, smelling and feeling signs and symptoms. These observations feed into the clinical reasoning process. How this process takes place is not clear. Part of the observations will probably be translated into verbal code, but experts will probably also use untranslated, or even to themselves untranslatable, information. In that case multiple representations are not themselves coordinated, but are used independently in a coordinated way.

A final topic: back to knowledge encapsulation. What do the present analyses say about knowledge encapsulation and the role of multiple representations in that process? The way the student dealt with her pictorial and verbal representations in order to coordinate normal biomedical and pathophysiological knowledge suggests that translation into a verbal representation is necessary to arrive at encapsulations. The use of the term "hyperandrogenic amenorrhea" by one expert, and the progressive deepening and back again by the other expert, also suggest that encapsulations can entail knowledge from different sources and with originally different representations. The active process requires a translation step of a mental model or a pictorial representation into a verbal one. It is not clear whether other kinds of translations (for instance from the verbal into the pictorial or the spatial platform) can also be used in this process. The use of think-aloud protocols and verbal explanations is not the most adequate kind of medium to display this kind of interaction. Analyses of drawings and gestures, and experiments in which the visuo-spatial sketchpad, the auditory loop system (Baddeley, 1997) and the supervisory attention system that controls the information flow and processing in working memory (Norman & Shallice, 1986) are loaded separately by dual tasks might shed further light on these questions.

Acknowledgement

The authors wish to express their thanks to Prof. Dr Henk G. Schmidt, Dr R. Rikers and Dr N. C. Schaper for the long disicussions we had about the concept of knowledge encapsulation, the processes underlying it and how encapsulation can be stimulated in medical students. The views presented in this chapter are explictly our own responsibility.

Appendix 1

Case 1

The patient is a 23-year-old woman. She works as a physiotherapist at a boarding-school for physically handicapped children. She has no long-established relationship, but has a new female friend every 2-3 months. She doesn't want any children. Now she turns to you because her menstruation has stayed away for 4 months.

Her menarche occurred at 13 years of age and after that she has irregular cycles (28-38 / 5-9). It has already happened a few times that her menstruation stayed away for 2-4 months, but now she has finally taken steps. Also because her menstruation mostly starts very unexpectedly the past year (35-56 / 2-9). She doesn't use contraceptives because she only courts with women.

There is no question of galactorrhea. She doesn't suffer from hot flushes. Her weight has been too high for many years, 79 kg at a height of 1.70 m.

Physical Examination

Slightly increased hairiness of the face; slightly increased hair growth on the upper lip and pronounced side-whiskers; slightly androgenic pubic hair growth; and increased hair growth on the upper legs.

Mamma: normal in size and consistency on both sides.

Vaginal Examination: Normal uterus in mobile anteversion/anteflexion; adnexa slightly enlarged on both sides (valued at 4 up to 5 cm in diameter)

Laboratory

TSH normal; prolactin normal; FSH low/normal; LH increased; LH/FSH ratio increased; oestradiol slightly increased; free testosterone increased; androstenedione increased; T4 normal; pregnosticon negative.

Case 2

The patient is a 27-year-old woman. She has a demanding job as an executive secretary at a medium-sized printing-office. She is married and has no children. Now she turns to you because her menstruation has stayed away for 4 months.

Her menarche occurred at 14 years of age, and after that she had irregular periods, (28-42 / 5-10). About 3 years ago her menstruation also stayed away for nearly 4 months. She was relieved she was not pregnant at that time. The last year her menstruation mostly has started completely unexpectedly (42-70 / 2-10). She uses condoms for contraceptives.

She does not suffer from hot flushes or galactorrhea.

Physical Examination

Slightly increased hairiness of the face; increased hair growth on the upper lip; and pronounced side-whiskers; acne on the face; obese impression: weight of 75 kg at a height of 1.65 m; slightly androgenic pubic hair growth and increased hair growth on the upper legs. Massive mamma development on both sides.

Vaginal Examination: Normal uterus in mobile retroversion; on both sides slightly enlarged adnexa (valued at 4 up to 5 cm in diameter).

Laboratory

TSH normal; prolactin normal; FSH low/normal; LH increased; LH/FSH ratio increased; oestradiol slightly increased; free testosterone increased; androstenedione slightly increased; T4 normal; pregnosticon negative.

Appendix 2

Fourth-year student's (pp 4.4) think-aloud protocol on the second case of PCOS.

case text	verbal response	code
A 27-year-old woman,	that's a reproductive period of life,	D
she has a demanding job as an executive secretary at a medium-sized printing-office.		
Married		
and no children.	Eh, no children, well that's not really spectacular,	D
	she probably uses some kind of contraceptive	D
	or unwantedly childless.	D

Case text	Think-aloud protocol	Code
Now she turns to you because her menstruation has stayed away for 4 months,	she may be **pregnant** now, just wait if that's the case,	C
	or a **hormonal disturbance**.	B
Her menarche occurred at 14 years of age,	that's, yes, a little above the average,	D
	but it is normal yet.	D
And after that irregular periods,	so as for the hormones the cycle didn't get under way very well,	B
(28-42 days / 5 over 10)	in fact normal or prolonged,	D
	so no shortened cycles.	D
About 3 years ago her menstruation also stayed away for nearly 4 months, she was relieved she was not pregnant at that time.	This makes me think of the possible **syndrome of the polycystic ovaries** once more,	C
	yes, let's see, **polycystic ovaries**, yes. that's it,	M
	she was relieved she was not pregnant,	Rep
	yes, so it was indeed possible.	D
The last year her menstruation mostly has started completely unexpectedly.	She doesn't feel it coming,	D
	so in the sense of that ,.....probably progesterone .....not enough build-up.	B
she uses condoms as contraceptive,	condoms are less reliable than the pill, but anyway.	D
She doesn't suffer from hot flushes or galactorrhea,	so as for hormones you don't expect her to be in her menopause,	C
	that's not really the question at that age.	C
Physical examination:		
slightly increased hairiness of the face;	so, that's hirsutism.	C
	That's because of a surplus of androgens,	B
increased hair growth on the upper lip; and pronounced side-whiskers;		
Acne on the face;	that's because of the androgens.	B
Obese impression,	yes, in case of **PCO** you also have too much fatty tissue,	C
weight of 75 kg at a height of 1.65 m,	I think that's overweight	D
	At a weight of 75 kg you normally ....., no, a weight of 55 kg is normal at 1.75 m,	B
	so, this clearly is overweight.	D
Slightly androgenic pubic hair growth and increased hair growth on the upper legs.	This all points to that surplus of androgens.	C
Massive mamma development on both sides,	yes	

Vaginal examination: normal uterus in mobile retroversion, on both sides slightly enlarged adnexa valued at 4 up to 5 cm,	normal is 3	B
	so this also points to that **polycystic ovary syndrome**,	C
	then those ovaries will also increase.	B
Laboratory:		
TSH normal		
and T4 normal,	so you wouldn't really think of the thyroid gland;	C
prolactin is normal;		
FSH low/normal;		
LH increased;		
LH/FSH increased;		
oestradiol increased;		
testosterone increased;		
androstenedione is increased too;	so the diagnosis will be the **polycystic ovary syndrome**.	C

Appendix 3

Think-aloud protocol of gynaecologist G.4

case text	**verbal response**	**code**
History:		
Patient is a 27-year-old woman.		
She has a demanding job as an executive secretary at a medium-sized printing-office.		
She is married		
and has no children.		
Now she turns to you because her menstruation has stayed away for 4 months.		
Her menarche occurred at 14 years of age,	that is normal	D
and after that irregular periods.		
(28–42 / 5–10)		
About 3 years ago her menstruation also stayed away for nearly 4 months.		
She was relieved she was not pregnant at that time.		
The last year her menstruation has mostly started completely unexpectedly		
(42–70 / 2–10).	So that has shifted to a considerable **oligomenorrhea**	C
	in that period of, let's see, it is not said exactly, the last year.	Rep
	She has always had a somewhat prolonged cycle.	D
She uses condoms for contraception.		
She doesn't suffer from hot flushes		
or galactorrhea.	As if to say that this direction excludes an endocrine pathophysiology.	C

Physical examination:		
slightly increased hairiness of the face;	That leads us to a **PCO-like picture**.	C
Slightly increased hair growth on the upper lip;		
and pronounced side-whiskers.		
Some acne on the face.		
Obese impression:		
weight 75 kg, height 1.65 m,	quite a high weight	D
Slightly androgenic pubic hair growth,	that clearly is some kind of diamond-shaped hair growth.	D
Increased hair growth on the upper legs.		
On both sides massive mamma development.	It is not mentioned if there is hair growth on the breasts.	PA
	This usually is a sign.	C
Vaginal examination: normal uterus in mobile retroversion; slightly enlarged adnexa on both sides, valued at 4 up to 5 cm in diameter,	so almost pathognomonic for a **PCO-like picture.**	C
Laboratory:		
TSH normal;		
prolactin normal;		
FSH low/normal;		
LH elevated;		
LH/FSH ratio elevated;		
free oestradiol slightly increased;		
free testosterone increased;		
androstenedione slightly increased;		
T4 normal;		
pregnosticon negative.	Well, I think this is almost a classical description	C
	as it is presented here, of a **PCO-like picture**,	C
	with an invariably irregular menstruation,	C
	but in the sense of an oligomenorrhea.	C
	It is not clear why that oligomenorrhea has increased so dramatically in the last years.	C
	Possibly, but this is not mentioned in the case,	C
	by weight increase.	C
	Sometimes the picture gets worse because of an increase of the weight,	C
	by which there is peripheral conversion of androgens,	B
	through which the free testosterone level increases even more.	B

Pathophysiological explanation by expert G.4

Yes, the exact pathophysiology of PCO is still unknown. If it was known, then you could do something about it. There are people who say there is a disturbance of the LH / FSH synthesis in the pituitary gland, well, it will probably not be like that. I highly doubt that. But there probably is a disorder in the ovary itself, with a disturbance in the conversion of androstene-dione into oestradiol, which causes a disorder in the receptor production in the ovary, receptors

which are sensitive to FSH, less to LH, by which there is a disturbance of the normal folliculogenesis as well and you will notice many follicles maturing and no normal ovulation occurring. That also explains the polycystic picture, an absent ovulation, in this case with abnormal follicle development, by which it is striking that hardly any dominant follicles develop. And then it would appear that all follicles are in the same stage.
That's why inducing an ovulation sets so many problems with this patient. When using inducing drugs there are significantly more overstimulation syndromes with this kind of patient. That's all.

Appendix 4

Think-aloud protocol of gynaecologist G.1

case text	verbal response	code
History:		
Patient is a 27-year-old woman.		
She has a demanding job as an executive secretary at a medium-sized printing-office.		
She is married		
and has no children.		
Now she turns to you because her menstruation has stayed away for 4 months.		
Her menarche occurred at 14 years of age, and after that irregular periods.		
(28–42 / 5–10)		
About 3 years ago her menstruation also stayed away for nearly 4 months.		
She was relieved she was not pregnant at that time.		
The last year her menstruation has mostly started completely unexpectedly		
(42–70 / 2–10).		
She uses condoms for contraception.	Those **oligomenorrheas** that start from the menarche,	C
	then you will also think of a **polycystic ovary syndrome**.	C
	And obviously it has always been like that with this patient, that it is irregular,	C
	and now even stays away for 4 months.	Rep
She doesn't suffer from hot flushes		
or galactorrhea.	This fits in.	C
Physical examination:		
slightly increased hairiness of the face;	which also matches a **PCO**;	C
Slightly increased hair growth on the upper lip;		
and pronounced side-whiskers.		
Some acne on the face.		
Obese impression:		
weight 75 kg, height 1.65 m,	Everything points in the same direction.	C
There is a slightly androgenic pubic hair growth		
and increased hair growth on the upper legs.		
On both sides massive mamma development.	which can be a result of the high oestrogens	B
	that go with **the syndrome**.	C

Vaginal examination: normal uterus in mobile retroversion;	and the adnexa will be slightly enlarged.	C
Yes, slightly enlarged, valued at 4 up to 5 cm	So everything still points to the **polycystic ovary syndrome**,	C
	which is a general term for ovary disorders that lead to **hyperandrogenic amenorrhea**.	C
Laboratory: shows a normal TSH level; a normal prolactin level; FSH is low/normal; LH is increased; so the ratio is increased; free oestradiol is slightly increased; free testosterone is increased; androstenedione is slightly increased; T4 is normal and pregnosticon is negative.	So this is a **primary disorder in the pituitary–ovarian axis** on the basis of the **polycystic ovary syndrome**.	B C

Pathophysiological explanation by expert G.1

So this is a primary disorder in the pituitary–ovarian axis on the basis of the polycystic ovary syndrome, which is characterized by an increased LH production and because of that the ovary produces more androgens. Those, eh, androgens in itself cause the symptoms of masculinism, at least the acne and the hirsutism. Because of the obesity, much of the presented androgen is converted into oestrogen, which, eh, explains the oestrogenic side-effects, the massive mamma development, and the increased oestrogen also evokes an increase of the FSH level, by a negative feedback on FSH, which disturbs the ratio between the two and then the LH will increase again, just like the androgens, and then it will all get worse and worse. Something like that, yes?

13

Competence-Related Differences in Problem Representations: A Study in Physics Problem Solving

Elwin R. Savelsbergh, Ton de Jong and Monica G. M. Ferguson-Hessler

1 Introduction

One reason why experts are better problem solvers than novices is that they have better mental representations of the problems they are dealing with. There is a considerable body of research demonstrating expertise-dependent differences between peoples' mental problem representations (Chi, Feltovich & Glaser, 1981; Chi & Bassok, 1989; De Jong & Ferguson-Hessler, 1991; Larkin, 1983; Di Sessa, 1993). In this chapter we present a structured overview of features that determine the quality of problem representations. We review features proposed in the literature and we discuss their relevance to physics problem-solving.

Proficient problem solving in physics, as in other complex domains, requires specific content-related abilities, either in a rule-based form (Anderson, 1983), or in schema-like forms (Chi & Bassok, 1989; De Jong & Ferguson-Hessler, 1991; Larkin, 1983). Moreover, unlike the board games and puzzles that have been studied extensively in early problem-solving research (De Groot, 1946; Newell & Simon, 1972; Polya, 1945), complex systems are often not completely accessible to perception. Machines are hidden in their housings, industrial plants are too large to oversee in a glance, and many conceptual problems are not perceivable even in principle. These characteristics make obvious that one needs some internal representation of the system to solve a problem about it, and that for such an internal representation a static snapshot is not sufficient. Most researchers seem to have some kind of internal representation at the core of their theories on problem-solving (Chi & Bassok, 1989; Elio & Scharf, 1990; Gentner & Stevens, 1983; Johnson-Laird, 1983; De Jong & Ferguson-Hessler, 1991; Plötzner & Spada, 1993).

Early recognition of the importance of the problem representation comes from Gestalt psychology. Duncker, for instance, conducted a series of well-known experiments using the "radiation problem":

> Given a human being with an inoperable stomach tumour, and rays which destroy organic tissue at sufficient intensity, by what procedure can one free him of the tumour by these rays and at the same time avoid destroying the healthy tissue which surrounds it? (Duncker, 1945: 1)

The following summarised fragment of a protocol by one of Duncker's experimental subjects gives an impression of the way subjects typically approached the problem:

1. Send rays through the oesophagus.
2. Desensitise the healthy tissue by means of a chemical injection.
3. Expose the tumour by operating.
4. One ought to decrease the intensity of the rays on their way; for example — would this work? — turn the rays on at full intensity only after the tumour has been reached. (Experimenter: False analogy; no injection is in question.)
5. One should swallow something inorganic (which would not allow the passage of the rays) to protect the healthy stomach walls. (E: it is not merely the stomach walls that are to be protected.)

   ⋮

13. Somehow divert ... diffuse rays ... disperse . . . stop! Send a broad and weak bundle of rays through a lens in such a way that the tumour lies at the focal point and thus receives intensive radiation. (Duncker, 1945: 2)

Apart from the idea of using a lens, which does not work with γ-rays, the final proposal comes close to a realistic solution. Duncker interpreted the impracticable solutions proposed as arising from an incorrect representation of the situation; the proposed solutions *would* work if the situation were the way it is understood by the subject. Therefore, he stressed the role of the developing problem representation, and of reformulating the goal in particular. According to Duncker, the process of reformulation leads to a sharper, more specific problem representation that gives rise to new solution approaches and other reformulations.

Nowadays, mental model approaches are among the most influential ways of describing mental problem representations. Two books have been seminal to the field of mental models research: one is a monograph by Johnson-Laird (1983), the other is a collection of papers edited by Gentner and Stevens (1983). In the latter volume, papers by Di Sessa, by Larkin, and by De Kleer and Brown are of particular interest in the present context. Both books make an attempt to define mental models. In the approach adopted by Johnson-Laird, mental models are applied mainly to syllogistic reasoning. In contrast, in the papers collected by Stevens and Gentner the emphasis is on physical systems developing in time. As a consequence, Johnson-Laird's mental model theory does not comprise time, whereas in Stevens and Gentner's theory time plays a central role. This might be the most profound difference between the two. It is our feeling that despite several significant differences between the two mental model notions, they have crucial aspects about the structure and semantics of mental models in common. The first point they agree upon is that a mental model is a structural analogue of the world. This implies that the elements of the mental model, and the relations between elements in the mental model, can be mapped in a one to one fashion to elements and relations in the world. A second point of correspondence between both books is how the mental model allows subjects to reason about the world. In Johnson-Laird's theory, the reasoning is about whether a proposition is compatible with the subject's model of the world. In Gentner and Stevens' theory, the reasoning is about how a physical system will behave in time.

In both theories reasoning is much like "seeing it happen in the mind's eye", as opposed to applying rules consciously.

In physics problem solving, the internal representation of the problem can have several functions. First, it guides further interpretation of information on the problem. This requires the mental representation to represent the linguistic structure of the story in order to interpret the internal references between parts of the problem description, and also the structure of the underlying situation in order to interpret various kinds of physical relations (like "above", for instance) (Johnson-Laird, 1983). Secondly, the mental representation can be the basis for a qualitative simulation that, based on knowledge about the properties of system components, predicts the properties or behaviour of the system. This requires that functional or causal relations between the elements are represented in such a form that the mind can simulate the physical process going on. The final use of the mental representation is that a particular representation may become associated with a particular solution procedure. If the representation matches the characteristics required for the solution procedure sufficiently well, the solution procedure may become an active structure that guides further reasoning. More generally, a mental representation might thus serve as a trigger for the relevant solution schema. This requires that the essential physics properties of the problem are represented in the problem representation, and these may go well beyond the information given in the original problem statement. It is important to recognise that in physics, many problems can be represented in several, formally equivalent, ways. Given the different roles the mental model plays, and accordingly the different kinds of information there are in the model, it is clear that a mental model is more than the single enumeration of a minimal set of propositions that would suffice to specify the problem formally. Instead there are different representations of a single physical situation and each of the representations may trigger particular conclusions that are not drawn straightforwardly from other representations. As Sloman (1995) puts it, two representations that have the same expressive power still may differ in heuristic power.

When a problem solver starts reading a problem, the mental model is still to be constructed, and the only thing available is a problem description on paper. The problem solver has to construct a meaning for the text some way, either by building a mental model from scratch or by retrieving information from long-term memory. De Kleer and Brown (1981, 1983) present a view of how the construction of mental models is achieved in subjects reasoning about familiar physical devices such as doorbells. Although their theory is not in the first place on psychological processes, but on qualitative simulation in computers, they try to extrapolate to human reasoning. They distinguish four "stages" in the construction of a mental model: constructing the device topology, envisioning, constructing the causal model and running the causal model. The device topology specifies the physical elements of the device and their static properties (geometry, connections, etc.). The envisioning process subsequently determines the possible states of the entire device from the component models of the device components. In the causal model, the causal interaction of the device components is described. Finally, the running stage executes a simulation based on the interactions represented in the causal model of the system. As we have discussed, this mental simulation is only an intermediate phase in the process of physics problem solving. McDermott and Larkin (1978) present a taxonomy that goes beyond the use of mental models for simulation. They describe a sequence of four types of representations that are involved in proficient physics problem solving. As an initial representation they propose a literal encoding of the problem statement. The second representation is a *naïve* representation, consisting of literal objects and their spatial relations, comparable with De Kleer and Brown's envisioning. The next type of representation is the *physics* or *scientific* representation that contains idealised objects and physical concepts, providing a link to the solution method. The final stage results when formulas are plugged in, leading to an *algebraic* representation.

As we said, if proper information can be found in long-term memory, the mental model does not need to be constructed from scratch. With increasing experience, memories of earlier problem-solving episodes are stored in the long-term memory, either as an episode or in a more or less generalised form. Consequently, parts of the mental model for a new problem can be retrieved from long-term memory rather than being constructed from scratch. Most researchers agree that there is a major difference between the two constructs, as a temporary mental model represents the current concrete problem, whereas the long-term memory representation is independent of the specific situation at hand and forms part of a larger knowledge structure. In the literature, the long-term memory representation is referred to as "schema" (Chi, Feltovich & Glaser, 1981; Rumelhart & Ortony, 1977), or "frame", or "script" (Schank & Abelson, 1977). Notwithstanding the differences, schemata and mental models are close relatives: both constructs describe the organisation of knowledge in clusters, and both types of theories give an account of how solution information is triggered when reading a problem description. Schema theories tend to emphasise the long-term structure in the mind whereas mental model theories in general are more focused on the structure that is built in the working memory during the problem-solving process. It is evident, however, that, to build a mental model, some form of long-term memory information is needed to construct the model from. Moreover, if the mental model is supposed to trigger a solution schema in the long-term memory, it can be expected that there is a close correspondence between the elements of the mental model and the knowledge of conditions that reside in the long-term memory. The latter type of knowledge is known by the term "situational knowledge" (De Jong & Ferguson-Hessler, 1996) or "conditional knowledge" (Alexander & Judy, 1988).

The different aspects of mental models will affect one's problem-solving competence in different ways. To the study of learning and the acquisition of expertise, differences between people reasoning about the same problems are particularly interesting when they can be ascribed to differences in expertise. A step in exploring the relations between qualities of problem representations and performance is to compare problem representations between more and less competent problem solvers. Indeed, experts and novices do seem to perceive different things when looking at a problem description. For example, they do not perceive the same kind of similarities when grouping problems with respect to similarity (Chi, Feltovich & Glaser, 1981; Ferguson-Hessler, 1989). Furthermore, Chi, Feltovich and Glaser (1981) found that, while experts and novices identify the same keywords in a problem description to be crucial to the solution of the problem, the experts "look behind" problem keywords, whereas novices do not. Based on these findings they propose that experts, unlike novices, can build deep problem representations. Such a deep problem representation should be some kind of a mental model structured around physics principles, unlike the novice mental model which is built of the objects present in the problem description. The concept of depth remains ambiguous, however. In the sense in which the word is used by Chi et al., depth includes both the amount of processing required to come to the representation and the physics relevance of the representation. Other authors have used the concept of depth to refer to the amount of processing only (Craik & Lockhart, 1972; Hammond, 1989), or to distinguish the token (surface) from the mental concept (deep) (Newell & Simon, 1972). A still different use can be found with Chandrasekaran and Mittal (1984), who treat depth as an inherent property of knowledge when they stress generality as a property of deep knowledge.

In the remainder of this chapter, we take a closer look at the quality of mental representations. We will try to break up complex concepts like *deep model* and *physics model* into their constituent properties. This will help us to describe the differences that exist between mental models in different people, and, in particular, the differences that exist between the experts' deep or physics models and novices' superficial or naïve models. The present discussion will be limited to differences that originate in expertise. However, these are not the only possible differences between people. It is plausible that there are other differences between people's mental representations that are more related to individual cognitive styles (Vermunt, 1991). Cognitive style may have profound implications for the way a person represents situations. Most notable here might be the "concrete" style for it could involve a more extensive use of examples than other, more theoretically, oriented cognitive styles do.

We will discuss the properties that have been ascribed to deep models and compare them with empirical results from our own study. In our discussion we distinguish four important aspects of mental representations. First, since a representation consists of information (the content), and this information is structured in some way, two problem solvers may differ with respect to the content of their problem representations and with respect to the structure of their representations. The content and structure of a mental representation will at least partially define how the mental representation can be used in reasoning, but the literature suggests that there may be other, qualitative, differences in the way people use their mental model of a situation. These differences are discussed in the section about dynamics. Finally, since proficient problem solving requires the use of multiple, changing, representations, a further difference might be that some problem solvers use these multiple representations more flexibly than others. The remainder of the chapter is structured as follows: first we briefly present the experimental approach we used. Then we discuss aspects that relate to the structure, the content, the dynamics and the flexibility of problem representations respectively. We conclude with a framework for the description of mental models that should be suitable to break up the concept of quality of problem representations in more concrete properties.

2 Brief Outline of the Experimental Method

The present chapter is partly based on a study in which we have empirically observed the problem representations of persons of different levels of expertise. The experimental procedure is outlined in detail in Savelsbergh, De Jong and Ferguson-Hessler (1996, submitted), here we will give only a brief outline of the procedure. We had two kinds of experts (lecturers and PhD candidates), and two kinds of novices, namely proficient and less proficient students who had just taken the course and had prepared themselves for the examination. As a stimulus we gave subjects a core formula from a physics domain together with a physics keyword and then we asked them to describe two different problems that could be solved using this information. The experimental method was based on the assumption that verbal descriptions of problems, as generated by the subjects, retain essential characteristics of the underlying mental representations. In order to prevent subjects from censoring their thoughts, we made the subjects verbalise their thoughts concurrently, like in think-aloud protocols. Although performance may be hampered, reflective thought, and thus censorship, is effectively inhibited by such a procedure (Ericsson & Simon, 1993).

An example of the stimulus material, together with a typical response, is shown in Figure 1. We put some effort into creating stimulus material that would evoke different types of problem representations, but still there was an apparent consensus among subjects about several aspects of the format of a problem description: problems were described mainly verbally, although sometimes a drawing was made in addition to clarify the geometry. A possible threat to the validity of the method is that verbal description of the problem representation could be disrupted by a lack of verbal ability. However, since the theory of the domain is in a propositional format, verbal reasoning can be assumed to play an important role in problem solving in the domain. Consequently, we may assume that poor verbal ability leads to a poor mental representation of the problem.

The subject matter we used is from a first-year university course on electrodynamics. The course covers electrostatics, magnetostatics and electrodynamics, all limited to systems in vacuum. The course starts with the treatment of Coulomb's law, Gauss' law, Biot-Savart's law and the like. The course ends with the derivation of Maxwell's equations in vacuum. The tree diagram in Figure 2 gives a structured overview of the concepts that occur in the types of problem situations the course deals with.

Since we had reason to believe that mental models may differ on several aspects, and that the differences may occur at several scale-sizes, we have analysed the problem descriptions generated by our subjects at three different scale levels: the level of single words, the level of single sentences, and, finally, the entire problem description. At the level of words we looked for group-dependent differences in the frequencies of the use of the words displayed in Figure 2. The words were classified according to different properties such as whether they referred to abstract or concrete concepts. On the basis of these classifications we determined whether a particular group used more words of a particular type. At the level of sentences, we coded each sentence according to the role it played in the problem description, then we analysed the frequencies of the different

$$F = \frac{1}{4\pi\varepsilon_0}\frac{q_1 \cdot q_2}{r^2}$$

Surface charge

a. What does the formula mean?
b. What does the keyword mean?
c. What kind of problem do you think of now?
d. Describe – in short – a situation in which both the formula and the keyword apply.
e. Describe – in short again – a second situation in which both the formula and the keyword apply, and which differs as much as possible from the first situation.
f. Explain what's different and what's similar in the two cases.

Response

The type of problem can be a charged sphere with charge at the edge only, a surface charge, that is because the charges exert force at each other, as a consequence they move as far away from each other as they can, this results in a surface charge, a capacitor that is charged, and grounded at one side, then we have a surface charge also, since because of the forces all charge is driven to one surface, difference is that the nature of the forces, well, yes the similarity in fact, is that there are forces that make the charge drift to the surface, difference is that one has a radial force, whereas with the capacitor all forces have the same orientation. (proficient student, id14)

Figure 1: One of the stimuli used in the experiment was Coulomb's law combined with the keyword "surface charge", below the stimulus, the response of a proficient student is given

types of sentences. Finally, at the level of the entire problem description, we assessed global properties like the coherence of the entire problem description.

In the following sections we discuss the various relevant properties of problem representations, and illustrate these discussions with fragments of the problem descriptions collected in the experiment.

3 Structure

The first aspect in our discussion is the structure of the mental model. We will focus on two aspects of structure. The first aspect is the amount of structure the model has. The second aspect relates to quality of the structure.

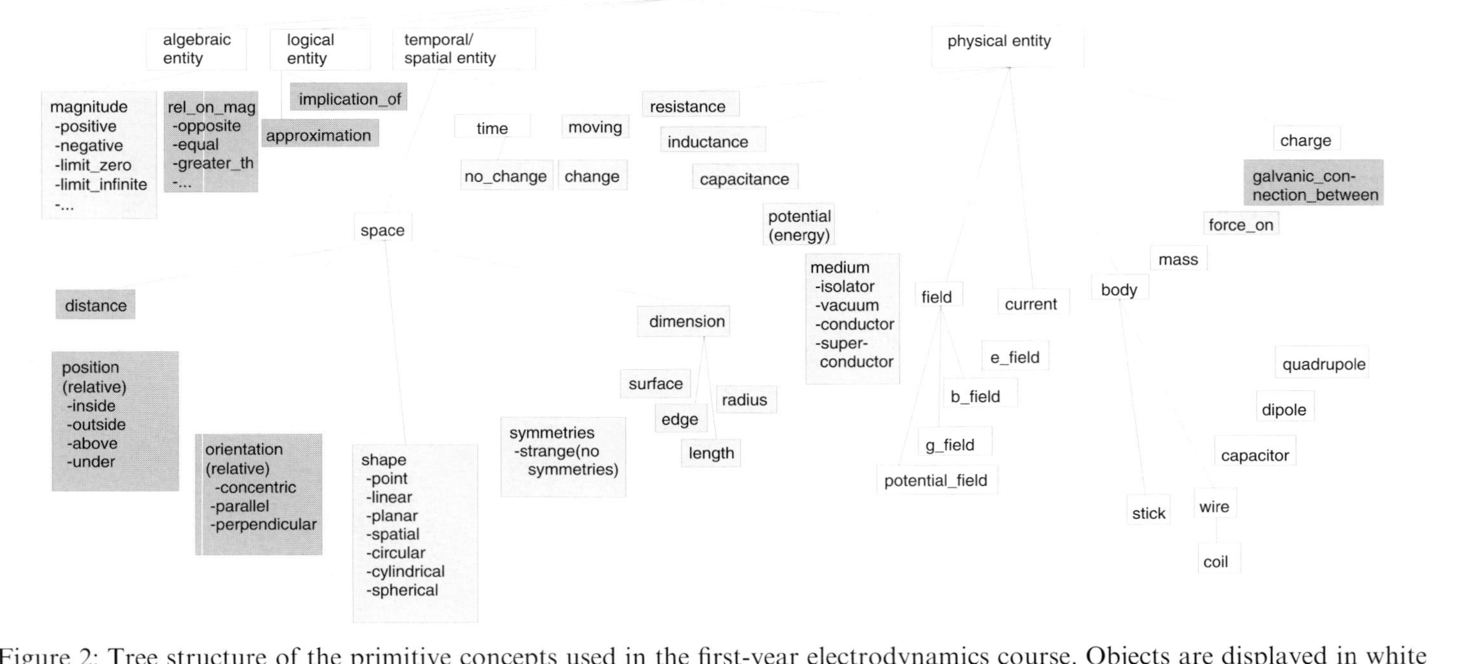

Figure 2: Tree structure of the primitive concepts used in the first-year electrodynamics course. Objects are displayed in white boxes, attributes are displayed in light shaded boxes, and relations are displayed in dark shaded boxes

With respect to the amount of structure, we distinguish between fragmentary and coherent models. From the above the suggestion may have risen that a problem representation is a (set of) coherent model(s) of the problem. This does not need to be the case, however. There may be representations of parts of the problem as well. In fragmentary models there are just isolated objects, whereas in a coherent model relations between the objects form part of the model. A fragmentary mental model is typically associated with novices in a field. As several other studies in different fields have shown, experts perceive larger chunks in a situation that is meaningful to them, and therefore are able to recall more of the relevant information (Chiesi, Spilich & Voss, 1979; Egan & Schwarz, 1979; De Groot, 1946; De Jong, 1986). In the most extreme case there may be no integrated conceptual model at all. Instead, in weak problem solvers, a particular problem-solving procedure might be directly triggered by the mere occurrence of a bit of lexical information in the description of the problem. As an example, consider problem solvers little proficient in arithmetic, who tend to apply a "minus" operation whenever a problem description contains the words "less than", whereas more proficient problem solvers apply the right operation independent of the exact wording of the problem (Hegarty, Mayer & Monk, 1995).

In the physics problem descriptions that we collected, we also found evidence that problem descriptions by novices are more fragmentary than those by experts, and, moreover, that those by weak novices are more fragmentary than those by good novices. The differences could be identified most clearly when the entire problem representation was taken into consideration rather than a smaller unit of analysis. Compare, for instance, the following two descriptions, both in reaction to Gauss' law with keyword "spherical symmetry".

- ***fragment 1:*** A sphere is not a field, but, I doubt whether that is relevant . . . There may be a field inside the sphere, but, well if there's no charge at least. (Experimenter: do you see a problem that employs both?) In fact I do but, yes, a sphere okay, but spherical symmetry . . . you may, a sphere, or uh, two spheres put together, where, if on one sphere you put this amount of charge, and on the other you put that amount, you can compute using the enclosed charge, you may compute the field, or when there is spatial charge inside the sphere, you may compute that as well . . . *(weak student, id18)*
- ***fragment 2:*** A charged sphere, for instance the electric field clearly is a spherical symmetry, and with a Gauss box, a sphere around it for example, so you can compute the field at a given distance of the sphere, and the other way round, with a given sphere or Gauss box, and you measure a certain field intensity at that distance, you can use the formula to compute the charge that is in the Gauss box, in the Gauss sphere *(proficient student, id16)*

The first fragment, a typical example of a problem description by a weak novice, illustrates the fragmentary nature of weak novice problem descriptions. The fragment is mainly an enumeration of isolated elements and some propositions. In the first few lines the student is apparently trying to get some grip on the terms used. Then, after an intervention by the experimenter, a situation is described. The wording remains fuzzy, and the relations between objects remain implicit or ambiguous. The spheres, for instance, are "put together" (probably concentric), and you have to use the "enclosed charge" (probably within a surface just outside the outer sphere), and finally a field is to be computed without specification of the position where it could be computed. The final sentence states that you might compute "that", without specifying what "that" is. Although in most cases there is a sensible interpretation to the student's statements, many of the statements, and especially the relations between statements, are unclear. In the second fragment, by a proficient student, there are far fewer ambiguities, and the relation between a statement and the previous statement is generally clear.

With respect to the quality of the structure, it has been claimed that all entities in a deep model of a physical device need to have *localised* properties in order to achieve the flexibility that is required for proficient reasoning (the "no function in structure principle" in Larkin, 1983; cf. De Kleer & Brown, 1981, 1983). According to this principle, none of the rules that specify the behaviour of any constituent part of the device refers to how the overall device works. De Kleer and Brown, who introduced the principle, claim that it would make mental models robust, that is, to make them insensitive to inconsistencies while running qualitative simulations. However, running qualitative simulations in the time domain is not so much associated with sophisticated expert models. On the contrary, as we will discuss further in the section on dynamics, these models are associated with constraint-based reasoning. The meaning and implications of the principle are ambiguous in domains where global constraints, like conservation of entropy, conservation of energy, or global spatial symmetries, play an essential role, since these constraints have their meaning only in the situation as a whole. Still, also in these cases, there is a difference between the novice's initial understanding of a concept and the more advanced understanding of a concept: in the following example by an expert, properties are clearly properties *of something* and relations are relations *between objects*.

- ***fragment 3:*** now we may think of the electric field of a point charge or of a sphere, . . . *[the first situation:]* a point charge *[the second situation:]* that is a charge distribution that has spherical symmetry, but now it is not in a point but distributed in space, so the symmetry remains the same, but the charge density is position dependent *(lecturer, id 8)*

This fragment provides a clear contrast to fragment 1 (by a weak novice) where a field is mentioned without direct reference to the charge that causes it, and spherical symmetry is mentioned as something associated with spheres but without clear reference to the symmetry of the entire situation. This difference may be described using the "levels of understanding" introduced by Van Hiele (1986), in his research on understanding geometry. According to the theory on levels of understanding, in the initial phase a subject recognises a concept as a whole, the properties that make the concept applicable still being inarticulate (this may be comparable to "non-local" in Larkin's terms). It is only when subjects come to a more advanced understanding that they are able to explain what properties in a situation make the concept apply. It requires an even more advanced level of understanding to be able to derive from the properties of a given situation whether a concept applies. Higher levels of understanding, according to Van Hiele, are built from the lower levels, and do not replace these lower levels. If deep models must specify attributes at different levels of understanding simultaneously, they will consist of a hierarchy of nested models, and therefore cannot satisfy the locality principle as it is defined by Larkin. The type of "non-locality" proposed by Van Hiele, in contrast, we could clearly demonstrate to be associated with weak novice problem descriptions. It is remarkable that, although words like field and force, that refer to a physical relation between concrete objects, apparently have a fuzzy meaning to the weak novices, the weak novices tend to use these words more frequently than subjects from the other competence groups do.

The emergent picture of the fragmentary problem representation of a weak novice is consonant with Di Sessa's (1983, 1993) theory on explanation:[1] Di Sessa holds that novice students do not have a coherent theory to explain for phenomena, but instead resort to a *fragmented* set of phenomenological primitives (p-prims). P-prims are the entities that are used as basic principles in an explanation. Di Sessa found that, with the increase of expertise, there is a gradual shift in the cueing probability of, and the confidence associated with, particular p-prims. In novices' explanations mostly naïve p-prims — closely related to direct observation — are used. With growing expertise naïve p-prims gradually lose their attractiveness, and become replaced with more abstract, physical, p-prims. This points to a further aspect that deserves attention, namely that also the content of problem representations may differ between expertise groups.

4 Content

Besides differences in the way a representation is structured, there may be differences in the content represented. Representations may differ for instance with respect to the amount of detail in the model, an aspect of which is the contrast between qualitative and quantitative representations. This distinction is clearly reflected in the model by McDermott and Larkin (1978) that, as we have seen, distinguishes between verbal representation, naïve representation, qualitative scientific representation, and quantitative scientific representation. Several researchers have claimed that the qualitative scientific representation is a missing link in novice problem solving (Chi, Feltovich & Glaser, 1981; Larkin, 1983). However, in our study we found no evidence of an expertise-related difference in the level of specification; neither experts nor novices gave quantitative descriptions, almost all problem descriptions were stated in qualitative terms, as can be seen from the various fragments reprinted in this chapter.

[1] It should be noted that Di Sessa's theory is mainly on explanation, as opposed to prediction. In novices at least these two types of knowledge can operate quite independently, causing them to make predictions about a system's behaviour that are inconsistent with their own subsequent explanations (Anderson et al., 1992; Chinn & Brewer, 1993; Kuhn, Schauble & Garcia-Mila, 1992).

In addition to the contrast between qualitative and quantitative representations, the four stages distinguished by McDermott and Larkin also reflect another property, namely the kind of entities the model refers to. In mechanics for instance, a force problem might be reformulated into an energy problem, without changing the level of specification. The abstractness of the concepts, however, is clearly different in both formulations, with the force formulation being closer to phenomenological entities and the energy formulation being more abstract. This distinction is particularly interesting because many powerful solution approaches in physics are applicable only when the problem is stated in abstract quantities rather than in phenomenological quantities. Moreover, concrete phenomenological quantities are often time dependent whereas the abstract quantities can be chosen in such a way that they remain constant. This difference may also lead to more profound differences in the dynamics of reasoning, as will be discussed in the following section. Because phenomenological quantities are closer to a naïve representation, and abstract quantities are important to expert reasoning, one may expect that experts will use relatively more abstract concepts. In our empirical data, we have tried to track differences in the concepts used by counting words. We had defined a group of words to refer to abstract concepts (such as *field*, *inductance*, *energy* and *flux*) and a group of words referring to concrete concepts (such as *plate*, *sphere*, *particle* and *charge*). There seems to be some difference in the proportion of abstract words used, with the weak beginners using relatively more concrete words but this is a slight difference only (for details see Savelsbergh, De Jong & Ferguson-Hessler, submitted).

A further difference related to the kinds of entities used lies in De Kleer and Brown's distinction between the device topology and the causal model of the device. In this case, however, it is not self evident what kind of model is more expert-like. According to De Kleer and Brown the device topology comes at the lowest level, and it is on the basis of this model that a causal model is constructed which employs functional concepts. However, since running a causal model has been associated typically with the novice way of thinking, it may be doubted whether the use of functional concepts really is a property of expertise. Instead it may be argued that constraint-based reasoning is more dependent on the geometrical properties of the problem situation (i.e. symmetry) than running a causal model is, and that therefore topological considerations may play an important role in expert mental models. This is especially true in the domain of electricity and magnetism, where symmetry plays a central role. A problem concerning a spatially extended charge distribution, for instance, could be transformed into a much simpler problem on a point charge, given that the charge distribution has spherical symmetry. As one of the experts participating in the experiment puts it, the "electricity and magnetism" course he teaches "is made or broken by the student's insight in symmetry". In this case again, we tried to track the differences by counting words. We had defined a set of topology words (such as *between*, *distance* and *outside*) and a set of function words (such as *force*, *tension* and *cause*). Our results indicate no differences between groups of subjects. For this reason and for the reasons mentioned above, we doubt whether the contrast is meaningful to the domain under study, and even whether it is applicable to university-level physics in general.

In summary, we found no strong evidence for advanced problem solvers using different concepts than novices. This may be due to the way in which we have operationalised these properties — namely at the level of single words. Still, we conclude that there may be more importance to the ways the different concepts are used and coordinated in reasoning. For instance, as Plötzner (1993) has demonstrated, there may be profound difficulties in the coordinated use of qualitative and quantitative representations in physics problem solving. These difficulties will recur in the sections about dynamics and flexibility.

5 Dynamic Behaviour

An important aspect the mental models of Stevens and Gentner and the mental models of Johnson-Laird have in common is that they both are simulations. This means that unlike in rule-based reasoning where a variable can be left open if it is not specified, in a mental model all variables get values. If a variable can have more values, there will be several alternative models. In Johnson-Laird's version, a mental model is a representation of the way the world would be if the premises were true. If the premises are not conclusive about some aspects, the world can be in two (or more) states, and there will be two (or more) alternative representations. Thus, mental models become more complex if the number of states that are consistent with the premises increases. This is true for the logic-inference, but it is equally true for simulations in the time domain. If a system is simple and straightforward, it is not too hard to simulate the system in the mind's eye. However, if there are several switches that can be in different states, or feedback loops that make the input dependent on the output of the system, it becomes very difficult to trace the development of the system. A powerful approach to such problems that is often used in physics, is to use a constraint or invariant to make the problem more tractable.

As an example of the power of invariants — with no time involved in this case — consider the following tiling problem: an 8×8 matrix is given with two cells in opposite corners removed (Figure 3a). The problem is to prove whether the figure can be filled with dominoes without leaving open spaces. Each domino piece covers exactly two (adjacent) squares. One could try to solve the problem by trial and error, but this approach fails if the surface cannot be covered with dominoes. If one transforms the situation to a mutilated chessboard as given in Figure 3b however, it is easy to recognise that a domino always covers one white square and one black square. This constant ratio of black and white squares covered is an invariant for all possible tilings. Because of this invariant property, and because the number of black squares exceeds the number white squares on the mutilated chessboard, a complete tiling can never be accomplished.

The association of "running simulations in the time domain" with novice problem solvers is in accordance with findings of Goei (1994) who found that novice mechanical engineers, when reasoning about shaping an object with a computer-controlled lathe, tended to trace the trajectory the cutting tool would follow, whereas experts did not. Boshuizen and Schmidt (1992) have demonstrated a similar effect in medicine,

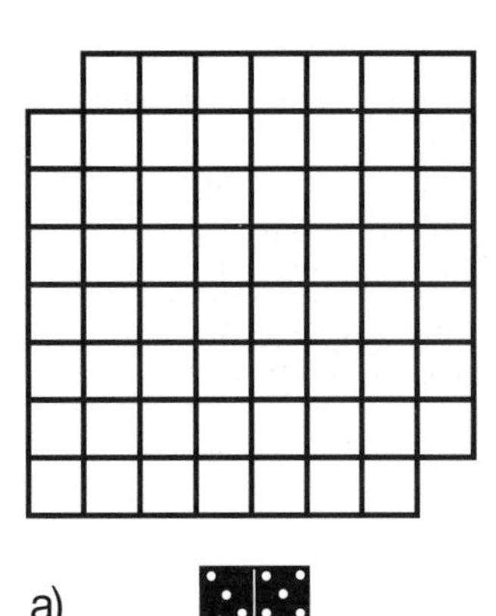

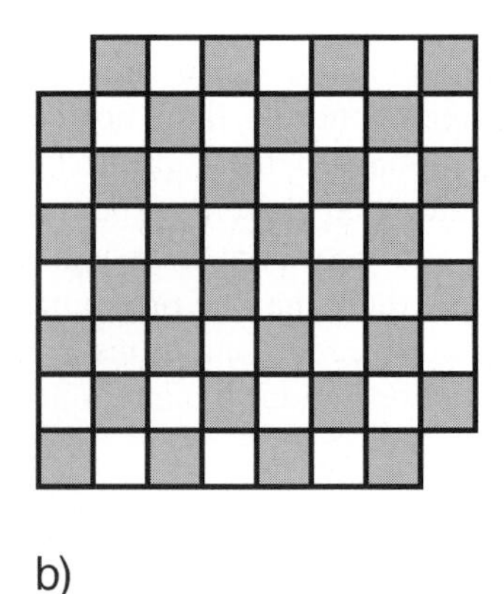

Figure 3: (a) Tiling problem with dominoes. (b) "Mutilated chessboard" representation.

where beginning clinicians construct and run elaborate physiological models, whereas experienced clinicians have constructed mental shortcuts that replace these overt simulations. In logic, finally, there is evidence that although lay persons may construct the "representative-sample" type of mental models as discussed by Johnson-Laird, experts in logic may reason about logic problems by applying abstract operators to the abstract formal representation of the problem. By using these abstract operators and representations, they avoid the problems that would arise if the problem were to be solved using the Johnson-Laird type of mental simulations (Stenning, 1992). Stenning refers to the first kind of reasoning (the straightforward mental simulation) as agglomerative reasoning and to the second kind of reasoning (where an appropriate abstracted representation is sought to simplify the problem) as analytic reasoning. This distinction is more or less synonymous with the contrast of time-based vs. constraint-based models, or vivid vs. static models. It will depend strongly on the kind of entities the model is built of whether it is analytic or agglomerative. In summary, we have shown that, in various domains, the distinction between agglomerative and analytic models may apply. In some domains the distinction may be reflected explicitly in the formulation of the problem, whereas in other domains, there is no explicit framework for expressing the analytic model, and the expert view will be based on tacit, intuitive knowledge.

In physics, the explicit formulation of the problem influences the course of reasoning in many ways; a few examples have been discussed in the section about content. In addition, tacit aspects of problem representations may influence one's thinking about problems in physics too. Di Sessa (1993), for instance, distinguishes between two naïve understandings of equilibrium: *dynamic* and *abstract balance*. Dynamic balance refers to situations where some forces are opposing each other, with none of them winning, so that the balance could be said to exist by accident. Abstract balance in contrast, applies to situations where balancing is considered an imperative. From a formal physics viewpoint, both are described by the formula $\Sigma F = 0$, and in both cases the same term "equilibrium" may be used in the problem description. Since our data are focused on the static aspects of problem representation, we did not expect to find any evidence on the tacit meaning behind concepts. This aspect could be better studied using think-aloud protocols or using extensive interviews as Di Sessa (1993) did. We did expect to find some evidence with respect to differences in the concepts used by different groups, but as we reported in the section on content, we found hardly any differences at the word level. Still, we found some evidence with regard to dynamics in the analysis of complete problem descriptions. We may distinguish between problems that involve changing situations and situations that relate to one static situation only. In the case of changing situations, both experts and novices gave script-like descriptions as in the following fragment, which was a response to the formula for Lorentz force plus the keyword "electric wire".

- ***fragment 4:*** . . . for example a, yes a bar with a rope attached, and a mass that drops down, and it pulls the bar forward like that, and then the bar rolls forward like that, and then here you have, square, so in fact you have a loop, and so the flux changes, and then you have, and then you can compute what the intensity of the magnetic field must be, to make it stand still, so that a Lorentz force *(PhD candidate, id11)*

In the descriptions of problems involving a static situation, we also found some descriptions that suggested a time course, but these were most frequently found in novice descriptions, as in the following example, which was a response to Coulomb's law with the keyword "surface charge".

- ***fragment 5:*** kind of problem can be a charged sphere, where the charge is found only at the edge, so surface charge, and that is because the charges exert a force on each other, which makes them try to get away from each other as far as possible, so that surface charge results *(proficient student, id14)*

This conception of the way charge is distributed in conductors resembles the notion of dynamic balance discussed by Di Sessa (1993).

A remarkable tendency that emerges more clearly from our data is the larger amount of "solution goals" stated by both good and weak beginners compared with experts. This tendency may be related to the difference between backward and forward thinking in problem solving. Beginners tend to start their reasoning from that which was "asked for", looking for a formula that has the desired entity as its outcome, whereas experts solve simple problems in a more forward fashion, starting from an analysis of the given situation (Sweller, Mawer & Ward, 1983). If weak beginners really are unable to construct a coherent representation of a problem, this sheds new light on the finding that beginners tend to skip the thorough analysis of the problem statement when they are trying to solve the problem (De Jong, 1986); since they are unable to construct a coherent representation, their best bet is to follow a trial-and-error approach, starting from something they need to know anyway: the entity that is being asked for. From that point of view it should be no surprise that strategy training gives so little results.

6 Flexibility

As a final factor we consider the flexibility of the representation. The flexibility of a model is largely determined by the possibility of switching between different problem representations. This possibility is closely related to the coherence of a problem representation, the way attributes are bound to objects (locality), and the consistency of the model. So most of the aspects that would determine flexibility have already been discussed in the paragraph on "structure". Still, there are some aspects of flexibility that are not accounted for by the structure of a single representation: the first property we will discuss is redundancy. One of the properties of Larkin's (1983) *physics models* would be redundancy. A redundant model accommodates several equivalent ways of representing, cq. solving, the problem. This redundancy enables the expert to find alternative approaches once the first approach fails to be successful, and moreover a redundant representation enables the problem solver to recover from errors in the representation.[2] The importance of redundancy has also been acknowledged in research on chess:

> "For example a single piece may belong to different categories mentioned by the subject in the protocol. The subject may mention (a) that the position is a Sicilian defence, variation Scheveningen, and (b) that the Pawn d6 is attacked twice when defended twice. This is redundant in that either of the two chunks of information would be enough to retrieve that the Black Pawn is located on d6. This is of course highly useful; the two characteristics of the position are linked together by the common element, Pawn d6, and its double determination counters forgetting. In terms of information theory, redundancy increases the likelihood of transmitting a message in a noisy channel" (De Groot & Gobet, 1996: 256).

The differences in the presence of redundant multiple representations are clearly among the more convincing differences in our data. As an example consider the following fragments, both in response to Gauss' law with keyword "surface charge", the first by a lecturer, the second by a weak student.

- ***fragment 6:*** when you think of surface charge, in this case that implies conductors, so in the inside the electrical intensity amounts to zero, as a consequence there is no spatial charge ... yes, I think of a problem that includes a cylinder, cylinders put together, and spheres put together, leading to charges at the inside and at the outside, in which case components cancel out due to symmetries, which makes the integral easier to evaluate ... *(lecturer, id7)*

[2]As Sloman (1995) puts it, we may distinguish between expressive redundancy, which has to do with the role a piece of information plays in the formal specification of the problem, and heuristic redundancy, which has to do with the role a piece of information may play in finding a solution to a problem. When we claim that redundant information is important to the flexibility of problem representations, we refer to information that is redundant from an expressive viewpoint but not necessarily from the heuristic viewpoint.

- ***fragment 7:*** two capacitor plates at nearby places, you have to compute electrical intensity, something like that, I can't think of a real difference since these problems are always concerning capacitors, yeah, either flat plates or cylinders put together ... *(weak student, id11)*

In the first fragment a redundant description is given with the redundant elaborations underlined. All elaborations refer to abstract properties of the conductor mentioned in the first clause. The second and third elaboration refer to the solution method in addition. The second fragment, in contrast, has no elaborations; all concepts mentioned are at the same level of abstraction, and there is no explicit reference to solution methods.

As became clear in fragment 6, in addition to connections between different types of descriptive elements, there may also be connections between problem representations and solution information. In the analysis of problem descriptions at the sentence level we found only weak evidence that good problem solvers have more solution information as part of their problem representations. However, we had also asked the subjects to describe a second problem situation that differed as much as possible from the first. Upon comparison of these two alternative problem situations, it became clear that, with increasing expertise, there is an increase in the number of problem descriptions with relevant alternatives mentioned. An alternative was considered to be relevant if the solution method was different or if the physics in the problem was qualitatively different, so a high percentage of problem descriptions with relevant alternatives mentioned suggests that the subject has an insight into the factors that influence the solution to the problem, and the way in which they do. A typical excerpt from a weak novice problem description illustrates that alternative problem descriptions generated by weak novices do not involve any qualitatively different physics.

- ***fragment 8:*** two charges placed at an axis at equal distance, that cancel out completely or partially, or several charges, and then compute the force of the third charge, yes the force that is exerted on the third charge, second situation, that you have several charges of the same, or of different, you have different kinds of charges ... that amplify each other proportionally, or amplify. (Experimenter: what exactly do you mean by that?) Both cause the same force in the same direction, difference is that in the first situation you had different, or two equally charged that cancelled out each other, in the other you had the same, or yes different charges that amplified each other *(weak student, id12)*

In this example the situations described are identical for both problems, only the relative magnitude of the charges involved changes, which in this case does not lead to any difference in the solution approach. The student did not give any signs that he was dissatisfied with the alternative he described. This would be different in protocols by more proficient subjects, who in general produced more relevant alternatives and, in case they had generated an irrelevant alternative they would be dissatisfied with it at least. We take this as evidence that, in the minds of more proficient subjects, solution methods are associated with problem representations. We had hoped to find further evidence on this point from the ways in which subjects explain the differences between the situations that they mentioned. The information on the differences is hard to interpret, however; the average scores on this aspect were low and several expert subjects were slightly irritated by being asked to summarise the differences between the problems they had described, since they felt the differences were obvious. Still, the findings regarding solution information indicate that in more advanced problem solvers the representation of the problem is more tightly interwoven with solution representations than it is in less proficient problem solvers. We conclude that the power of expert models is not merely in the use of one particular deep model but rather in the ability to translate a given problem description to the proper mental representations.

7 Discussion and Conclusion

This chapter set out to specify the characteristics that determine the quality or depth, of a problem representation used in physics problem solving. The characteristics that we have introduced are summarised in Table 1. Features listed in the left-hand column are related to the novice model or the so-called "naïve" model. In the right-hand column features that are related to expert models or "deep" models are listed. For some contrasts it may be doubted what is the experts' side; in these cases we have followed the author who introduced the concept. Although the dimensions and groups of dimensions as mentioned in Table 1 have no claim on being independent of each other, they can help to bring some order into the multitude of factors.

Table 1: Characteristics of mental models summarised

Mainly novice		Mainly expert
structure		
fragmentary	vs.	*coherent**
diffused	vs.	*localised properties** *(in Van Hiele's sense)*
content		
numerical	vs.	*qualitative specifications*†
phenomenological	vs.	*abstract entities*†
topological	vs	*functional relations*†
dynamic behaviour		
agglomerative	vs.	*analytic model*
time-based	vs.	*constraint based*
flexibility		
single	vs.	*multiple redundant representations**
inconsistent	vs.	*consistent representation*
problem and solution are separate	vs.	*integrated solution information**

Note. Contrasts that were confirmed by our data are marked with a "*". Contrasts that we looked for but that could not be confirmed are marked with a "†".

Our experimental results suggest that there is no single factor that accounts for the depth of a representation. Instead, depth appears to be a composite concept that comprises several characteristics that can be identified individually. Part of these characteristics can be traced back to the level of words used, or to the level of single sentences. It appeared, however, that at the level of the entire description there are more pregnant differences that could not be captured at a smaller scale level. The most important gain from an analysis at different scale levels is that it gave us the possibility to triangulate our analyses, and to interpret findings at one level by the use of findings at another level.

The properties that we found to differ between expertise groups can be summarised by two broad factors that determine the quality of a problem representation. The first factor is the internal structure in a single representation, which will be designated by the term "coherence". The second factor is the intertwinedness between multiple representations of different kinds, such as that between abstract representation and concrete representation or between the representation of the situation and the representation of solution information, which will be termed "flexibility". The two factors are interrelated in that an incoherent representation is supposed to be more brittle and more susceptible to inconsistencies than a coherent representation is. In case an inconsistency occurs, or in case the reasoning process comes to a halt, multiple connected representations give a better chance for recovery than a single mental model does. From our findings it seems that earlier accounts of "depth", that have stressed a particular set of concepts and a particular content for the mental model as deep knowledge, should be adapted.

The findings could be interpreted in the context of cognitive flexibility theory. According to the definition by Jacobson and Spiro, physics knowledge at an advanced level could be said to be ill-structured knowledge:

> "Ill-structured knowledge domains are characterised as follows: (a) there are many relevant concepts in a typical knowledge application situation and, (b) the application of a concept or combination of concepts can vary widely across different case situations". (Jacobson & Spiro, 1995: 302)

Although, from the viewpoint of an expert, there are just few theoretical concepts in the domain, in the eyes of a beginner, the domain has a less orderly outlook. Moreover, although an expert will see the number of methods as limited, the expert also has to deal with a wide range of situations in which the knowledge can be applied. At the level of superficial situational knowledge, we may conclude that there are numerous concepts. So, when it comes to applications, physics knowledge is not so different from knowledge in domains that have been recognised previously as ill-structured (medicine for instance). In ill-structured domains cognitive flexibility is an important determinant of problem-solving performance. Multiple representations of a situation clearly enhance the cognitive flexibility of the problem solver. In cognitive flexibility theory, there has been an emphasis on creating different interrelated perspectives by "criss-crossing the conceptual landscape". In the context of the present domain, several combinations of perspectives can be relevant, such as the coupling between concrete and abstract representations, between semantic and episodic information (Tulving, 1983), between situation and solution representations, and between visual and verbal representations (Paivio, 1986). In earlier research De Jong and Ferguson-Hessler (1991) found evidence that in good beginners the latter two modalities of problem representation are more tightly coupled than they are in weak beginners. In the current experimental set-up, subjects apparently found little incentive to describe episodic memories, or to describe a problem in a visual format. Still the data show evidence of different levels of "abstraction" mutually connected within the verbal mode.

In this chapter we have primarily focused on problem representations in electrodynamics problem solving. The domain of electrodynamics shares several essential characteristics with other physics domains: first, it is a conceptual domain, which implies that many of the concepts in the domain are not directly perceivable; they are more or less abstract objects. Secondly, it is a formal domain, which means there is a limited set of well-defined concepts and laws that can be applied in a limited and well defined range of problems. The third characteristic shared by electrodynamics and other physics domains is that it supports different ways of describing a problem. Although these different representations can be mapped onto each other without loss of information, one way may be more apt in a given situation than another. The existence of explicit alternative formulations distinguishes formal physics from domains such as medicine, mechanical engineering or computer programming. In these fields multiple representations will be related to perceptual or conceptual chunking, instead of being an explicit reformulation of the systems description (cf. research in chess by De Groot (1946); in machine programming by Goei (1994); in medicine by Boshuizen & Schmidt (1992)). The notion of deep models as introduced in the foregoing thus seems to be restricted to domains that have explicit theories or metaphors associated with them.

To conclude, from the present study it appeared that weak beginners in physics problem solving have mostly single problem representations that are rather incoherent. Apparently, successful and more advanced problem solvers elaborate on the initial elements to create a richer and more coherent representation of the problem. The finding that weak beginners have so many incoherences in their problem representation is indicative for the hypothesis that the fundamental entity in a beginner's problem representation is not the entire problem situation, but rather a smaller unit. In further research we will pursue this line of thought, and we will examine the effect that the availability of extra, redundant, information about the problem situation has on the reasoning process of good and weak novices.

Acknowledgements

This research was carried out in a joint project of the Faculty of Technology Management, Eindhoven Technical University, The Netherlands, and the Faculty of Educational Science and Technology, University of Twente, The Netherlands.

We thank Jules Pieters for the useful discussions we had on setting up this research. Correspondence should be addressed to the first author.

14

A Utility-Based Approach to Speedup Learning with Multiple Representations

Maarten W. van Someren, Pietro Torasso and Jerzy Surma

1 Introduction

In general learning is the improvement of performance on certain tasks as a result of external influences. Performance can improve along different dimensions. Here we shall use terminology from problem solving and talk about "problems" and "solutions". Three major dimensions of problem-solving performance are generality (problems can be solved that could not be solved before), accuracy (more correct solutions are found than before learning) and speed (the same solutions are found as before but in less time). These dimensions are independent in the sense that improvement along one dimension need not affect performance along another dimension. In this study we consider speedup learning. This is a form of learning in which performance becomes faster but the generality or the accuracy is not changed. The problem solver finds the same solutions for problems before and after learning.

It is well known that human performance on tasks that involve cognitive skills becomes faster over time as a result of practice. The cognitive processes involved adapt to the task. If we want to understand how this adaptation is achieved, a suitable approach is that of "rational task analysis" (see Simon (1981) and Marr (1982) for the fundamental ideas of this approach, Anderson (1991) for an brief argument in the context of cognitive psychology and Van Someren, Barnard and Sandberg (1993) for an elaboration in the context of the think aloud method). This approach consists of two steps: (1) design an optimal method for performing a task, within weak assumptions about the human cognitive architecture and the knowledge that may be available to people; (2) refine this rational model from empirical data to find a stronger model. The first step amounts to a close analysis of the task and the environment in which it is performed. Here we analyse a learning task: reducing the solution time in problem solving that involves multiple representations. A representation is associated with a component in the problem-solving system. Such a component can be viewed as a system in its own right. It can perform problem-solving (sub-)tasks using its own methods, its own knowledge and its own representation language. A prime example is visual representations vs. verbal representations (e.g. Kosslyn, 1975; Larkin & Simon, 1987). We argue that it is plausible that in the human cognitive system such components can be distinguished.

Our analysis is based on a distinction between levels in the architecture of problem solvers. These levels differ in "permanency": the lowest level is the fixed underlying cognitive architecture. This corresponds to Pylyshyn's concept of "cognitive architecture" (see Pylyshyn, 1989). This part of a system is not accessible and cannot be changed. Pylyshyn makes a distinction between this cognitive architecture and "knowledge", which is accessible and modifiable.

In the context of problem solving it is useful to make another distinction between problem data and knowledge. Knowledge is more permanent than problem data, although it can be modified. Our analysis involves another distinction, between "problem-solving architecture" and "knowledge". The problem-solving architecture is a structure of components that is more permanent than the knowledge that is represented in the components of the problem-solving architecture. In the example of a problem-solving system that we shall study in some detail below, two components are a "case-based reasoning" component and a "model-based reasoning" component. The first consists of a memory of specific cases that is used to retrieve old similar experiences and use these to find a solution for a problem, and the second consists of a causal model that is used to find causal explanations for observations. The architectural levels of this problem-solving system can be viewed as follows:

- *"knowledge":* the knowledge inside the case-based reasoning component (for example old solved problems);
- *"problem-solving architecture":* the structure of case-based and model-based reasoning components and how these are connected;
- *"cognitive architecture":* the system in which the problem-solving architecture is realised (in this case a programming language which in turn runs on a computer).

We do not want to argue that the problem-solving architecture of this system corresponds to the human cognitive architecture (although a claim similar to this has been made in cognitive psychology, for example by Tulving (1972) who postulates a distinction between episodic and semantic memory at this level). Rather we believe that human memory behaves as if it consists of two components because it tends to reason at one time from specific cases and at other times from mental models. These processes are accessible (e.g. a person can report that he/she is constructing a causal explanation) and modifiable (e.g. he/she may decide to stop doing this and try to retrieve a relevant case). There is a strong analogy here with the structure of computer systems. The hardware of a computer system provides basic functionality such as memory and a set of basic operations. A computer system cannot modify its own hardware. Programs and data are stored in the memory of the system and they can be executed by the basic instructions. With respect to the hardware, a program and its data are all data for the basic operations in the hardware. The hardware operations are able to inspect and modify the software. However, a program that resides permanently in the system and that is run with different sets of data can be viewed as an architecture in its own right. Our analysis only assumes an architecture for problem solving with components that are organised in a control structure and that process data. To avoid confusion with Pylyshyn's notion of cognitive architecture, we shall use the term "problem-solving architecture" and in the rest of this chapter this will be the default meaning of the term "architecture".

Different components in an architecture may have different representation forms, for example specific cases, causal models, visual or verbal forms. It is possible that knowledge that is the same in function (the input and output of using the knowledge is the same) exists in different representations. For example, we can have a visual representation and a verbal representation that produce the same solution for a particular problem. Here we assume that representations are associated with components: different components may have different representation forms. These different representation forms determine performance characteristics of the component. A particular problem may be solved much more quickly by one component than by another. This means that the overall performance of a system can be improved by "moving" knowledge between components.

Many empirical studies have found reduced solution times as an effect of practice on a task and several computational models present tentative explanations of the phenomenon (for an overview see VanLehn, 1989). Examples of methods for optimising within a component are chunking (Laird, Rosenbloom & Newell, 1986) and explanation-based learning (De Jong & Mooney, 1986; Mitchell, Keller & Kedar-Cabelli 1986). Architectures with multiple components allow an additional form of learning: translation of knowledge between components and thereby between representations. Although there are a number of examples in the literature that illustrate the drastic effect of differences in performance between different representations of a problem (for example Amarel, 1968; Simon, 1975), less attention has been paid to the process of "rerepresenting" knowledge to improve efficiency. Yet this is likely to be an important form of learning.

An original study of this process was carried out by Anzai and Simon (1979) who model shifts between problem-solving strategies on the basis of a think-aloud protocol. An interesting aspect of these protocols (and of the model) is the introduction of new concepts in which the problem is represented. These concepts allow increasingly efficient strategies to be expressed. In this model the conceptual changes are not modelled completely, however, and each change immediately gives a substantial improvement in efficiency of the problem-solving process. Boshuizen et al. (1992; this volume, chapter 12) describe the concept "knowledge encapsulation". This refers to declarative knowledge from different domains that becomes integrated as part of procedural knowledge for diagnostic problem solving. On the basis of various types of data Boshuizen et al. argue that experts and novices in medical domains differ in the extent to which knowledge is encapsulated. Experts develop both procedural shortcuts and more compact declarative ways to express their knowledge. These new representations co-exist with the original ones. Under certain conditions, experts fall back on the original representations, for example for difficult problems or when asked to explain the solution to a particular problem. This suggests that their problem-solving consists of first applying their shortcut skills or compact explanations. If this is not enough, they use their full problem-solving knowledge. The shortcut knowledge and compact explanations are most likely derived from the initial more complete knowledge. In a problem-solving architecture with multiple components that have different representations there are at least two ways to reduce the solution time: optimising knowledge *within* a component or optimising *between* components.

In this chapter we address the question of how solution times can be reduced in problem-solving (and learning) architectures that involve multiple components. These components can have different representational formats with their own performance characteristics. This has been observed by a number of authors. A well-known example is the difference between visual and verbal information. Larkin and Simon (1987) compare in detail the costs of accessing information that is represented in diagrammatic form with the same information represented in "sentential" descriptions. According to Larkin and Simon some operations are very much "cheaper" when applied to diagrammatic information data than to "sentential" information. This is mainly because diagrams group information together on the basis of meaningful relations while "sentential" descriptions may group information by different criteria (for example formulae vs. data). The effect is that it is easier to find relevant information in diagrams than in sentential form. Because perceptual operations are relatively "cheap" compared with operations on sentential representations, diagrammatic representations are (sometimes) more efficient than their sentential counterparts. However, as Larkin and Simon's analysis implies and as demonstrated by Schijf and Simon (chapter 11), Ainsworth, Bibby & Wood (chapter 7) and Savelsbergh, De Jong & Ferguson-Hessler (chapter 13, all this volume), and others this is not true for all representations, all operations (and all persons). Furthermore, it is sometimes possible to translate diagrams into descriptions that have a similar organisation, or to draw diagrams that contain the information of a sentential description. This has implications for solving single problems and for performing tasks that consist of a class of related problems. When solving a single problem it may be good to translate the problem information into a particular representation to facilitate the reasoning process. Acquiring skill in performing a particular cognitive task raises the more general choice of in which component to store knowledge: what is the optimal distribution of knowledge over representations?

Optimising this with respect to a task is a problem in its own right. A person who is using graphical representations to solve a problem may get lost in the diagrammatic representation and decide to stop using it. ((Tabachneck-)Schijf and Simon (chapter 11, this volume) quote a domain expert who claims that diagrams are redundant and possibly confusing.) A more subtle approach, characterising experts, is to use different representations selectively. This requires a kind of "meta knowledge" about when a particular representation is best. As we shall demonstrate below, the optimal solution will depend on the distribution of problems.

In section 2 we discuss the main difficulty in speedup learning: the utility problem. In section 3 we demonstrate the utility problem in the context of a relatively simple problem-solving architecture that combines two components with different representations and different performance characteristics.

2 Task Analysis of Speedup Learning: The Utility Problem

How can solution time on a problem-solving task be reduced given an underlying cognitive architecture, an initial problem-solving architecture, initial knowledge and experience in problem-solving? Time reduction is possible because within a single cognitive architecture the same solutions can be found in different ways. Knowledge can be "moved" between components in the given problem-solving architecture and in principle it is also possible to change the problem-solving architecture. Speedup learning means that the same cognitive functionality is realised by different knowledge (in the same architecture) which leads to shorter solution times. Analogous to Langley (1996) we can distinguish between problem-solving architectures that include different forms of multiple representations. The representation formats and the procedures in the architecture that operate on the knowledge structures determine the computational costs that are associated with a particular component and representation. How components are integrated in the problem-solving architecture is another factor. There are different types of problem solving architectures. "*Competitive*" architectures have a number of components that are applied in parallel. Each component can have its own internal representation. Each component in the representation has a "strength" parameter and the "strongest" applicable component is used by the problem solver. In such a competitive architecture, the "strength" parameter depends on the frequency of use and success in achieving goals. A different type of problem-solving architecture (corresponding to "decision lists" in inductive learning) is "*default control*". The knowledge is organised in a number of components. The components are ordered and there is a fixed toplevel control structure of the form shown in Figure 1.

TRY component$_1$
IF solution found THEN stop
ELSE TRY component$_2$
IF solution found THEN stop
ELSE TRY component$_2$
......

Figure 1: Schema of default control architectures

In comparison, a "competitive" architecture would look like Figure 2.

UNTIL solution found DO:
CONSIDER applying component$_1$ ... component$_n$ TO current state;
FIND BEST component/state combination;
APPLY best component TO state

Figure 2: Schema of competitive control architectures

Here all components are considered at each cycle in the computation. Their strengths are compared and the best one is selected and applied.

A third type of architecture involves "control knowledge": (modifiable) knowledge is used to decide how to continue reasoning (Figure 3).

CLASSIFY problem → problem class
IF problem class = class1 THEN $\mathbf{component_1}$
IF problem class = class2 THEN $\mathbf{component_2}$
..

Figure 3: Schema of control knowledge architectures

Within some problem-solving architectures it is possible to represent knowledge that is "functionally equivalent" in different components. For example, consider the schema in Figure 1. The knowledge that "if a patient is unconscious then conclude that his nervous system is affected" can be included in $\mathbf{component_1}$ but also in $\mathbf{component_2}$. If $\mathbf{component_2}$ has a different representation language than $\mathbf{component_1}$ then the knowledge will take a different form but it may well be functionally the same in both cases. In the simple and small example above this is unlikely but many experiments have shown that inferring the same conclusion from the same data is sometimes done much faster in, for example, a visual representation than in a verbal representation. If there are indeed alternative representations for the same knowledge (functionally!) then this raises the question where to represent a particular piece of knowledge. Before we discuss this question we need to look at transformations between representations that make it possible to represent knowledge in different ways.

Which (functionality preserving) transformations are possible depends on the underlying cognitive architecture. The best known computational cognitive architectures are SOAR and ACT-R. These are both "competitive" in the sense defined above. The SOAR architecture creates new productions out of solved problems (and "impasses") and *adds* these to the production memory. The new operator avoids the search that took place before it was constructed and it will be used by the architecture if a similar (sub)problem is encountered again. This results in reduced solution times. This is the "utility problem". The SOAR approach (Tambe, Newell & Rosenbloom, 1990) assumes that chunking only happens in reaction to an "impasse" in problem solving. An impasse occurs if a solution cannot be found within one problem space. Hence learning to reduce search does not occur when problem solving takes place within a single problem space. The effect of chunking is that new knowledge is constructed that competes with the other knowledge in a problem space.

The ACT-R architecture explains speedup learning as the effect of several processes. One process resembles the chunking process of SOAR. As a result of overcoming an "impasse" a new structure is added. In ACT-R, however, the structure is added to the "production memory" where overcoming the impasse takes place with knowledge in "declarative memory". A second process is that "activation" weights of productions and declarative memory structures are acquired and modified. These scores represent the frequency with which productions and declarative knowledge structures have been used successfully. Declarative structures with high scores match faster than structures with lower scores. This means that acquiring the correct weights for a task will result in speedup. Note that ACT-R does not make the assumption that SOAR makes: learning is impasse driven.

Both these architectures are susceptible to the "utility problem". This problem was studied in the area of machine learning in the context of explanation-based learning (De Jong & Mooney, 1986; Mitchell, Keller & Kedar-Cabelli, 1986). Explanation-based learning is a technique for constructing a direct association between problem data and solution that is more general than the complete problem data and that follows directly from the knowledge in the system. For example, suppose that we get a description of a patient and are asked to find a diagnosis. With much problem solving and reasoning we can construct an argument why this patient must have a particular disease *D*: the "explanation" of the solution. Explanation-based learning (EBL) now constructs the minimal (most general) conditions under which this derivation holds. It is possible that certain patient data are not needed to conclude that the patient must have disease *D*. The minimal set of conditions is associated with the conclusion of disease *D*. This new knowledge is of course more direct than the original knowledge that was used for the full problem-solving process. If the same patient would be presented again, the new knowledge would immediately, without reasoning, find the right solution. If a patient is presented that differs from the first only in properties that are not needed to conclude *D* this will also be done directly.

A complete EBL system consists of two components: (a) a problem solver that reasons with domain knowledge to find a solution; and (b) a component that directly recognises the solution, using shortcut knowledge that was constructed by EBL. These two components are integrated in a default control architecture: first the fast recognition component is tried and if that fails, the system falls back on its complete domain knowledge. In a sense (logically) the shortcut recognition knowledge is redundant because the same solutions can be found by reasoning with the domain knowledge. The only *raison d'être* of the shortcut knowledge is that it increases speed. EBL is also the mechanism that SOAR uses to construct chunks.

However, as demonstrated by Keller (1988) and Minton (1990) for EBL, and Francis and Ram (1985) for CBR, the shortcut rules do not always lead to reduced solution time. When more shortcut rules are added, trying all these rules on problems takes a growing amount of time. To some extent this is compensated because ever more problems can be solved directly by shortcuts. However, empirical studies show that simply adding all shortcuts that are constructed by the EBL component at some point results in a slowing down of performance rather than a speeding up. This raises the so-called *utility problem*: when do new knowledge structures that achieve a local improvement also improve global performance?

SOAR and ACT-R incorporate different solutions to this problem. The approach taken by SOAR is to restrict the language for productions to a relatively simple one that allows efficient indexing (see Tambe, Newell & Rosenbloom, 1990). The effect is that retrieval times for productions grow relatively slowly. Because SOAR also uses a form of generalisation (like EBL) it does not encounter the utility problem in most cases. However, this is clearly not a principled solution to the problem.

ACT-R uses a "storage restriction" approach in combination with its numerical competitive problem-solving architecture. Productions and declarative memory structures that have scores below a threshold are deleted. The effect is that the size of production memory and declarative memory are restricted and this keeps memory size relatively low thereby reducing the problem of a growing number of memory structures. Like the SOAR approach, this is also not a principled approach to the utility problem. Solution times may still go up instead of down.

Optimising the knowledge in a given architecture with respect to solution time means transforming knowledge in such a way that the result is an overall optimum. The optimal distribution of knowledge will depend on the architecture (and the performance characteristics of the components) and also on the distribution of the problems that are to be solved. Transforming knowledge between components will affect the solution times of both components and this can have a complex effect on the overall solution times. Another complication is that it may be difficult to predict the effect of changes in the distribution of knowledge over components. Experience with a representation must produce the necessary data on the solution times. In the next section we introduce a relatively simple architecture that makes it possible to study these issues in detail.

3 A Default Control Problem-Solving Architecture

To understand the task of optimising learning in an architecture with multiple components that have different representations we study a relatively simple architecture. This can be characterised as a "default control" architecture where one component uses case-based reasoning (CBR) and the other model-based reasoning (MBR). CBR means that a problem is solved by retrieving an old similar problem of which the solution is known and using this to find a solution to the new problem. MBR means that an explanation of problem data (faulty behaviour of a system, symptoms) is constructed from a causal model of a system and assumptions about possible faults. In the problem-solving architecture the MBR component is used if the CBR component cannot solve the problem. The CBR component is always tried first. This architecture is suitable for this analysis for several reasons. The main reason is that it is a clear example of an architecture with different components that have different performance characteristics for pieces of knowledge that are functionally equivalent. In architectures with many small components the issue becomes more complicated. The second reason is that this architecture has at least some relevance for human learning. It is similar to the default control architecture proposed by Rasmussen, that includes skill-based, rule-based and knowledge-based problem solving. Automatic skills are tried first and if this fails to produce a solution then control is passed one level up to rules and only if this also fails is reasoning from explicit knowledge then used. Although the content of the levels is different both Rasmussen's model and ADAPtER are "default control" architectures.

The CBR–MBR architecture presented in this section is essentially the one of the ADAPtER system (Abductive Diagnosis through Adaptation of Past Episodes for Re-use; see Portinale & Torasso, 1995), a diagnostic system which integrates a case-based reasoning component with a model-based component. In the following we just sketch the structure of the system. The MBR component (see Torasso et al., 1993) uses an abductive procedure for finding an explanation for the observations and uses consistency constraints for disregarding the tentative solutions which predict manifestations that are in conflict with observations. The MBR component uses a *Causal Model* of the behaviour of the system to be diagnosed that includes initial causes for representing possible faults and malfunctions, internal states for representing influence of faults on the different subparts or components of the system, manifestations for representing observable consequences of the behaviour of the system and contexts for representing "exogenous" observable parameters. The problem data for ADAPtER consist of symptoms (behaviour or properties that cannot be explained from normal functioning) and other so-called "exogenous" observable parameters: properties that are not really symptoms but that play a role in an explanation of the symptoms. For example, a person who has a headache may have recently been bitten by an insect, or a car that does not start may not have been used for 3 months. These are not symptoms but it is information that is relevant for explaining the actual symptoms.

We illustrate the type of reasoning of ADAPtER with an example. To avoid technical terminology we present the example in a form that is not the same as in the actual system. Suppose that we have a causal model (relating causes to effects), possible faults of a system, problem data that consist of symptoms and "contextual observations" and suppose that we search for a diagnosis that explains the symptoms.

Causal model:

power on battery & light switch on → power on front lights
power on battery & ignition key on → power on ignition
power on ignition → ignition
ignition & gas → car starts
power on front lights → front lights on
battery not loaded → not (power on battery)
car used → battery loaded

Possible faults:

battery broken: "not (battery loaded)"
light broken: "not (front lights on)"

Problem data:

Symptoms: not (car starts); not (front lights on)
Context: car used

The causal model explains properties of a car under normal conditions. If unusual symptoms are observed these may be explained from the normal causal model. If that succeeds they simply follow from how the car was constructed and how it works. If they cannot be explained, ADAPtER can assume one or more possible faults. These describe a malfunction of the car. ADAPtER searches for an explanation of the symptoms, assuming a minimal number of faults that explain, together with the rest of the causal model, the observed behaviour, including "contextual" information.

When a solution is found, ADAPtER can store the observations, the "fault(s)" and the derivation of the observations from the faults and the causal model as a case. In the example above we would get the following derivation:

car used & battery broken: "not (battery loaded)" $\rightarrow$
not (power on front lights) $\rightarrow$ *not (front lights on)*

car used & battery broken: "not (battery loaded)" $\rightarrow$
not (power on ignition) $\rightarrow$ *not ignition* $\rightarrow$ *not (car starts)*

This explains the two symptoms. Note that the explanation does not involve the fault assumption that the front lights are broken (because it is redundant) and relies on the fact that the car was used (because otherwise this would have explained the symptoms). ADAPtER can now store the symptoms, the context, the fault that was found and the explanation in *Case Memory*. When it is given a new problem of car that does not start and of which the front lights are off then Case Memory Manager (RETRIEVE) retrieves cases from the *Case Memory*. RETRIEVE evaluates the degree of match between the current case to be solved and the retrieved ones using a heuristic function which estimates the adaptation effort between the current case and the retrieved one. RETRIEVE thus retrieves the most promising case. It uses a threshold and thus can fail when there is no similar case in the *Case Memory*. The component (OK-SOLUTION) checks if the old solution still holds for the new problem by replaying its derivation. The retrieved solution is used together with the contextual data of the case under examination (potentially slightly different from the contextual data of the retrieved case) and the causal model to recompute all the possible consequences and therefore to recompute a causal chain from initial causes to manifestations. This step is done to evaluate what the retrieved solution actually predicts for the new case. If there is a conflict for at least one of the observations the replayed solution is not a solution for the case under examination and OK-SOLUTION fails. If all manifestations are covered, OK-SOLUTION succeeds and the replayed solution is a solution for the case under examination according to the definition of diagnosis, otherwise OK-SOLUTION fails. In this case the replayed solution is passed on to the ADAPTATION component.

This component attempts to prove that the old solution still holds for the new problem, even if the old derivation does not apply. ADAPTATION can make limited adaptations to the old derivation using the *Causal Model* that is also used by MBR. In particular, ADAPTATION tries to find an explanation in terms of initial causes for the observations which were not explained by the old solution. The toplevel problem solving architecture of ADAPtER is shown in Figure 4.

```
ADAPtER(new-case,Case-Memory,Causal-Model):

IF NOT RETRIEVE(new-case, Case-Memory, retrieve-solution)
  THEN MBR(new-case, Causal-Model)
  ELSE IF OK-SOLUTION(new-case, retrieve-solution, Causal-Model,
          replayed-solution)
    THEN return replayed-solution
  ELSE IF ADAPTATION(new-case, replayed-solution, Causal-Model,
               adapted-solution)
    THEN return adapted-solution
    ELSE MBR(new-case, Causal-Model)
```

Figure 4: Toplevel architecture for ADAPtER

Figure 4 gives an instance of the default control architecture schema. ADAPtER always tries CBR first and within CBR there is another default structure. CBR always tries OK-SOLUTION first and only if that fails, does it try ADAPTATION. IF CBR as a whole fails then ADAPtER tries MBR. The ADAPTATION component is quite expensive from a computational point of view. As shown in Portinale and Torasso (1996), in the worst case ADAPTATION is as expensive as MBR itself. As we saw in the context of EBL, one can imagine adding all solved problems (along with the derivation of their solutions) to *Case Memory*. Like EBL, however, this will lead to the utility problem (see Francis & Ram, 1985). The costs of retrieving cases, checking their solutions and trying to adapt the derivations will increase when more cases are added and at some point this becomes counterproductive and adding cases will begin to slow performance down instead of improving it. In other words, like EBL, this method suffers from the utility problem. See Portinale and Torasso (1996) for a discussion of this.

4 The Learning Process

Speedup learning means that the knowledge in the components of the problem-solving architecture, or the problem-solving architecture itself is changed in such a way that the solution time decreases without changing the function of the knowledge (the solutions that it finds for problems). For example, when repairing a machine that is broken we may first check whether it is a frequently occurring problem with a particular part. If that fails then we may follow the checklist that is supplied with the machine more systematically and if that fails we may start to think about possible causes of the problem in terms of the structure of the machine and the way in which it works. This is an example of a default problem-solving architecture that is itself learned. Note that both the "subprocedures" (checking parts, following the checklist, reasoning about the machine) and the overall structure are learned. In this case the representations may not differ dramatically but suppose that reasoning about the machine involves constructing a mental image of the internal structure of the machine to explain its behaviour. Under which conditions will this pay back the reasoning "costs"? More generally, the question is: how can performance in such complex tasks be optimised? Even starting from relatively simple initial knowledge there are many ways in which this knowledge can be modified to give faster problem solving.

Here we briefly describe some preliminary experiments with various aspects of the ADAPtER architecture. These experiments assume an underlying cognitive architecture that supports nested "default control structures" and some more specific functions for performing actual reasoning steps. We shall not discuss these in detail here. The experiments explore the effect of several factors on problem-solving performance, in particular speed. The experiment described in section 4.1 concerns the effect of the size of Case Memory, section 4.2 addresses changes in the problem-solving architecture and section 4.3 reviews an experiment with dynamic learning the content of case memory. All these experiments involve learning operations that preserve the functionality of the problem-solving system: after learning it will still find the same solutions as before learning.

4.1 The Effect of the Size of Case Memory

This experiment concerns the size of *Case Memory*. In ADAPtER *Case Memory* and the CBR component were added to the original MBR component of ADAPtER to improve its efficiency. One parameter of the ADAPtER architecture is the size of this *Case Memory*. To assess the effect of the size of *Case Memory*, experiments were performed with causal models and cases from three domains, car faults, leprosy (a medical diagnosis domain) and a technical diagnosis domain. The experiments were designed as follows. Case memory was filled with sets of cases of different sizes. The resulting system was run on a test set of 100 problems. If *Case Memory* is empty then all cases are solved by MBR.

Tables 1–3 show results of experiments with different numbers of cases in *Case Memory*. In the tables below the second column gives the average solution times, standard deviations and probability of success (PS) with a case base of 50 cases, the second for a case base with 75 cases and the third with 100 cases.

Table 1: Experiments with "faulty cars" with 50, 75 and 100 cases in Case Memory

	50 cases		**75 cases**		**100 cases**	
Component	*Av. cost (SD)*	*Ps*	*Av. cost (SD)*	*Ps*	*Av. cost (SD)*	*Ps*
RETRIEVE	551 (208)	1	873 (325)	1	1186 (527)	1
OK-SOLUTION	77 (12)	0.75	77 (12)	0.83	78 (13)	0.83
ADAPTATION	222 (71)	0.82	239 (73)	0.75	266 (92)	0.66
MBR	10,679 (50,346)	1	10,679 (50,346)	1	10,679 (50,346)	1
ADAPtER	814 (496)	1	1084 (553)	1	1518 (714)	1

Table 2: Experiments with leprosy: 50, 75 and 100 cases in Case Memory

	50 cases		**75 cases**		**100 cases**	
Component	*Av. cost (SD)*	*Ps*	*Av. cost (SD)*	*Ps*	*Av. cost (SD)*	*Ps*
RETRIEVE	581 (102)	1	901 (165)	1	1212 (228)	1
OK-SOLUTION	88 (25)	0.86	92 (25)	0.88	93 (27)	0.92
ADAPTATION	213 (35)	0.82	212 (41)	0.75	205 (47)	1
MBR	19,795 (112,233)	1	19,795 (112,233)	1	19,795 (112,233)	1
ADAPtER	701 (152)	1	1050 (208)	1	1347 (262)	1

Table 3: Experiments with technical diagnosis with 100, 150 and 200 cases in Case Memory

	100 cases		**150 cases**		**200 cases**	
Component	*Av. cost (SD)*	*Ps*	*Av. cost (SD)*	*Ps*	*Av. cost (SD)*	*Ps*
RETRIEVE	2982 (1118)	1	4746 (1692)	1	6890 (2337)	1
OK-SOLUTION	111 (18)	0.78	111 (18)	0.83	111 (18)	0.89
ADAPTATION	393 (96)	0.95	421 (118)	0.95	405 (81)	0.92
MBR	556 (825)	1	556 (825)	1	556 (825)	1
ADAPtER	3383 (1055)	1	5281 (1539)	1	7372 (2093)	1

We note that in car faults and leprosy, even with 100 cases in *Case Memory*, complete ADAPtER is still much faster than MBR alone. In these two domains it is quite apparent that the CBR component is a really useful addition to MBR. However, as already pointed out in Portinale and Torasso (1996) the benefits only hold on average, since in many specific cases MBR could be better than ADAPtER because the cost of MBR is relatively small. It is worth noting the extremely large standard deviation of cost (MBR) which makes clear that even in the same domain diagnostic problems can require very different amounts of time to be solved. This is not always the case. In the technical diagnosis domain, the costs of MBR are relatively low compared with CBR and we see that empirically, ADAPtER has higher costs (longer solution times) than MBR alone. This suggests that in this domain the entire CBR component can be dropped from the problem-solving architecture.

For example in the technical domain, cost (MBR) is relatively constant and of the same magnitude as cost (RETRIEVE). In such a domain the addition of a CBR component to MBR is questionable. We also observe that increasing the size of the *Case Memory* is not always beneficial. In fact there is an increase in Ps (OK-SOLUTION) but this increase does not fully compensate the increase in cost (RETRIEVE). Therefore, the overall solution times of the entire system increase quite steeply, which shows that on average, caching makes the system slower rather than faster: the "marginal" effect of new cases in *Case Memory* (the difference in solution time for each new case) becomes negative.

Next consider the *marginal utility* of subprocedures of the CBR component: the contribution of the utility of a subprocedure to the overall utility. (In a default control architecture components can be dropped without changing the overall functionality because many components are functionally redundant.) In the leprosy domain we see that ADAPTATION is very useful because it is always successful and has much smaller costs than MBR. In the car fault domain, ADAPTATION solves most of the cases it tries also at much smaller costs than would be accrued if problems were solved directly by MBR. In both cases ADAPTATION has a positive marginal utility.

This experiment illustrates several general points:

- The size of *Case Memory* does indeed affect performance.
- The effect of the size of *Case Memory* is not the same for all tasks.
- A single architecture may not be optimal for all tasks. This means that there can be useful learning processes that can modify the entire architecture.
- Some effects of the size of *Case Memory* can be predicted without making observations (for example the cost of RETRIEVE) and others can be estimated by extrapolating a general pattern (for example the cost of OK-SOLUTION is relatively independent of the size of *Case Memory*).
- In principle it is possible to automatically optimise the size of *Case Memory*. However, as our experiments show, this involves substantial experimentation and it is complicated by the fact that changes are not independent. Changing the problem solving architecture may change the effect of the size of *Case Memory*.

We return to these issues below. First we look at other possible changes to ADAPtER.

4.2 Learning by Changing the Problem-Solving Architecture

The data above suggest that, beside improving the representation within a component and moving knowledge between components of a problem-solving architecture, another way to improve performance is by changing the problem-solving architecture itself.[1] As we noted above, for some domains, one or more components of the architecture may have a negative utility and it would be better to drop them altogether. For example, in the technical domain above, ADAPTATION was not a useful step on average. The average costs of ADAPTATION did not outweigh its benefits. It is possible to apply our approach at different levels: to keeping or dropping single *cases* or to keeping or dropping complete *components*. In the technical domain a rational

[1]Obviously there is an underlying cognitive architecture that cannot be changed. In the context of our software architecture, the meaning of terms in the underlying language (IF ... THEN ... ELSE, REPEAT ... UNTIL ...) cannot be changed and similarly, people cannot change certain aspects of their cognitive architecture.

decision would be to completely drop the ADAPTATION component. Such changes in problem-solving architecture are quite plausible in human learning. For example, the studies by Savelsbergh, De Jong and Ferguson-Hessler (chapter 13) by Boshuizen and Van de Wiel (chapter 12), and by (Tabachneck-)Schijf and Simon (chapter 11) show that experts switch between representations more frequently than novices. This suggests that they shift from coarse-grained default control to either more fine-grained default control or to control knowledge (knowing which representation to use for which (sub) problem).

4.3 Learning by Changing the Content of Case Memory

As we mentioned above, for some problem-solving architectures it is possible to "move" knowledge between components with different representation forms. ADAPtER is a prime example because a solved problem can simply be added to *Case Memory* without substantial translation or transformation. As we mentioned above, the frequency of use and of the probability of success of cases is needed to decide if adding an individual case to *Case Memory* will be useful. This information can only be collected by keeping the case around for some time to collect this information. In other words, "forgetting" cases is probably a better approach than selectively storing cases. In a follow-up study (Portinale & Torasso, 1998) the first results of experiments in this line are reported. Two strategies for changing memory content are compared. The first is to simply add all cases that are solved by MBR to *Case Memory*. The second is to replace a case $Case_1$ by a new case $Case_2$ if ADAPTATION of $Case_1$ failed, MBR time of $Case_2$ is above MBR time of $Case_1$ and the estimated ADAPTATION time of $Case_2$ lower than that for $Case_1$ by a certain threshold. (Note that the cost of adapting $Case_1$ to $Case_2$ can be different from the cost of adapting $Case_2$ to $Case_1$.) These two strategies were compared by running ADAPtER with both strategies on 5000 problems. The experiments show that for the add-only strategy the solution times grow approximately linearly with the number of problems, with a factor of about 25 ms per 1000 solved problems to an average of 180 ms after 5000 cases. This is a significant improvement with respect to the 855 ms that are needed when only MBR is used. This again demonstrates the utility of the CBR component for this domain. The solution times of the replace strategy, however, increase very little, only about 1 ms per 1000 cases. The size of *Case Memory* increases approximately linear to a memory of 140 cases after 5000 problems with the add-only strategy but the replace strategy increases slowly and levels off at about 75 cases after 5000 problems. This demonstrates that the replace strategy is much more effective than the add-only strategy.

Another approach to this is elaborated by Anderson (1990) who uses assumptions about the distribution of problems to estimate the probability that a problem will be encountered again. For example, the assumption that symptoms follow a Poisson distribution makes it possible to estimate the parameters of the distribution from a sample and use this to estimate the probability that a case will be used again after N new problems.

5 Using Cost Models to Reduce the Utility Problem

Above we indicated some possible changes to the ADAPtER system that can possibly improve its performance. There are other ways to improve the solution time of ADAPtER that we have not explored. For example, we did not consider ordering cases in *Case Memory* or ordering the *Causal Model*, nor did we consider moving to an architecture with *control knowledge* that uses domain- or task-specific knowledge to decide which component is to be used. For completeness we mention the possibility to construct new components in the problem solving architecture or to construct new representation forms.

Even without these types of learning operations, the number of variations of ADAPtER that all preserve its functionality is very large.

Optimising the knowledge is difficult because there are many possible improvements but it is complicated by the fact that the effects of these improvements interact. The overall expected solution time depends on the costs of components of the system and the frequency with which they are used. This in turn depends on the control flow because that determines under which conditions a component will be used. Therefore, optimising the structure of the system will interact with improving the components.

The ultimate goal of a system is to perform its problem-solving task as efficiently as possible. Learning is instrumental and brings its own costs. The costs of learning are in the collection of data and in the costs of finding the optimal problem solving knowledge. It takes time to collect data about the frequency with which problems are encountered, components are applied and the costs of these applications. When something is changed in the system this will affect all these properties and therefore many new data must be collected. This means that simple learning methods are likely to be expensive.

Fortunately, it is possible to optimise knowledge more efficiently by constructing and using a model of the "cost structure" of the system. This can be used to predict the costs of changes to a system without collecting a full new set of performance data. The model and the data are then used to predict the effect of changes. Changes take the form of adding or deleting a single case, a set of cases but also a complete component.

5.1 Constructing a Cost Model of the Control Structure

The cost model expresses the cost of a compound system in terms of its components by considering branches in the control flow. The control flow is laid out as a graph consisting of nodes that represent choice point in the control structure and arcs that represent procedures. Associated with an arc is also a probability that it will be followed from the choice point from which it starts and the average costs of applying the procedure. For example, the control structure:

IF $Proc_1$ THEN $Proc_2$ ELSE ($Proc_3$; $Proc_4$)

branches after executing $Proc_1$. After $Proc_1$, either $Proc_2$ or $Proc_3$ is executed, depending on the outcome of $Proc_1$: success or failure. These occur with a certain probability. This gives the following control graph:

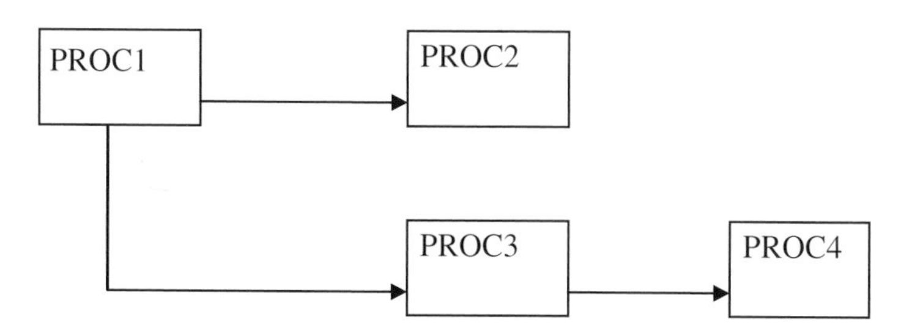

The cost model expresses the control expression in terms of the expected costs of its component procedures. The control graph above corresponds to the following formula for the expected costs of the compound procedure (**Ps** refers to probability of success and **Pf** to probability of failure) (Figure 5).

cost(IF $Proc_1$ THEN $Proc_2$ ELSE $Proc_3$) =
cost($Proc_1$) +
(Ps($Proc_1$)* cost($Proc_2$)) +
(Pf($Proc_1$) * (cost($Proc_3$) + cost($Proc_4$)))

Figure 5: cost model of conditional

Analogous formulae can be derived for other control structures. The cost model in Straatman and Beys (1995) gives formulae for sequence, if-then-else and while-do. Now we construct a model of the total costs of ADAPtER as a function of the costs of its components. This model will then be used to predict the effect of changes in the costs and probabilities of success that result from learning. This in turn will be used to control the learning process.

The cost model is derived from the problem solving architecture of ADAPtER given in Figure 4. We can express the expected costs of the entire system ADAPtER in terms of the costs of the basic procedures and the probabilities of success and failure as given in Figure 6.

cost(ADAPtER) =
cost(RETRIEVE) +
Pf(RETRIEVE) * cost(MBR) +
Ps(RETRIEVE) * [cost(OK-SOLUTION) +
Pf(OK-SOLUTION) * (cost(ADAPTATION) +
Pf(ADAPTATION) * cost(MBR))]

Figure 6: cost model of the toplevel architecture of ADAPtER

If the parameters of this model (Pf, Ps and expected costs for branches and components) are known then this model can be used to predict the costs of ADAPtER. There are basically two ways to find the values of these parameters: collecting performance data to measure them or deriving them from a combination of analytical knowledge about the system and performance data. We illustrate this with ADAPtER.

5.2 Predicting the Effect of Learning Operations in ADAPtER

We can use the cost model to predict the effect of changing the problem solving architecture from the probabilities and costs of the basic components. In the experiment described below we obtained these values from measurements, assumptions and analytical arguments. We use these to evaluate the difference in performance of two alternative architectures.

Consider the following example in the domain of car faults. Suppose that we have measurements on the performance of ADAPtER and its components with case memories of 50 and 75 cases. What will be the utility of a *Case Memory* of 100 cases? We shall use a combination of assumptions and empirical data to evaluate the effect:

Assumption 1: Retrieve time grows linearly with number of cases

If we extrapolate the data from 50 and 75 cases then we estimate the solution time for 100 cases at 1120. (The observed value is 1186.)

Assumption 2: Cost of OK-SOLUTION is constant

This is based on general experience with OK-SOLUTION.

Assumption 3: Cost of ADAPTATION is constant

This is a more questionable assumption.

Assumption 4a (pessimistic): adding cases do not improve success of components

This gives:

Ps100(OK-SOLUTION) = Ps75(OK-SOLUTION)
Ps100(ADAPTATION) = Ps75(ADAPTATION)

Assumption 4b (optimistic): adding cases increases success of OK-SOLUTION

Ps100(OK-SOLUTION) = 1

Now performance of ADAPtER with a *Case Memory* of 100 cases can be estimated from the cost model. This gives:

cost(ADAPtER) = 1197 (optimistic)
cost(ADAPtER) = 1677 (pessimistic)

The actual value (measured) with *Case Memory* of 100 cases is: 1518. If we compare this with the expected utility with a *Case Memory* of 75 cases, which was 1084 we see that adding another 25 cases to *Case Memory* is probably not a good idea. Note that this prediction is made using the cost model and reasonable assumptions about the architecture and the domain, and so without actually experimenting with this new situation. The latter approach would incur substantial learning costs.

If the model is used to compare different versions of the architecture or to predict the effect of changes in components then it is enough to know the difference between the values before and after changes. These can often be estimated or reasonable assumptions can be made. For example, the effect of adding cases to memory on retrieval can be predicted if we know how retrieval time is a function of the number of cases in memory. If the *Case Memory* Manager performs a linear search, the retrieval costs will grow linearly with the size of the *Case Memory*.

What are the implications of this rational analysis for human learning? The main conclusion is that optimising a problem-solving architecture is an inherently complex task. There are many different ways to organise the knowledge for performing a problem-solving task, even if we do not consider the accuracy of performance. This means that even for a person who has achieved a certain competence on a task, there is no *a priori* uniform best problem-solving architecture and which is best depends on characteristics of the task such as the problem distribution and performance characteristics of the components of the architecture.

Optimising the problem-solving architecture within a reasonable number of experiences and without incurring enormous extra costs for storing and using these experiences (and their performance data) will have to be based on general "cost knowledge" of the kind that is expressed in cost models. For example, a person may know that it is a good idea to remember very "expensive" problems (that took a long time to solve) explicitly, to represent complex relations that all involve two numerical variables as a graph, to intensively train a skill that is used very frequently, to first reason with a diagrammatic representation and only after that with a mathematical model. These decisions are based on the kind of meta-knowledge about the mental effort or load of cognitive processes that is captured in the cost models.

One implication of our analysis is that it is useful to experiment with new representations to avoid "local maxima" in performance. A person who never draws a diagram will not collect performance data to know if or when diagrams are efficient.

6 Conclusions and Discussion

In systems with an architecture that makes it possible to represent knowledge in different ways that are functionally closely related (for example one representation is subsumed by another) and to translate or move knowledge between components, the learner faces the task of optimising the distribution of knowledge. We analysed this task in the context of a "default control" problem-solving and learning architecture. This is analogous to a decision list (see for example Langley (1996)) or an "if-then-else" control structure: first a problem is given to one component. If this component fails to solve it then the problem is given to the next component. This architecture can be optimised by using the first component(s) for quick solutions that apply only for certain problems. If this fails then the system falls back on slower but more general knowledge. Although popular models of human cognition have a "competitive" architecture, the "default" architecture is relevant for understanding learning with multiple representations because it is less likely that people apply all these representations in parallel, "competitive" style. Reasoning within one representation seems to allow little direct parallel "competition" from knowledge in different representation forms.

Our analysis of problem solving and learning in an architecture with multiple representations and the possibility to move knowledge between components with different representations shows that:

- A form of "forgetting" (undoing translations) is necessary because there are costs associated with the size of the knowledge.
- Optimising the control structure requires statistical data about performance. Exploration is needed to collect performance data about the effect of possible improvements in the knowledge. This is necessary at each level at which changes may be made. In our example we focused on operations on *Case Memory* but we noted that the same method could be applied to complete components of the architecture (and even to the structure of the architecture).
- Fine-grained iterative improvement by "moving" knowledge between components is possible and is shown to be effective in at least one domain. A simple hill climbing strategy based on replacing cases in *Case Memory* produces sustained improvement in performance.
- "Competitive" architectures will learn relatively slowly because optimisation involves substantial information about performance characteristics of the system for the current task. Acquiring this will need substantial exploration and memory. Models of the architecture of human memory (e.g. Anderson, 1993) view the distinction between cases and models not at the architectural level. These are different contents but the architecture treats them in the same way. This means that solved problems are stored along with the deep knowledge and both are subject to gradual processes of activation. If knowledge is not (successfully) used then it gradually becomes deactivated which reduces its probability of being used during problem solving.

We can compare this analysis with the work by Dobson (chapter 5). He studies solving syllogisms with different representations. The computational costs of problem solving in these representations vary and it also depends on the type of problem. A learner could find the optimal representation for this task in the same way as ADAPtER. One of the implications of our model is, however, that a careful analysis of the computational costs (or "cognitive load") is needed and also the distribution of problems that the learner encounters. This may explain the results described in Dobson (chapter 5).

This means that the complete architecture should include a component for "exploration" that is able to "forget" or somehow "undo" earlier changes. Here we considered only the costs of reasoning. Many tasks include other aspects such as (motor) actions, perception or in general some form of interaction with the environment. In this case, the analysis can be extended with the costs of interaction processes and possibly even with more indirect costs, such as (expected) costs of the effect of actions. The study by Ainsworth, Bibby and Wood (chapter 7) is an example in which interaction costs play an important role. Different representations of (functionally) equivalent knowledge leads to different "perception" and "action" costs and subjects seem to optimise these costs.

In competitive control architectures, the utility problem is somewhat different. Competitive learning involves two types of learning operators. The first type is strengthening/weakening of (pieces of) knowledge (and thereby changing the probability that it is used) and creating new pieces of knowledge. There are two main differences between the architecture that we study here and the competitive architectures: ADAPtER only has two learning operators that create new knowledge items: (1) adding a case (that was solved by MBR) to *Case Memory*, and (2) dropping a case from *Case Memory*. The second difference is that ADAPtER does not have the continuous strengthening/weakening mechanism that characterises competitive learning. Finally, the goal of learning in other architectures not only reduces the solution times but also avoids performance errors. Here we assume that MBR contains adequate knowledge for problem solving. Reasoning with the CBR component should therefore not change this and give the same solutions as the MBR component.[2] In competitive architectures a component can be strengthened because it is *fast* but also because it is *correct*.

It seems likely that a continuous competitive model is adequate for learning within a representation and the more discrete model for learning between components with different representations.

Note that we did not consider accuracy of solutions (the proportion correct solutions). Different pieces of knowledge may differ in solution time but also in correctness. This gives a different argument for using and constructing different components. For example, if the solution to a problem is found by *combining* solutions of different components then using more cases may well improve the accuracy of the solutions. This leads to a more complex problem in which the knowledge should be optimised

[2]In fact this is not guaranteed in ADAPtER. If the ADAPTATION component is used then the solution may differ from the solution that would be found by MBR.

with respect to both solution time and accuracy. A version of this problem was described earlier in pattern recognition as the "edited Nearest Neighbour" (Dasarathy, 1991). Recently machine learning research offers a variety of different deletion policies in order to preserve accuracy in CBR systems while reducing solution time. One of the most influential strategies is the "Remembering to Forget" approach (Smyth & Keane, 1995). In this approach it is assumed that in case memory some cases contribute mainly to the accuracy of the system (pivotal cases) and others may predominantly contribute to its solution time (auxiliary cases). The deletion of a pivotal case directly reduces the competence of a system. On the other hand the auxiliary cases do not effect competence at all, and those cases are selected as the candidates for deletion. This strategy might be applied in ADAPtER relatively easy and consequently improve its performance.

Another related line of research is reinforcement learning. Reinforcement learning is normally concerned with accuracy of performance, but one can view our approach to speedup learning as a form of reinforcement learning that: (a) optimises speed instead of accuracy, and (b) involves memory structures other than performance parameters of knowledge elements (the cost model).

Acknowledgement

The authors wish to thank Peter Reimann for his comments on an earlier version of this chapter which led to substantial clarification.

PART 3

General Issues and Implications for Education

15

Multiple Representations and Their Implications for Learning

Alan Lesgold

1 Introduction

The chapters in this book describe some of the representation forms that are helpful to various practices[1] in the world. A strong case has been made, both in earlier literature and in the present book, that experts are more successful in their practice than novices because they have different and more useful representations that they can build and use in practice. The present book includes chapters showing that a particular problem faced by experts as well as novices is the coordination of multiple representations that each contribute to successful practice. In this brief chapter, I will try to suggest some ways of looking at the coordination problem.

A starting point is to consider some of the representational forms or systems of knowledge that have proven useful in past research. I then consider some of the research efforts aimed at tying various representational forms together. After this, I suggest some approaches to thinking about the ways in which multiple representations can, in principle, work together, and I propose further researches aimed at understanding which forms of representation linking can be productive and how one might learn to better do this linking.

2 Systems of Knowledge and the Spaces Between Them

There is a long history of research on representational variations, upon which the chapters of this book build. A number of categorisations of representation forms have been suggested in these chapters. In addition, other literature suggests several categorisations or contrasts among representational forms that should be considered. I focus briefly on the latter, since the authors of this book have described their own views better than I can, and these are readily accessible to the reader.

2.1 Rasmussen's Levels

To many of us in the cognitive engineering world, the strongest multiple representation approach in practical use is that of Jens Rasmussen (Rasmussen, Pejtersen & Goodstein, 1994). Rasmussen referred to three levels of cognitive function: skill-

[1]By a practice, I mean carrying out activities, solving problems, and otherwise acting in a domain in ways that are socially valued and personally adaptive. In part of the literature, "problem solving" might be substituted for this use of "practice", while in other parts, "activity" might be substituted. I use the term practice to avoid commitment to any one of the more specific terms that have been used.

based, rule-based and knowledge-based. Skill-based activity is essentially perceptual-motor, driven by the characteristics of an unfolding situation. Specific recognitions trigger direct, real-time actions. Rule-based activity involves a more complex process. The situation, or one's ongoing mental activity, triggers various rules or schemas that in turn generate specific mental or physical actions. Finally, there is a knowledge-based level of processing, in which relatively weak (i.e. general) methods are used to infer appropriate actions from abstracted representations of the current situation. These three stages can be illustrated with a simple medical example. If a patient goes into shock, emergency physicians or technicians carry out a number of actions very automatically. They may only be able to describe what happened and how they responded after the fact. However, if a patient is not *in extremis* but shows a striking set of symptoms, the physician tries to match the symptoms he is seeing to one or another schema from his medical knowledge. Once a schema is triggered (Rumelhart & Norman, 1981), the physician knows a set of rules that can be applied to refine the diagnosis and develop a treatment plan. On occasion, however, a case does not match any of the schemas the physician has learned (such cases are often featured in medical grand rounds and in the sections of medical journals devoted to hard cases). In such cases, the physician is reduced to reasoning from his knowledge of body function and disease process in order to develop an explanation for the symptoms. From this explanation, a treatment plan can be developed by reasoning about what mechanisms are available to ameliorate the disease process.

Rasmussen's idea is that control of practice jumps from a lower level to a higher level in the skill–rules–knowledge progression whenever the lower level is unable to sustain practice sufficiently. In situations where actions and decisions have life-and-death implications, such as some areas of medicine, it appears that experts work at multiple levels partly to verify their solutions and not simply in response to lower-level impasses (cf. Lesgold, 1984; Patel & Kaufman, 1995).

2.2 Concrete vs. Abstract Knowledge

A related but older distinction among representation types is the concrete–abstract dimension. From Piaget and other developmental psychologists, we get a picture of abstraction as layered on top of a more concrete level of representation. The distinction made by the Piagetians was between operations that were bound to things that could be enacted directly in the physical situation vs. operations that were more purely symbolic. For example, on the one hand, many simple arithmetic problems that are assigned in schools can be solved in a concrete manner, and some children who can do arithmetic concretely may not be able to execute school-taught algorithms (cf. Carraher & Schliemann, 1988). On the other hand, some physics problems require algebraic computations, the steps of which are not bound to images of direct enactment in the problem situation.

Our information age environment is full of objects that fall somewhere in between the concrete and the abstract. For example, computer programs have some properties of both the abstract and the concrete. In principle, a computer program to control a robot might be considered concrete, because it directly refers to observable events in the world ("turn", "move forward", etc.). However, many activities within a computer program are quite abstracted from the physical (e.g. generating a random number, sorting, combining rules into chunked rules, etc.). The informatics world blurs the distinction between the abstract and the concrete.

In addition, certain symbolic activities become so common as to make them concrete. For example, many programmers are described as "hackers". In part, this means that they produce programs — which are at least partly abstract — by writing fragments of code and then thinking about how to stretch them or interconnect them. As their activity seems bound to their direct representations of these code fragments, it seems sensible to speak of it as concrete, especially if they do not use certain broader abstractions that are common to high-quality programming work.

At the same time, computer programs, by their nature, are not concrete things but rather formal descriptions of concrete processes, and hence somewhat abstract. The information age has produced numerous cases of process descriptions that become objects of discussion and hence partly abstract. Further, things like programs, which are instructions to manipulate symbols, are inherently more abstract than things like cooking recipes, which are instructions to manipulate physical objects/substances.

2.3 Professional vs. Technical Knowledge

Another way of viewing this distinction is by looking at professional vs. technical knowledge. I currently have a joint project with Intel Corporation, in which we are developing training for technicians who maintain and repair chip-making equipment. As we do our work, we often find that certain kinds of knowledge, generally at Rasmussen's knowledge-based level, are provided to process engineers but not to technicians, while rule-based training is focused on the technicians. It is not quite that simple, of course. Technicians in the high technology world are generally provided a lot of explanatory general knowledge as well as specific rules, but the knowledge is usually treated as being somewhat in the background. Moreover, less of it is provided. Often absent as well are specific links between the rule-based and the knowledge-based level of activity. It is assumed that the technician will "use his head" to solve problems based on what rules he has been taught and what systemic understanding he may have.

While technicians often are assumed to have minimal abstracted knowledge or representation capability, engineers are perceived differently. It is assumed that they have the theoretical background, and need to acquire the ability to bring their theoretical knowledge to bear on practical problem situations. Technicians generally feel that engineers do not really understand the specific systems with which the technical staff works, while engineers often feel that technicians fail to use theoretical knowledge intelligently. Overall, this suggests first that technicians and engineers use different kinds of representations and second that the ability to use multiple representations simultaneously is difficult, as several chapters of this volume also suggest.

2.4 Education vs. Training

More generally, our education systems seem to make strong distinctions between education and training. It is assumed that education provides broad understanding that can be brought to bear on whatever situations are encountered in life. Training, however, is seen to be very specific and procedural. Some of this distinction is snobbishness. The core arts and sciences faculties of universities like to feel that the education they provide is superior to that of professional schools and technical schools. However, the continued existence of the latter attests to a broad social perception that general education is not sufficient preparation for complex professions and occupations. Still, general education programmes, even those in specific subject matters, also have a good survival record, and it is worth trying to consider what these programmes teach and why it is needed.

As will become evident below, there is an interesting distinction to be made: between skills of learning and actual domain content knowledge and skill. However, it is not clear that skills of learning can be acquired except by experience with rich contexts of performance. That is, the skills of learning in school may not be the skills needed to learn even the conceptual backings for required job performances. To learn school knowledge, skills such as rote memorisation can be very important. More powerful forms of school learning include the use of personally meaningful scaffolding knowledge, even if that knowledge is totally unrelated to the specific content being learned. On the other hand, in demanding work situations, it makes more sense to use work domain experiences to scaffold the acquisition of domain knowledge. If nothing else, this should help decrease the "knowledge inertia" problem pointed out by Whitehead.

2.5 Summary

Two kinds of distinctions among representations seem important for learning. First, there is variation in level of abstraction. Some representations are made up of components that point very directly to objects in the immediate environment. Others are made up of abstract components whose links to the environment require significant inference. Secondly, and overlapping partly, different representations rest on different kinds of systemic characteristics. For example, when a medical case is represented in terms of physiological functions, there are causal aspects to the representation. On the other hand, a representation organised around the signs and symptoms that relate to various diseases may have interconnections among its elements that are probabilistic rather than causal. Overall, then, representations can be distinguished in part by the immediacy of links from their components to the task environment and in part by the qualitative character of the links among their components.

3 Cognitive Approaches to Learning Across Knowledge Domains

Learning to be competent in a particular environment (or environment family) therefore requires not only acquiring the ability to represent the environment in multiple ways. It also requires learning how to interconnect the multiple representations and use them in combinations to accomplish cognitive goals. Several very different approaches have been taken to these learning requirements. I discuss these below.

3.1 Multiple Analogies

Spiro and colleagues (Spiro et al., 1989) put forward a rather interesting view of complex knowledge that bears on the multiple representation issue. They pointed out that the schemas by which medical experts represent various disease possibilities are sufficiently complex that any one analogy fails to capture all of what is going on. Yet, in order to help students learn complex medical concepts, analogies often are used. The solution they proposed was to present the various analogies that represent views on a medical knowledge schema and then create learning situations in which the student is forced to deal with multiple analogies simultaneously. That is, the situation must be so constrained that superficial application of only one of the analogies prompts ideas that conflict with others of the analogies used initially. Learning then becomes a dialectical process in which the various analogical views of the overall schema or task situation must somehow be reconciled. Completing the reconciliation is equivalent to completing the learning.

This provides us with a model for teaching using multiple representations:

1. present all the representations (or analogies) that are needed to convey the full content of a domain; and then
2. provide a variety of learning experiences in which the student must reconcile multiple representations in order to be successful.

One can imagine various forms of this approach, including both single- and multiple-student variations. The latter would permit the dialectic to be turned explicitly into a conversation, in which different participants start from the positions of different representations and then negotiate a unifying view.

The idea of collaborative dialectical approaches to learning is certainly not new. At least in elite education, it has been common to have seminars and other discussions in which multiple positions were put forth and efforts made to reconcile them. In recent years, collaboration has been seen as an important approach to teaching hard-to-learn concepts, partly because it was felt that through conversations among learners, views of a domain that seem initially to be inconsistent could eventually be reconciled (or "negotiated," to use a currently popular term). For a particularly good exploration of this possibility, see Roschelle (1992).

3.2 Multiple Bodies of Knowledge Reflected in Multiple Representations

In this book, Boshuizen and Van der Wiel (chapter 12) discuss their views about medical education, a domain in which multiple bodies of knowledge are important — with each body of knowledge reflected in a separate representational capability. Their new work derives in part from earlier work by Boshuizen and Schmidt (1992) which suggested that clinical cases are possible vehicles for bridging among the various forms of knowledge (with their associated representations) that make up medical expertise.

Boshuizen and Schmidt suggest that there are at least three forms of knowledge that come into play during medical diagnosis, and that many diagnoses require use of multiple knowledge forms, perhaps in multiple representations. Their three candidate forms are basic biomedical knowledge of normal function (e.g. physiology and anatomy), pathophysiological knowledge and clinical knowledge. To this, we might add, in the spirit of Rasmussen (Rasmussen, Pejtersen & Goodstein, 1994), a rather automated capability for dealing with the routine day-to-day stream of colds, juvenile ear aches, and the aches and pains of getting old. In Rasmussen's terms, the skill level consists of the automated handling of these routine conditions, the rule level reflects clinical knowledge, and the schema level reflects biomedical knowledge of normal function and pathophysiology.

Representation of the patient will be quite different at these different levels. At the skill level, much of the representation might be almost social in nature, since no treatment might be merited. At the clinical level, a disease schema might be triggered that includes both rules for establishing the details of the patient's disorder as well as rules for assuring that the wrong schema has not been triggered. When clinical knowledge does not suffice to find and confirm a diagnosis, then the two higher levels of normal and pathological medical knowledge are brought into play, each with its own representation. Normal physiology can be represented by a system model, in which the system components are interconnected by paths that permit them to influence each other. Pathophysiology, however, requires the ability to capture a variety of possible failures that might occur in the system. The failures themselves are unique to each patient, but the potentials for failures can be thought of as links from an individual organ in a physiological representation to a model of a form of dysfunction.

Boshuizen and Schmidt (1992) pointed out that real clinical cases are the primary source of information about which linkages among multiple representations are useful or important to expertise. Hence, it is certainly adaptive for medical students to use real cases as a base for acquiring the new knowledge structures or connecting links among the knowledge bases they already have. While forming these links, there is, in principle, also a possibility for developing new representational processes that can permit more integrated representations (over multiple representation forms) of future cases.

All of this is very consistent with some earlier general ideas about cognitive learning and performance. Rasmussen et al. (1994) and Rasmussen's earlier writings speak of a problem-solving process that escalates through the skill, rule and knowledge layers as required to reach a solution. And, Newell (1990) proposed a theory of learning in which multiple actions, potentially in multiple representations, were then "chunked" into an integrated action rule. Anderson's ACT theory (1993) also has a form of chunking mechanism that can attain the same sorts of effects. So, we have good general theory about how the learning might occur, but we still need more work on how to stage the experiences that will allow cross-knowledge-base and cross-representation learning to occur.

3.3 Complex Learning by Doing

Lesgold (Gott, Lesgold & Kane, 1997; Lesgold et al., 1992) introduced a training regimen that relates to the present discussion, which he has called *coached apprenticeship with reflection*. The approach is being employed primarily for training complex technical skills, although it is probably extensible to other domains. The core of the approach is a simulated work environment in which realistic problems can be presented. These problems are selected to be the hardest problems encountered in real life, problems the trainee may not be able to solve unassisted. However, whenever the trainee reaches an impasse, he or she can ask for help, which is provided by an intelligent computer-based coaching system. In the US, we sometimes refer to a strategy called "sink or swim". Lesgold's scheme extends that strategy by providing enough coaching so that no one sinks. Every trainee successfully solves every problem, often with the assistance of the on-line coach.

Lesgold's systems provide a lot of forms of coaching, at a number of levels of detail. However, studies now being conducted by Katz (1997) suggest that certain forms of coaching are not referenced very much while a problem is being solved. In particular, trainees seek "*what* to do" information rather than "*why* to do it" information. This is consistent with the suggestions by Sweller (1994) that mental load during problem solving is too high for the formation of broad, adaptive knowledge.

Lesgold and his associates dealt with this problem by adding a reflection phase after each problem is solved. Upon completion of a problem, the immediate memory of the problem context and the actions taken for solution are very fresh. They can be reinstated readily through "instant replays" of the solution process. The reflection period in Lesgold's systems consists of several different affordances: critique of the trainee's performance, ability to replay the trainee's performance, and availability of an expert solution to the problem. Furthermore, all of the "why" coaching is available as these different forms are used. Katz's data suggest that trainees do indeed reference explanatory coaching during their replays and as they follow up critical comments on their just-completed performance.

Because problems of great difficulty are used, it is generally the case that this entire technology serves substantially to provide opportunities for combining multiple representations in order to solve problems. Indeed, in the most recent venture of the Lesgold group (a joint project with Intel that is not yet documented), the multiple representations needed to solve problems are represented explicitly. The task involves repairing faults in machines that make computer chips, and putting a layer onto a wafer of chips involves delivering a variety of gases into a chamber, providing appropriate heat and RF power, and having the wafer mechanically in the right position. The tools provided to trainees are all accessed through a scheme in which these multiple representations (chemical, electrical and mechanical) are used explicitly as separate indexing schemes to access schematic diagrams, possible data-gathering and corrective actions, and coaching. The problems are sufficiently difficult that many require integration of knowledge over these multiple representation forms.

We would like to support learning of both the multiple representations that experts find useful and some heuristics for linking these representations. Indeed, our fundamental challenge has been to integrate the multiple representations and the linking heuristics into a single training technology. In the version tested so far, we have not completed this integration, but we believe we have learned a lot about how to do it.

For example, we have observed that technicians seem quite able to integrate over portions of multiple representations when they can see, at one time, some physical objects that correspond to components of the different representations. If visual proximity is a good way to promote a link, then we might be able to help the technicians connect their representations by highlighting components that would be salient in multiple representations of an expert. This would be a simple use of the visual display capabilities of the kinds of training systems we currently build. Certain tabular displays can also be useful. For example, we have, in our work with Intel, developed what we call a "process explorer". This shows, in one table, the extent to which each of a collection of processes might play a role in the problem the technician is trying to solve. To the extent that different representations are required for these processes (e.g. when a failure of a system may be due to poor placement of a wafer by a robot or a leak in a vacuum seal), such tabular displays should help technicians link representations (e.g. robotic movement vs. vacuum containment). We are exploring this possibility over the coming year. And, we hope to develop and test other approaches as well.

3.4 Experientially Grounded vs. Abstract Knowledge Integration

The previous section raises the important question of how different trainees accomplish the cross-representational integrations that are required. Specifically, how concrete must the situation and the representations be before most trainees can integrate different knowledge forms? The traditional view has been that a small elite think abstractly while the vast bulk of workers think very concretely. In the strong statement of this distinction, Lévi-Strauss (1974) spoke of the average worker as a bricoleur, one who cobbles together solutions on the fly, heavily influenced by memories triggered by direct contact with objects in the work environment. For the bricoleur, materials which are at hand "suggest" adaptive courses of action (Chandler, 1995). The process is very "bottom-up," in the sense that plan components emerge only through looking at the problem situation directly.

In contrast, Lévi-Strauss thought of the engineer as one who develops solutions to problems in his mind and through formalisms rather than through direct contact with the physical objects involved in a problem situation. The engineer, in Lévi-Strauss's formulation, works more "top-down," using mental representations and formalisms to produce plans that then are carried out in the problem situation. This whole formulation resembles Piaget's distinction between a concrete operations stage and a formal operations stage in part, although it can be seen as even more insulting to the modern worker.

Today, the objects of many work environments are themselves abstract — or at least not visible as physical objects. A simple caricature of the calls for new levels of work force education is that that today everyone must be more than a bricoleur, that they need more formal problem-solving skills, and that these skills are provided through standard schooling. That is, while in the past, only the top students made it through the education system into engineering, today all students need to get at least through the secondary education path that would lead them toward such careers as engineering. Programs such as Goals 2000 (an American plan for improving schooling) implicitly reflect this kind of strategy.

However, it is not clear that this approach, of getting more people through the same old educational system but at higher levels of achievement, is really sufficient or, in that specific formulation, necessary. On the one hand, there may be levels of competence that are important but that are more in the realm of an extended view of bricolage. If so, then some alterations to standard, highly verbal–symbolic education may be helpful, if they provide opportunities to combine what is learned through standard schooling with opportunities for practising bricolage in a variety of settings. On the other hand, there may be skills needed by modern workers that historically have not been taught either to potential future bricoleurs or to budding engineers.

These possibilities have major implications for education. It is important to establish more clearly whether existing and proposed standards are sufficient to assure a work force that can be sufficiently flexible and adaptive. Furthermore, it is at least worth exploring whether existing approaches to schooling will provide, by themselves, the needed competence for future work. Norton Grubb (1996) has argued that at least some students need schooling that is more integrated with work and work-related experiences.

We suspect that some aspects of the skills needed for high-performance work require practice involving a combination of school-type abstract knowledge and real world bricolage-type experiences. A simple thought experiment might help make this more apparent. Suppose that a high-performance work team were assigned the task of making elevators run more efficiently in a government or university building. If they attack this purely as a formal queuing problem, they may miss certain subtleties of real elevator life. For example, most people, when asked whether it is possible for elevators to arrive repeatedly headed upward and almost never headed downward, will say that this cannot happen. However, there are real-life circumstances when it can. To be able to plan or tune the operations of an elevator control system, a worker must be able not only to view the world theoretically but also to couple the theory with the results of direct observations.[2]

Another example in which theoretical and bricolage activity are combined, but this time by teaming different people, is the design of the wing on the Boeing 777. Engineers came up with a great design for a composite-intensive wing. Fortunately, they discussed their ideas with mechanics who pointed out that the wing would be impossible to repair, since no airport in the world had an oven big enough to bake the larger composite panels that were required. The theoretical work of the engineers required a bricolage supplement from the mechanics in this case. In principle, engineers could have some systematic database-searching activities for checking whether aircraft parts will be able to be repaired in current facilities, and they do. However, it seems unlikely that the rules for checking those databases will ever be complete enough to permit designs to be fielded without some explicit checking by bricoleurs.

Situations abound in the modern work world in which workers at all levels need to combine bricolage and abstract activity. In order to capture the requirements for such modern jobs, we will need to build a better theory of the interactions between concrete and abstract thought. It is extremely unlikely that incumbent workers, left to their own devices, will be able to sort out these distinctions, important as they might be. For one thing, in the modern world of relatively abstract entities, we can see bricolage involving the manipulation of rather abstract entities. An example of this is the computer hacker, who cobbles together subroutines but cannot lay out an overall design in advance.

The concrete/abstract distinction is central to understanding and measuring worker skill, both for immediate performance and for future learning. The information age is, in great part, the age of information objects, entities that are concrete in the sense that we routinely talk about them and send them to each other but abstract in the sense that they have no essential physical existence. From one side, this means that even concrete work is abstract now. From the other, it means that traditional abstractions and the viewpoints of traditional abstract thinkers, like educators, may still be outside of the everyday work world, in which abstractions are perceived and handled as objects. Both workers and subject-matter specialists need a vocabulary and a conceptual structure for distinguishing everyday information objects from extended flights of abstracted reasoning or new mental representations.

With respect to the issue of multiple representations, we instructional designers must not be too "schoolish" in deciding which aspects of expert representations are practically important and hence important to teach. Rather, we must attend explicitly not only to the content of those representations in terms of their connections with the real world but also with the forms of reasoning, knowledge extension, and representational process that experts and non-experts apply when using and extending their

[2]In the case of the elevator, when a large group is waiting for an elevator, once one stops, the others can sneak by while the first is loading, leaving those who are still waiting with the experience of seeing more elevators going one way than are coming back the other way. Depending on the number of elevators, number of people waiting, and number of floors the elevators must cover, this might or might not be a significant factor. It is unlikely even to enter the consciousness of an elevator worker unless he/she actually observes and thinks about real problem situations in elevator scheduling.

multiple representations. That is, we need to be sure that we develop accurate accounts of what is represented in a task situation, how the representation is built, and how tightly it is bound to the situation.

4 Some Thoughts on the Ways in which Multirepresentational Learning Can Be Fostered

Having discussed the need for cross-representational linkage in fostering expertise, I turn to a brief consideration of the ways in which such cross linkages might be stimulated. These approaches fall into two categories: scaffolding by bringing multiple representations into consciousness simultaneously and directly teaching skills of cross-representational linkage. We will consider these in turn, concentrating primarily on the former.

4.1 Memory for Cases Can Link Representations

As Boshuizen suggests, cases are an ideal form for linking across representations. If a given case can be solved only through the use of multiple representations, then memory for the case might act as a link among representations of future cases. This will more likely occur, I suspect, if the linking case is understood by the learner to be useful linking information for the future. That is, I think the links among cases that are exemplified by the case need to be explicitly reflected upon, along with an explanation for the importance of multiple representations in solving that case. If this is done, then it will be more probable that the case will be recalled in future situations for which it is relevant and that the specific ways in which it might be relevant will be more apparent.

Note that tutorial learning in which this kind of scheme is used will, of necessity, require some level of conversational modelling in order to work well. That is, some of the tutorial comments will need to contain references to prior cases: *"remember the case we had where.... Doesn't this case seem similar in important ways? For example..."*. Human tutors seem to be able to do things like this — at least some of them can. It will be worthwhile to develop conversational models explicitly for this kind of tutoring, and these models, in turn, might make it easier to train other human tutors as well as to develop machine tutors with this kind of capability.

4.2 The Current Situation Anchors Multiple Representations

Another related way to link representations is to develop situations in which crossing from one representation to another is a central part of the demands of the situation. Of course, medical or technical diagnosis cases are one special form of this, but other forms might also exist. One could imagine creating some artificial machine environments in which it is only possible to manipulate the machine to do something interesting by attending simultaneously to multiple representations. A simple example might be a complex toy electric train arrangement, in which the electrical circuit representation of the system must be combined with the mechanical switching activity of the system in order to discover how the system is controlled. A little experience of the right form might be sufficient to prepare people to modify the track setup to obtain various results, and that experience would almost inevitably have to involve combined attention to both electrical and the mechanical/switching representations for the system.

4.3 One Representation Focuses the Attention of Another

Sometimes, one representation can serve as an anchor for another. Consider the toy train example just given. Quite conceivably, a child might have a vision of a complex train layout. That vision, and the corresponding track layout representation for that vision will be relatively strong in the active memory of the child — he will be able to "see it in his mind." However, the electrical representation will be both more abstract and less familiar to the child. Even so, the requirements of the track representation provide a strong constraint on the electrical representation. In that sense, the one representation scaffolds the other. It seems feasible to work explicitly to foster such anchorings. For example, we might ask the child to sketch the overall arrangement of track he wants. This should make it easy to start thinking about the layout more completely, e.g. including the necessary switches and patterns of switching. That, in turn, lays the scaffolding for thinking about the electrical representation of the wiring needed to operate and control the various switches.

4.4 A Learned Regimen Disciplines the Jumps Among Representations

The ultimate goal, often, is for the learner to acquire his own strategies for interconnecting multiple representations and for using them to solve problems. This has not been an explicit focus of learning at this level of generality, although work such as that by Dunbar and Klahr (1989) certainly aims in this direction. Dunbar and Klahr have focused specifically on the coordination of hypothesis representations (what they call the hypothesis space) and data representations (what they call the experiment space). Having studied individual differences in coordinating experiment and hypothesis spaces, Klahr is now turning attention to explicitly training the problem space coordination skills he believes are important for learning to handle complex tasks (E. Toth, personal communication).

I can imagine a variety of ways in which people could learn to coordinate complex representations and have been involved in one such effort, the Belvedere project (Suthers, Erdošne Toth & Weiner, 1997). The purpose of that project was to help students learn to use a graphical scaffolding tool to gather and coordinate complex data involving multiple representations. So far, the work has focused on public policy and science theory argumentation. Argumentation seems the right space to be in for this purpose, as the dialectical activities of arguments are exactly what we are after (I consider a dialectic to be, in part, the coordination of understanding among multiple representations).

What Belvedere does is to provide a graphical space for noting components of different representations and for explicitly linking among those representations. Efforts to date indicate that students can indeed learn to do this kind of explicit representation. What remains to be done is to develop schemes for projecting the complex graphical networks that eventually result from such an exercise into simplified views that highlight particular kinds of interconnections. Future work will include explicit consideration of what those simplifications should look like to be maximally useful and how the simplifications can most easily be generated. While we see this as partly a machine intelligence problem, I expect some possibilities for human teaching and learning from this work as well.

Overall, the chapters in this book give us a lot to think about as we develop additional research efforts. A particularly important locus for such research must be the emerging needs for people to be able to adapt quickly to changes in the domains in which they work and often to become proficient quickly in new domains. This adaptability is central to economic well being. This is especially the case for emerging areas of work for which there are not yet any experts altogether.

Acknowledgements

I thank Maarten van Someren for insightful comments on an earlier draft. This chapter was supported by contracts with the United States Department of Labor and Intel Corporation, which do not necessarily endorse the views presented herein.

16

Representation and Conceptualisation in Educational Communication

Keith Stenning

1 Introduction

Cognitive science was, to a great degree, inspired by what could be achieved in the way of modelling behaviour by reasoning *internal* to systems of rules. Grammars were perhaps the initial favourite systems, but logics and expert systems played similar roles in studying reasoning. In a number of fields there appears to be a realisation of the limits of system–internal reasoning, and a search towards frameworks which can treat mental processes as combinations of external reasoning *about* systems, with reasoning *within* them. Observations of the importance of multiple representations in teaching and learning are one example area which invites this theoretical approach. If it is possible to transition between representation systems, then there may be dry ground between them, and a perspective from that dry ground from which the range of alternatives can be surveyed, and new ones constructed.

There are, of course, other practical reasons why multiple representations may have surfaced in the literature now. The technology has liberated us from the puritanism of the ASCII terminal, and we are just discovering the extent of the labyrinth in which we have been turned loose. We have the facilities for previously undreamt of combinations of display media, along with the nightmare that their "obvious" benefits may melt away at the first signs of serious evaluation.

A group of colleagues in Edinburgh have been developing a theory of the cognitive efficacy of different modalities of information presentation, focusing on the use of graphics and sentential languages in teaching reasoning. Reading the early drafts of the chapters of this book, I was struck by the need to distinguish two alternative perspectives on representational systems: what I would provisionally call *representational* systems in contrast to *conceptual* systems. I do not intend this distinction to be a loaded one. Conceptual systems can usefully be regarded as representation systems but they are representation systems seen from a different perspective. The fact that I choose to retain the term "representational" for the former perspective marks nothing more than the direction from which I have come to this topic (and perhaps my lack of terminological imagination).

The contrast I have in mind is between, on the one hand, the sense of multiple representation illustrated by the distinction between say graphical vs. sentential treatments of the syllogism; and, on the other, between say medieval vs. particulate theories of matter. The former are distinguished by their "surfaces" — their representing objects (which is not, of course to suppose that the differences are cognitively superficial). Clearly the pair of physical theories just mentioned could each be represented (in this surface sense) by different constructions of graphics and language — their contrast is at a conceptual level. Very often contrasts at this conceptual level are only accompanied by verbal (or possibly mathematical) differences at the surface level. This raises acute issues about when different purely *verbal* treatments constitute multiple representations. If these conceptual contrasts are to be treated as representational differences too, we need to be clear how. It may be that the two kinds of representational differences (surface and conceptual) need to be associated with each other in an optimal teaching practice. Perhaps surface differences help students to detect hidden conceptual differences? If so there is all the more reason to be careful to distinguish them at the outset.

So struck was I by the need for this distinction between aspects of representation, that I came to the conclusion that if this chapter could do nothing other than clarify this distinction, it might thereby provide a useful framework for advancing empirical studies in this area. Obviously the program has to be an empirical program, but perhaps it need not be one prosecuted in a conceptual minefield. There are perhaps now enough examples of attempts to interpret real data that it is time to stand back and analyse some of the components of a theory of the impact of multiplicity of representations on learning?

Teaching, at least of the kind treated here, is about teaching concepts and new conceptual systems. Learning these novel conceptual systems certainly has to be connected with the operations and skills required to put them to use in controlling the physical or social world, but it is quite hard to give any account of say physics learning at this level which sees its core as anything other than conceptual learning. If we are to think of conceptual systems (such as the medieval or particulate theories of matter) as representations, then representations are very much the *objects* of this teaching. On this view teaching *is* teaching representation systems, and therefore necessarily involves *multiple* representation systems.

To the extent that novel representational surfaces (e.g. new diagrammatic systems) function to aid this conceptual learning, they are *tools* of learning. To the extent that novel representational surfaces frequently transfer beyond the domain they are originally taught in (e.g. maths learnt in association with one domain may be applied to another), these representational *tools* often become educational *goals* in themselves. Much of mathematics teaching can usefully be seen as learning hopefully portable representational skills of providing new representational surfaces for both old and new conceptual systems. Some representation systems may be more goals than tools; some more tools than goals. As several chapters in the book remind us, learning novel representational systems can itself be hard work — an investment that has to be measured against the work they can be expected to perform.

Frameworks, just like representation systems, ought to do some work in proportion to their trespass on the readers' patience. The specific piece of work I will attempt here is to see what light a framework for thinking about representations can throw on the relation between learning physics and learning logic, and on the use of multiple representations in both activities. The two subjects are united in their difficulty; but divided by at least contemporary views of their utility. I hope to show something about what they share, and something of how they contrast. A long-term goal is the rehabilitation of the latter's reputation by the strategy of redefining what it ought to be seen to teach, and showing just how central that goal is to providing students with representational skills. On this view, logic is the study of representational systems, and so should play a major role in our understanding of the didactic effects of representations in teaching and learning. But my immediate goal is to show that distinguishing between representations as surface, and representations as conceptual systems is necessary for understanding either the learning of physics or of logic, and the role of multiple representations in that learning.

The next section lays out a framework for understanding representation and conceptual systems. The third section applies the framework to the comparison of learning logic with learning physics. The concluding section draws out some consequences for the use of representations in teaching.

2 Representation Systems: Their Anatomy and Family Life

Representation systems consist of represent*ing* objects (sentences, diagrams, animations, etc.) and represent*ed* objects (the entities these sentences, diagrams and animations are about). The mapping between the two is called an *interpretation* — mathematically it is a function from representing to represented objects, although in interesting cases, a function with much internal structure. While we are on the list of ingredients, we should remember that representation systems also have *transformations* which turn one member representation within the system into another, with specified results — a clean example is provided by formal "rules of inference" to which we will return later. The transformations are what make these systems systematic. There may well be representations which do not belong to systems, but if so they do not belong in this chapter.

The transformation rules of a system of representation are also what often sets such systems off from conceptual systems *tout court*. Conceptual systems are often thought of as classifications of reality which are prior to any reasoning about reality. I would suggest that we should downplay this way of differentiating representational systems from conceptual systems. I do this because closer inspection of the cases critical in education reveals that the conceptual systems do come with systems of transformation (indeed they are only accessible through those systems). More of this below.

Representation systems come in families. Because they bear two faces, representation systems can be classified by properties of either face and by the mappings between faces — that is by kinds of interpretation. Let us take as the most concrete centre of the space, a fully specified representing system — a completely defined "surface" which is fully interpreted, for example, an artificial language interpreted on a particular domain. We beg the reader's patience with a formal example, but matters are so much clearer here. We return to look at some of the problems of applying what we develop in the artificial case to natural languages later on.

3 Representational Surfaces

One can abstract over surfaces, or over the things represented. An example of the extreme of abstraction over alternative domains of objects to be represented is an uninterpreted formal "language" such as propositional calculus. The propositional variables (Ps, Qs . . .) are completely uninterpreted: they may denote propositions about numbers or hiccups, or anything in between. But even in such apparently completely uninterpreted systems, a few core signs are at least constrained in the kind of interpretation they may receive. For example, the connectives (the so-called logical *constants*) are open to a restricted range of different interpretations (e.g. truth functions over *n*-valued truth values).

One can alternatively abstract over what systems represent: numbers, hiccups, cabbages, kings. It is not of much (logical or representational) interest whether the things denoted are cabbages or kings, but there is a style of mapping between the logical "constants" and the domain which *is* of the essence. These patterns of mapping are identifiably invariant over the domains of objects (vegetables and monarchs). And indeed, there are families of similar but not identical such mappings, as well as families of augmentations of the core systems. In other words, the important classifications are by *kinds of interpretation.*

So a many truth-valued interpretation of the propositional connectives or an augmentation with modal operators (for necessity and possibility) yields different representational systems, yet ones whose family relations we would not want to forget. For one thing, the one system may be a fragment of another, and particular discourses may be conducted within fragments. It then becomes a moot point what representational systems is "in play" for its user — a point of great significance in considering how to understand alternative conceptual systems as representations, even when they are all accessed only through an apparently homogeneous language.

With artificial languages it is clear that this classification by kind of interpretation is the dominant method of classification. First order predicate calculus (FOP) is FOP whether it is in Russell notation, Polish notation, Frege notation, or even Pierce's diagrammatic notation. In other words, whatever surface it is given. What makes FOP, FOP is its interpretation — any syntax (or in the case of graphical treatments we should perhaps say "form") which will sustain the essential interpretation of the logical meanings constitutes an acceptable "surface" for FOP. Some may be more user friendly for (some) users, but semantic sufficiency can be rigorously defined. Notice that the mapping of the "variable" vocabulary is quite inessential to defining the identity of the language; whether we use some fragment to reason about natural numbers or hyenas and bankers, we use the same FOP. So the essence of the language is the existence of some surface mapped onto the critical semantic functions of the representation system.

The fact that the core classification is of semantic interpretations should not lead us to assume that surfaces are not important. This illusion could easily arise because logic is almost always pursued using sentential surfaces. For sentential systems, choice of alternative syntax is generally a matter of little more than convenience. But surfaces become important when they either constrain interpretation (e.g. when they are not sufficiently structured to sustain the required interpretations) or when the device that is going to process them makes the choice more than conventional. For example, Polish notation is particularly congenial to calculators; and Pierce's graphical inclusion notation for FOP is not congenial for any known processor.[1] These questions about alternative surfaces are precisely the kinds of question that arise about alternative graphical and sentential presentations of domains in teaching. They usually interact with questions about what kinds of interpretation the surfaces are given.

To round off the discussion of how artificial representational system's arrange themselves in space, we must note that there is yet another dimension of relevant variation even for these "simple" artificial systems — that is the dimension of what I called above transformation systems. The whole purpose of an artificial language is to allow a uniform definition of patterns of inference. Propositional calculus can be provided with many different proof theories (e.g. natural deduction systems, axiomatic systems, mental models notation, proof table checking algorithms). Algorithmic or heuristic theorem provers are then built on top of them, to guide the search for proofs in the mazes which these proof theories provide. Most importantly we should remember two things about theorem provers. First they are not strictly part of the logic itself — but equally, without them, nothing happens. These are therefore the features of representation systems which are most directly related to *skills* of representation, and to the meta-systematic perspective which is so important to understanding multiple representation. We return to some empirical evidence on this point below.

We noted above that if conceptual systems are to be thought of as representation systems, then their associated transformational apparatus must be included with them. This simple, almost terminological, point takes on central significance when we realise that transformational operations are of the essence in conceptual learning both in physics and in logic because they are the only access students have to the new conceptualisations. This is the point to which we return in the next section.

I have used a logical framework to structure our problem. But how does this apply to natural languages rather than artificial ones? There is, of course, a sizeable subdiscipline wholly devoted to answering this question. Large parts of linguistics and computational linguistics are given over to using the artificial languages of logic to analyse the natural languages we speak. The critical point for our current purposes is that natural languages (viewed at the level of English, French, . . .) are only (very) *partially* interpreted systems. Students do not have the relevant interpretation of words such as "volume" before they learn physics.[2] But physics is not taught in language schools. And when children do acquire the physicist's concept, they do not generally acquire any new surface for it — no new word. Do they thereby have a new representation? The answer I will pursue here is that inasmuch as they have connected the concept to a system of concepts and their transformations, they may well be usefully considered to have acquired a new representational system. So are

[1]Even if some argue that the propositional version is useful to some students in some circumstances.
[2]Lots of 8-year-olds have an interpretation in which it denotes only the knob on the radio.

natural languages single or multiple representation systems, with single or multiple surfaces?

One kind of example of the multiplicity of representational systems hidden beneath the apparently homogeneous surface of natural languages is the case of analogies. Analogies may or may not involve the introduction of new words, but they systematically transform the concepts attached to whole fields of old worlds in ways that are usefully conceived of as creating new representation systems. When we learn that atoms are like billiard balls we may learn a new word "atom", but we also transform our old understanding of words like "pressure", "volume", etc. Analogies are, of course, rather closely related to theories. They are therefore particularly relevant examples of how partially interpreted natural languages are, and how interpretations change in learning. Taking this view seriously we realise that natural languages are not single representation systems at all but enormous aggregations of interpretations each of which might constitute something like an interpreted representation system in our sense. When we count modalities we may think of graphics as one and language as another, but this arithmetic is highly misleading if we begin to think of modalities as representation systems.

4 Conceptual Depths

In illustrating how an artificial language "surface" can be arbitrarily mapped onto monarchs and menageries, and how it is the "style" of this mapping that defines the interpretations that these systems are classified by, I may well have eloquently reminded the reader why they do not believe that logical classification of representational systems can help with understanding learning. Learning physics (substitute most subjects of intensive conceptual learning as you will) is about learning particular concepts, and in serious learning (as opposed to frivolous logic) it matters what these concepts are: mass and volume, not hiccup and cabbage, or worse still, concepts defined by randomly assorted objects.

If this serves to remind us that logic has nothing to say about which concepts are interesting and which ones not, that is all to the good. I shall still argue that logic nevertheless provides the only framework on which we might construct an understanding of how conceptual systems work. Much else has to be added to this framework, but without its foundation, we cannot start our construction.

Concepts, from a logical point of view are ways of classifying things. Concepts are, therefore, abstract things. Abstract, but nonetheless real. They are part of the world (along with many other abstract things like numbers or differences). The psychological problem that they pose, and especially pointedly for theories of conceptual learning, is how people know them, or come to know them. And of course there is the more practical but just as profound question, especially pointed for theories of conceptual teaching, of just how we know whether others do possess them.

For some kinds of concept, it is plausible that we base our answer to these questions in terms of abilities to "sort" the furniture of our environment. Being able to sort cabbages from cauliflowers (and other things) is plausibly what we possess when we have acquired the concept of "cabbage", and plausible evidence for teacher of us possessing the concept. There is trouble with "unicorn" and with "mass", but even before that trouble, there is trouble with "cabbage". At least, plausibility must not be taken for sufficiency. In order to know what a cabbage is we need to know what a plant is, and what a species is, and what a living thing is, let alone an object. Our tutorial in the vegetable garden only works because we share with our teacher at least approximations to these concepts, and a great deal of perceptual experience about the range of variability of exemplars. Only sharing all this can we learn to apply cabbage and cauliflower correctly and robustly.

Even sufficiency is not the whole problem. If we do possess the concepts of vegetable, plant, species and object, then knowing that a cabbage is a species of plant gives us most of the concept before we have any grasp on sorting the vegetable garden. Sorting ability isn't necessary or sufficient for possession, even though it may be, for these concepts, an indispensable aid to finding out what concepts students possess.

So we should not overestimate the role of sorting as criterion of knowledge, and we shall see below that for many "abstract" concepts it is not even a useable guide to checking for possession of concepts. Logical discussions, focused as they are on the logical vocabulary and its associated transformations systems, love to use the plausibly sortable cases of concepts as examples (what have been called the middle-sized white goods of the universe) because it has problems enough, and conceptual acquisition and testing is not the problem it focuses on.

To know a concept we have to have at least some mental representation of it, and its conceptual nexus. These mental representations may not be sufficient for conceptual deployment without a supporting environment, so we should remember that the exact distribution of representation between mind and environment is an important issue. There are those prepared to argue against *any* representational theories of mind, but I think few here, and so the arguments perhaps need not distract us. So possessing concepts requires that we have some representational surface (some combination of mental/public surface) with which to compute about our concepts and their manifestations in the world. What these representations are like, and how some account of their processing can be given that allows escape from the regresses they invite, are age-old questions. Study of the internal representations of concepts has made some progress in the last century of psychology — but it has been slow. This is one place where the "practical" question how can we establish with some reasonable assurance that we share a concept with another (say our student) is the only approach we have to the "theoretical" question about what it is that either of us has. We may not be able to compare our internal representations yet, except by using external ones to communicate with each other. These external ones used in communication manifest as surfaces.

So how is it that we classify conceptual systems? Not by their surfaces but by their content. Conceptual systems are related because they attempt to classify the same parts of the world, although they do it in competing ways. The medieval and the particulate theories of matter are related because they attempt to categorise some of the same phenomena. They also share relations with the pre-theoretical conceptualisations (systematic or not) of the same phenomena. And all these three conceptualisations (or perhaps many more if there are competing "pre-theoretical" cases) are built on shared background concepts such as "thing", "time" and "causality". If the foreground theories (medieval vs. particulate) do not contrast in their treatment of these concepts, they may remain submerged throughout, just as the "ontologies" of knowledge engineers often remain submerged while the details of knowledge of domains are examined.

We approach our educational communication problem with public access to the world that is being classified and to the public surfaces of our representational systems. We must also share a great deal of our internal mental apparatus, and must assume so to get communication started. With this much capital, we have to try to align the concepts that we command. Teaching proceeds by systematising the pre-systematic concepts that students bring. This happens by differentiating concepts. With some concepts, sorting situations are readily available. As we shall illustrate with physics and with logic, in many abstract fields they are not. Then we have to resort to access to concepts through their transformational relations. The central pedagogical question of this volume is when is adding explicitly new surfaces for conceptual systems helpful and when is it not? What logic has to contribute is some techniques for understanding the relations between surfaces and conceptual systems that determine the complexity of reasoning resulting from resurfacing, and some grasp of the metalogical properties of systems that are critical in communicating about new surfaces.

How can any of this study of representational surfaces or conceptual depths help with understanding how multiple representation systems affect students' learning as exemplified in the chapters of this book? At the very least, it serves to illustrate how complex is the full process of coming to understand or to create a representational system. But my hope is that we can do rather more. It suggests that at the finer grains of distinction between representational systems, all learning involves multiple representation systems, both as educational tools and as educational goals. It allows us to refine what we think is significant multiplicity that will avoid trivialising the concept. And it suggests a bifurcation into two general kinds of significant multiplicity which might provide a handle on our topic — multiple surfaces and multiple conceptual systems.

5 Physics and Logic — Accessing Concepts Through Transformations

5.1 The Example of Physics

Many of the chapters in this book deal with scientific learning about the conflicts of appearance with underlying reality. Theories which operate with tightly coupled sets of new concepts have to be acquired, and conflicts between them and prior systems of concepts (often labelled by the same words) have to be resolved, or at least lived with — old conceptual scheme alongside new. All of this has to be achieved by applying what is very indirect evidence as to the nature of the concepts to be acquired. It is the nature of this evidence and its relation to the task of understanding the interpretation of the new conceptual system that we must understand if we are to understand learning with multiple representations.

In Piaget's theory, the crucial conceptual reorganisations of physics knowledge come in two stages.[3] At the concrete operational stage, the child has to differentiate a set of concepts — amount, weight, volume, density, mass, length. The pre-operational child has concepts already attached to at least some of these words. It is quite hard to state exactly what these concepts are because they are not totally systematic and are largely evidenced by judgements made in a variety of perceptual situations. But whatever the pre-operational concepts are, they are different from the post-operational ones which are *systematic*. So the pre-operational child's concept of weight is certainly connected with some of the weighing operations with which it is associated after acquisition of the abstract physics concept, as well as with lots of primitive perceptual experiences.

The learning problem is that the abstract concept of weight is only indirectly connected with *any* perceptual experience. Appearance and reality have to be separated further than they were previously. The new abstract concepts are only accessible through learning which physical transformations preserve which abstract quantities and which change them. Compression of a fixed mass preserves weight but changes volume and density. Compression by introduction of more mass may preserve volume but increases weight. The concepts of mass and weight can only be separated by the transformation of moving to a different gravitational field. Without a trip to the moon, such separation can only be done indirectly through a grasp of theory.

Which brings us to Piaget's second transition of development — formal operations. Here the abstract concepts acquired through their relations to concrete operations have to be connected together into systems by theories enabling genuine hypothetical thinking. The particulate theory of matter is an excellent example of how theory is required to connect concepts in a way that allows understanding of situations which are even more complexly and indirectly connected to perceptual experience (see Rohr and Reimann, chapter 3). In order to understand that pulling a syringe to increase its internal volume lowers internal pressure, and atmospheric pressure causes an inflow of air, requires not only coming to appreciate the properties of the invisible medium in which we live, rather in the manner of a goldfish recognising water, but also requires the child to dismantle its common notion of causality. According to this conception, the efforts of the agent pulling the syringe are the cause of whatever happens as a result of those efforts. This is, of course, not merely the child's common conception of causality, but it is also our adult conception. It is not something we abandon when we learn that physics has an agentless understanding of causality in terms of systems achieving equilibria. Nor is it a "wrong" conception of causality (or agency if one prefers), but it certainly is different.

Even further backgrounded, but equally important in the large picture of learning, are the unique social arrangements which underpin scientific theories. The student is not just learning that a particular abstract theory can mediate between appearance

[3] I will not be much concerned here with issues about stages in acquisition — but Piaget's specification of the concepts to be learnt seems the most useful starting place.

and reality, but that a definite shift in the social authorisation of knowledge occurred with the creation of the first such theories. They learn that Galileo dropped stuff off the tower in Pisa and got into lots of trouble with the inquisition because of his views about theory, experiment and authority.

How are teacher and student to communicate about this novel conceptualisation? If concepts are ways of classifying things, then pointing out which are which would be an approach with some appeal in the case of discriminating cabbages and kings. With volumes, weights, masses and amounts this approach has less appeal for the simple reason that when we point to one, we thereby point to all the rest. The tasks of conceptual acquisition involved in learning physics require a learner to access new concepts indirectly through transformations, and the consequences of transformation for representations. What does this mean for the role of representation systems with contrasting surfaces in this learning/teaching process?

In the case of learning a theory such as the particulate theory of gasses, a variety of surface representations often are used in teaching. These range from experiments with example real objects (perhaps the limiting case of representation), through animations of syringes containing representations of magnified atoms in motion and collision, analogies such as the "billiard ball analogy" to diagrams of syringes, verbal descriptions of physical situations, and mathematical equations. There are evidently going to be cost functions in learning to interpret these new surfaces. We will return to these when we have compared and contrasted the situation in logic.

5.2 The Example of Logic

What is the situation with learning logic? Here it is immediately controversial what it is to teach logic and what has been taught when we succeed, in a way that these questions are uncontroversial about physics. Logic may be formal or informal. It may be thought of as a narrow set of techniques, or a broad study of representational systems; the construction of arguments within them, and the social relations of authority which surround them. What I want to do here is to look at the teaching of a formal logical system and ask what novel representational surfaces and what conceptualisations are involved in the teaching of that system, in order to compare this case with physics.

In teaching say first-order predicate calculus, the conspicuously novel representational surface which is involved is the artificial language itself, along with its proof systems and proof strategies. This is sometimes taken by outsiders to be the goal of teaching — a new set of operations for reasoning. Of course, even this view requires that there also be some teaching of ways to connect the new formalism with material to be reasoned about. Minimally this might be thought of as teaching some translation into and out of the formalism from the students' existing natural language.

This minimal model of logic teaching is quite prevalent among non-logicians, and is usually, unsurprisingly, associated with the belief that this occupation is a waste of student and teacher's time. It is essentially the view that logic is a novel representational surface or surfaces. But this view completely ignores the conceptual content of logic. So what is this conceptual content?

The usual place to start is with the concept of validity — valid arguments preserve truth. Validity is what logic is about — it defines what it is that is preserved by the transformations which figure in particular logical systems. A valid argument's conclusion is true whenever its premisses are true. The concepts *premiss*, *conclusion* and *whenever* constitute much of what remains of the fundamentals of logic. To understand what differentiates premisses and conclusions is to understand the distinctive social relations in argument which they mark. Conclusions are what can be drawn "unilaterally", by any party to the argument, by applying transformation rules to premisses. Premisses are what cannot be changed without agreement on the part of all parties. Understanding what *whenever* means requires us to understand what is meant by a representation system (usually but not necessarily a sentential language), and *all* its possible interpretations (a rather condensed tutorial appeared above in section 2). Understanding all that requires a grasp of meta-representational properties of such systems — properties such as soundness, completeness and decidability. There is of course more, but here are the main entrypoints to the network of concepts which constitutes the topic of logic.

So far from logic being about some specific new surface of Ps and Qs, it is about general properties of representation systems which hold whatever surface is assigned a given kind of interpretation. The situation is just like physics in that none of these concepts are entirely unrelated to concepts which students have before they get into logic classes — concepts for the most part implicit in their communicative behaviour. However, as with physics the implicit concepts are at least undifferentiated from, and sometimes downright in conflict with, the logical concepts which are the object of teaching. As with physics, these "primitive" concepts will endure alongside the new differentiated conceptual apparatus after learning logic, and indeed will need to if the student is to continue his/her communicative life.

The situation in logic is also like that in physics in that the ostensive approach of pointing out which things are cabbages and which are kings won't work, and for the same reasons. For example, premisses and conclusions are sentences, but that statement is as helpful for the learner as the statement that masses are volumes and weights. It is the role that sentences play in arguments that makes them premiss or conclusion. The roles that premiss and conclusion play in the transformations of reasoning are the only route the learner has into the new conceptual system that is being taught.

With physical knowledge, the "commonsensical" ability to get around in the world that pre-dates teaching, will still be the basis of the students' everyday dealings with the physical world after learning physics. It is a good question what impact there is on behaviour. There may be some, but outside professional practices, these may be surprisingly limited. Do we teach physics at school level in order to change behaviour? To change conceptual understanding? Or to give an example of how a developed abstract scientific theory can mediate between sensory experience and underlying reality? Or even to teach about the changes in the social arrangements for knowledge that come with the introduction of experimental science? Or all of the above?

In logic there is a similar relation between the pre-theoretical concepts which allow us to communicate and reason, and the post-theoretical system of concepts which constitute logic. We do not give up our intuitive grasp of language because we have learned logic. It is a deep and didactically important question to what extent reasoning and communicative behaviour is changed by learning the theoretical concepts. It is my suspicion that this impact may be greater for more students than is the case in physics, just because more people are faced by reasoning problems where the concepts are applicable, and where their intuitions are palpably not enough. Be that as it may, the analogy raises the same questions about why we might teach logic as we just raised about physics. It is wildly implausible that we should teach logic for a specific logical system and its reasoning transformations — say propositional calculus (unless perhaps we have in mind some very specific application, say electrical circuit design). More plausible is the idea that we teach logic because it is a deep abstract theory of the surfaces of representation systems and how they are interpreted that can connect the complexities of the sensory experiences involved to an underlying communicative reality. Of course the critical didactic questions are how to do this. Whether a novel formal surface aids in teaching, or whether it is better to stick with structures implicit in students' natural language. This battle rages between camps — informal logicians, critical thinkers, argumentation theorists, formal folk. . . . There does not have to be one answer for all students with all educational goals. But there are some serious empirical questions to which we turn below.

If these analogies between physics and logic are well conceived, then we can observe that the two subjects share the fact that they are taught for the underlying conceptual systems that they embody, not the surfaces of the representations that must be used to teach them. When we turn to the question how the two subjects differ, we can return from our extended detour to the question what role different surface representations may play in the teaching of these (and any other) subjects.

6 The Costs and Benefits of Resurfacing the Road to Knowledge

After our travels, we can now restate one question that this book asks. When teaching new conceptual systems through their attendant transformation systems, when is it advantageous to explicitly introduce novel surfaces for the representation systems involved, and to explicitly teach about their equivalences?

Resurfacing has obvious costs. However expert we are at the arts of representation, some new cost is involved. This will be less if there is an old use of the surface familiar to the student for which they already have an interpretation. Transfer of bits of mathematics is often like this, although there is still generally considerable cost in adapting the interpretation to the new domain.

One important consideration in judging whether this cost is worthwhile will be whether the representational insights gained thereby are likely to be useful elsewhere. Mike Dobson's comparison of Euler and Venn's graphical systems for teaching syllogistic reasoning are an apt example here.[4] Euler is the more constrained system. It has important meta-properties which make it less complex (in a mathematically precise sense) to reason with, and may make its reasoning properties easier to understand. It is for example, self-consistent. One cannot diagram inconsistent sets of premisses in the Euler system. Venn is more expressive (it can be used to teach elementary set theory — concepts of union, intersection, etc. beyond the syllogism) and this is the reason why it is almost always the circle system which is seen in modern texts. But its expressiveness costs the system Euler's self-consistency. One can represent say "All Xs are Ys and some Xs are not Ys" in the same Venn diagram.

Which system is best? How does absolute mathematical complexity translate into classroom experience? Well that will depend on what we are trying to teach, and why. Are we trying to get students to draw the right conclusions from syllogisms? Are we trying to introduce a new representational surface which we might want to use to teach set theory? Are we only interested in teaching the syllogism because it is one of the smallest fragments of logic for which we can introduce the study of basic logical invariants like validity, and through them meta-representational properties such as soundness, completeness and decidability — a thumbnail appreciation of the study of representation systems? Are we interested in teaching the social arrangements for argumentation?

The at least implicit answer suggested here is that we should take seriously the explicit teaching of the business of representation, and that fragments of logic like the syllogism are of scant interest in themselves, but may be of considerable use in teaching about representations (which is, after all, what logic is about). I am not pretending that we can answer these questions without empirical experiment to find out what actually happens in the classroom. What I am suggesting is that without this kind of refinement of the questions, we will not understand our experimental results.

I will finish by mentioning two empirical lines of research which address some specific cases of these questions in the area of logic teaching. The first is Van der Pal's (1995) study of teaching conditional reasoning in which he showed that teaching the formalism of propositional calculus (in a graphical environment — Tarski's World) did have advantages over teaching in the students' natural language (in the same graphical environment). Apparently, the formalism does help to make explicit, structures which are both complex and loaded with extraneous functions in the natural language (at least for those students with this background assessed in this way).

The second line of research is our Edinburgh work comparing graphical and sentential teaching of first order logic using Hyperproof (Oberlandeer et al., 1996; Stenning, Cox & Oberlander, 1995). This work reveals very large individual differences between students in their response to heterogeneous graphical + sentential as opposed to conventional purely sentential presentations. Analysis of the underlying basis for these aptitude treatment interactions reveals that the students who benefit most from the heterogeneous environment do so because they are adept at knowing when to translate information in which direction between sentences and graphics. The

[4]See Stenning (in press) for a more extended discussion of the issues in applying the theory of expressiveness to interpreting Dobson's experimental results.

students who have trouble with the heterogeneous environment do so because they tend to translate into the graphical mode, and fail to translate in the reverse direction at the strategic points. This result illustrates nicely the centrality of the skills underlying strategic transformation of multiple representations in the determination of learning outcomes.

Acknowledgements

I would like to thank the other authors (and especially the editor) of this volume whose chapters' early drafts provoked the writing of this chapter: and my colleagues in Edinburgh with whom much of the thinking set down here went on. The support of the Economic and Social Research Council UK (ESRC) is gratefully acknowledged. The work was partly supported by the ESRC funded Human Communication Research Centre. We also wish to acknowledge partial support from the following grants: Grant \# L127251023 from ESRC; Cognitive Engineering and Grant \# GRL 86930 from the EPSRC.

References

Chapter 2

Alessi, S.M. (1995, April). Dynamic vs. static fidelity in a procedural simulation. *Paper presented at the Annual Meeting of the American Educational Research Association* (San Francisco, CA).

Bertin, J. (1981). *Graphics and Graphic Information-processing*. (W.J. Berg, & P. Scott, Trans.). Berlin: Walter de Gruyter.

Bertin, J. (1983). *Semiology of Graphics: Diagrams Networks Maps*. Madison: The University of Wisconsin Press.

Brown, J.S., & Burton, P.R. (1978). Diagnostic models for procedural bugs in basic mathematical skills. *Cognitive Science*, *2*, 155–191.

Chi, M.T.H., Slotta, J.D., & de Leeuw, N. (1994). From things to processes: A theory of conceptual change for learning science concepts. *Learning and Instruction*, *4*, 27–43.

Coleman, T.G., & Randall, J.E. (1986). *HUMAN-PC: A Comprehensive Physiological Model* (computer software). Jackson: University of Mississippi Medical Center.

Cognition and Technology Group at Vanderbilt (1992). The Jasper series as an example of anchored instruction: theory, program, description, and assessment data. *Educational Psychologist*, *27*, 291–315.

Davis, R., Shrobe, H., & Szolovits, P. (1993). What is a knowledge representation. *AI Magazin*, *14*, 17–33.

de Jong, T., & Ferguson-Hessler, M.G.M. (1991). Knowledge of problem situations in physics: a comparison of good and poor performers. *Learning and Instruction*, *1*, 289–302.

de Jong, T., & van Joolingen, W.R. (1995). The SMISLE environment: Learning and design of integrated simulation learning environments. In P. Held, & W.F. Kugemann (Eds.). *Telematics for Education and Training* (pp. 173–187). Amsterdam: IOS Press.

de Jong, T., & van Joolingen, W.R. (1996). Discovery learning with computer simulations of conceptual domains. University of Twente, Faculty of Educational Science and Technology. Internal Memo 96-02.

de Jong, T., van Joolingen, W.R., Swaak, J., Veermans, K., Limbach, R., King, S., & Gureghian, D. (1998). Self-directed learning in simulation-based discovery environments. *Journal of Computer Assisted Learning*, *14*, 235–246.

de Jong, T., Martin, E., Zamarro J-M., Esquembre, F., Swaak, J., & van Joolingen, W.R. (1995, April). Support for simulation-based learning; the effects of assignments and model progression in learning about collisions. *Paper presented at the Annual Meeting of the American Educational Research Association*. (San Francisco, CA).

de Jong, T., Härtel, H., Swaak. J., & van Joolingen, W. (1996). Support for simulation-based learning; the effects of assignments in learning about transmission lines. In A. Díaz de Ilarazza Sánchez, & I. Fernández de Castro (Eds.). *Computer Aided Learning and Instruction in Science and Engineering* (pp. 9–27). Berlin: Springer-Verlag.

de Kleer, J., & Brown, J.S. (1984). A qualitative physics based on confluences. *Artificial Intelligence*, *24*, 7–83.

Dibble, E., & Shaklee, H. (1992, April). Graph interpretation: a translation problem? *Paper presented at the annual meeting of the American Educational Research Association.* (San Francisco, CA).

Dobson, M.W. (1996). Specificity and learning with graphics. *Proceedings of the IEE Colloquium Thinking with Diagrams.* London: IEE.

Dobson, M.W. (in press). Information enforcement and learning with interactive graphics. In M. Dobson (Guest Ed.). *Learning with Interactive Graphics. Special Edition of Learning and Instruction.* Oxford: Permagon.

Forbus, K.D. (1984). Qualitative process theory. *Artificial Intelligence, 24,* 85–158.

Gentner, D., & Gentner, D.R. (1983). Flowing waters or teeming crowds: Mental models of electricity. In D. Gentner, & A.L. Stevens (Eds.). *Mental Models* (pp. 99–130). Hillsdale, NJ: Erlbaum.

Gruber, H., Graf, M., Mandl, H., Renkl, A., & Stark, R. (1995, August). Fostering applicable knowledge by multiple perspectives and guided problem solving. *Paper presented at the conference of the European Association for Research on Learning and Instruction.* (Nijmegen, The Netherlands).

Huff, D. (1954). *How to Lie with Statistics.* New York: W.W. Norton.

Janvier, C. (1987a). *Problems of Representation in the Teaching and Learning of Mathematics.* Hillsdale, NJ: Erlbaum.

Janvier, C. (1987b). Translation processes in mathematics education. In C. Janvier (Ed.). *Problems of Representation in the Teaching and Learning of Mathematics* (pp. 27–32). Hillsdale, NJ: Erlbaum.

Kaput, J.J. (1992). Technology and mathematics education. In D.A. Grouws (Ed.). *Handbook of Teaching and Learning Mathematics* (pp. 515–556). New York: Macmillan.

Koedinger, K.R., & Anderson, J.R. (1994). Reifying implicit planning in geometry: Guidelines for model-based intelligent tutoring system design. In S.P. Lajoie, & S.J. Derry (Eds.). *Computers as Cognitive Tools* (pp. 15–46). Hillsdale, NJ: Erlbaum.

Kosslyn, S.M. (1989). Understanding charts and graphs. *Applied Cognitive Psychology, 3,* 185–226.

Kozma, B., & Russell, J. (1997). Multimedia and understanding: Expert and novice responses to different representations of chemical phenomena. *Journal of Research in Science Teaching, 34,* 949–968.

Kozma, B., Russell, J., Jones, T., Marx, N., & Davis, J. (1996). The use of multiple, linked representations to facilitate science understanding. In S. Vosniadou, E. De Corte, R. Glaser, & H. Mandl (Eds.). *International Perspectives on the Design of Technology Supported Learning Environments* (pp. 41–61). Hillsdale, NJ: Erlbaum.

Kuipers, B. (1984). Common-sense reasoning about causality. *Artificial Intelligence, 24,* 169–203.

Kulpa, Z. (1995). Diagrammatic representation and reasoning. *Machine Graphics and Vision, 3,* 77–103.

Larkin, J.H., & Simon, H.A. (1987). Why a diagram is (sometimes) worth ten thousand words. *Cognitive Science, 11,* 65–99.

Mayer, R., & Anderson, R. (1992). The instructive animation: Helping students build connections between worlds and pictures in multimedia learning. *Journal of Educational Psychology, 84,* 444–452.

Mayer, R.E., & Sims, V.K. (1994). For whom is a picture worth a thousand words? Extensions of dual-coding theory of multimedia learning. *Journal of Educational Psychology, 86,* 389–401.

McKenzie, D.L., & Padilla, M.J. (1986). The construction and validation of the test of graphing in science. *Journal of Research in Science Teaching. 23*, 571–579.

National Council of Teachers of Mathematics (1989). *Curriculum and Evaluation Standards for School Mathematics*. Reston, VA: NCTM.

Newell, A. (1982). The knowledge level. *Artificial Intelligence, 18*, 87–127.

Ohlsson, S., & Langley, P. (1984). *P.R.I.S.M.: Tutorial, Manual, and Documentation* (Technical Report of the Robotic Institute). Pittsburgh, PA: Carnegie-Mellon University.

Opwis, K., Spada, H., Beller, J., & Schweizer, P. (1994). Kognitive Modellierung als Individualdiagnostik: Qualitatives und quantitatives physikalisches Wissen. *Zeitschrift für Differentielle und Diagnostische Psychologie, 15*, 93–111.

Padilla, M.J., McKenzie, D.L., & Shaw, E. (1986). An examination of the line graphing ability of students in grades seven through twelve. *School Science and Mathematics, 86*(1), 20–25.

Paivio, A. (1986) *Mental Representation: A Dual Coding Approach*. New York: Oxford University Press.

Pinker, S. (1990). A theory of graph comprehension. In R. Freedle (Ed.). *Artificial Intelligence and the Future of Testing* (pp. 73–126). Hillsdale, NJ: Erlbaum.

Plötzner, R., Spada, H., Stumpf, M., & Opwis, K. (1990). Learning qualitative and quantitative reasoning in a microworld for elastic impacts. *European Journal of Psychology of Education, 4*, 501–516.

Quinn, J., & Alessi, S. (1994). The effects of simulation complexity and hypothesis generation strategy on learning. *Journal of Research on Computing in Education, 27*, 75–91.

Rieber, L.P. (1990). Using computer animated graphics in science instruction with children. *Journal of Educational Psychology, 82,* 135–140.

Rieber, L.P., & Parmley, M.W. (1995). To teach or not to teach? Comparing the use of computer-based simulations in deductive versus inductive approaches to learning with adults in science. *Journal of Educational Computing Research, 14*, 359–374.

Rieber, L.P., Boyce, M.J., & Assad, C. (1990). The effects of computer animations on adult learning and retrieval tasks. *Journal of Computer Based Instruction, 17,* 46–52.

Scardamalia, M., & Bereiter, C. (1990). An architecture for collaborative knowledge building. In E. de Corte, M. Linn, H. Mandl, & L. Verschaffel (Eds.). *Computer-based Learning Environments and Problem Solving* (NATO ASI series F: Computer and Systems Series) (pp. 41–66). Berlin: Springer-Verlag.

Sime, J-A. (1994). Model switching in intelligent training systems. Unpublished PhD thesis, Heriot-Watt University, Edinburgh, UK.

Sime, J-A. (1995). Model progressions and cognitive flexibility theory. In J. Greer (Ed.). *Proceedings of the AI-Ed 95, the 7th World Conference on Artificial Intelligence in Education* (pp. 493–500). Charlottesville, VA: AACE.

Sime, J-A. (1996). An investigation into teaching and assessment of qualitative knowledge in engineering. In P. Brna, A. Paiva, & J. Self (Eds.). *Proceedings of EuroAIED* (pp. 240–246). (Lisbon, Portugal).

Sime, J-A. (1998). Model switching in a learning environment based on multiple models. *Journal of Interactive Learning Environments, 5*, 109–125.

Sime, J-A., & Leitch, R.R. (1992a). A learning environment based on multiple qualitative models. In C. Frasson, G. Gauthier, & G.I. McCalla (Eds.). *Intelligent Tutoring Systems. Proceedings of Second International Conference* (ITS'92) (pp. 116–123). Berlin: Springer-Verlag.

Sime, J-A., & Leitch, R.R. (1992b). Multiple models in intelligent training. In *Proceedings of the 1st International Conference on Intelligent Systems Engineering (ISE'92) Conference Publication 360*, 263–268. London: Institute of Electrical Engineers.

Spada, H., Reimann, R., & Häusler, B. (1983). Hypothesenerarbeitung und Wissensaufbau beim Schüler. In L. Kötter, & H. Mandl (Eds.). *Kognitive Prozesse und Unterricht. Jahrbuch für empirische Erziehungswissenschaft* (pp. 139–167). Düsseldorf: Schwann.

Spada, H., Stumpf, M., & Opwis, K. (1989). The constructive process of knowledge acquisition: Student modeling. In H. Maurer (Ed.). *Computer-assisted Learning* (pp. 486-499). Berlin: Springer-Verlag.

Spiro, R.J., & Jehng, J-C. (1990). Cognitive flexibility and hypertext: Theory and technology for the nonlinear and multi-dimensional traversal of complex subject matter. In D. Nix, & R.J. Spiro (Eds.). *Cognition, Education and Multi-media: Exploring Ideas in High Technology* (pp. 163–205). Hillsdale, NJ: Erlbaum.

Stenning, K., & Oberlander, J. (1995). A cognitive theory of graphical and linguistic reasoning: logic and implementation. *Cognitive Science*, *19*, 97–140.

Stumpf, M. (1990). Adaptivität durch Flexibilität in Repräsentationsform und Konstrollstruktur im System DiBi-MR. In R. Reuter (Ed.). *Gesellschaft für Informatik — 20. Jahrestagung* (pp. 377–385). Berlin: Springer-Verlag.

Suthers, D., Weiner, A., Connelly, J., & Paolucci, M. (1995). Belvedere: Engaging students in critical discussion of science and public policy issues. In J. Greer (Ed.). *Proceedings of the AI-Ed 95, the 7th World Conference on Artificial Intelligence in Education* (pp. 266–273). Charlottesville, VA: AACE.

Swaak, J., & de Jong, T. (1996). Measuring intuitive knowledge in science: The development of the WHAT-IF test. *Studies of Education Evaluation, 22,* 341–362.

Swaak, J., van Joolingen, W.R., & de Jong, T. (1998). Support for simulation-based learning; the effects of model progression and assignments on learning definitional and intuitive knowledge. *Learning and Instruction*, *8*, 235–253.

Tabachneck, H.J.M., Leonardo, A.M., & Simon, H.A. (1994). How does an expert use a graph? A model of visual and verbal inferencing in economics. In A. Ram, & K. Eiselt (Eds.). *16th Annual Conference of the Cognitive Science Society* (pp. 842–847). Hillsdale, NJ: Erlbaum.

Tergan, S. (1997). Multiple views, contexts and symbol systems in learning with hypertext/hypermedia. *Educational Technology*, *37*(4), 5–18.

Tessler, S., Iwasaki, Y., & Law, K.H. (1995). Qualitative structural analysis through mixed diagrammatic and symbolic reasoning. In B. Chandrasekaran, J. Glasgow, & N.H. Narayanan (Eds.). *Diagrammatic Reasoning; Cognitive and Computational Perspectives* (pp. 711–729). Cambridge, MA: AAAI and MIT Press.

Thompson, P.W. (1992). Notations, conventions and constraints: contributions to effective uses of concrete materials in elementary mathematics. *Journal for Research in Mathematics Education*, *23*, 123–147.

Tukey, J.W. (1977). *Exploratory Data Analysis*. Reading, MA: Addison-Wesley.

van der Hulst, A. (1996). Cognitive tools. Two exercises in non-directive support for exploratory learning. Ph.D. thesis, University of Amsterdam, Amsterdam.

van Joolingen, W.R. (1995). QMaPS: qualitative reasoning for intelligent simulation learning environments. *Journal of Artificial Intelligence in Education, 6,* 67–89.

Wainer, H. (1984) How to display data badly. *The American Statistician*, *38*, 137–147.

Weld, D.S., & de Kleer, J. (Eds.) (1990). *Readings in Qualitative Reasoning about Physical Systems*. San Mateo, CA: Morgan Kaufmann.

White, B.Y. (1984). Designing computer games to help physics students understand Newton's laws of motion. *Cognition and Instruction*, *1*, 69–108.

White, B.Y. (1993). Thinker tools: Causal models, conceptual change, and science education. *Cognition and Instruction*, *10*, 1–100.

White, B.Y., & Frederiksen, J.R. (1986). *Progressions of Qualitative Models as a Foundation for Intelligent Learning Environments*. BBN Laboratories Report No 6277, May 1986.

White, B.Y., & Frederiksen, J.R. (1989). Causal models as intelligent learning environments for science and engineering education. *Applied Artificial Intelligence*, *3*, 83–106.

White, B.Y., & Frederiksen, J.R. (1990). Causal model progressions as a foundation for intelligent learning environments. *Artificial Intelligence*, *42*, 99–157.

Woolf, B., Blegen, D., Jansen, J.H., & Verloop, A. (1987). Teaching a complex industrial process. In R.W. Lawler, & M. Yazdani (Eds.). *Artificial Intelligence and Education. Volume 1. Learning Environments and Tutoring Systems* (pp. 413–427). Norwood, NJ: Ablex Publishing.

Yerushamly, M. (1991). Student perceptions of aspects of algebraic function using multiple representation software. *Journal of Computer Assisted Learning*, *7*, 42–57.

Chapter 3

Benson, D.L., Wittrock, M.C., & Baur, M.E. (1993). Student's preconceptions about the nature of gases. *Journal of Research in Science Teaching*, *30*, 587–597.

Ben-Zvi, R., Eylon, B.-S., & Silberstein, J. (1986). Is an atom of copper malleable? *Journal of Chemical Education*, *63*, 64–66.

Brook, A., & Driver, R. (1989). *Progression in Science: The Development of Pupils' Understanding of Physical Characteristics of Air Across the Age Range 5–16 Years*. University of Leeds: Children's Learning in Science Project.

Brook, A., Briggs, H., & Driver, R. (1984). *Aspects of Secondary Students' Understanding of the Particulate Model of Matter*. University of Leeds: Children's Learning in Science Project.

Buck, P. (1990). Jumping to the atoms: The introduction of atoms via nesting systems. In P.L. Lijnse, P. Licht, W. de Vos, & A.J. Waarlo (Eds.). *Relating Macroscopic Phenomena to Microscopic Particles*. Utrecht: Centre for Science and Mathematics Education.

Burstein, M.H. (1986). Concept formation by incremental analogical reasoning and debugging. In R.S. Michalski, J.G. Carbonell, & T.M. Mitchell (Eds.). *Machine Learning: An Artificial Intelligence Approach* (pp. 351–378). San Mateo, CA: Morgan Kaufmann.

Chi, M.T.H., Feltovich, P., & Glaser, R. (1981). Categorization and representation of physics problems by experts and novices. *Cognitive Science*, *5*, 121–152.

Chi, M.T.H., Bassok, M., Lewis, M., Reimann, P., & Glaser, R. (1989). Self-explanations: how students study and use examples in learning to solve problems. *Cognitive Science*, *13*, 145–182.

Chi, M.T.H., Slotta, J.D., & de Leeuw, N. (1994). From things to processes: A theory of conceptual change for learning science concepts. *Learning and Instruction*, *4*, 27–43.

Chinn, C.A., & Brewer, W.F. (1993). The role of anomalous data in knowledge acquisition: A theoretical framework and implications for science instruction. *Review of Educational Research*, *63*, 1–49.

Clement, J. (1994). Use of physical intuition and imagistic simulation in expert problem solving. In D. Tirosh (Ed.). *Implicit and Explicit Knowledge: An Educational Approach* (pp. 204–244). Norwood, NJ: Ablex.

Davis, R., Shrobe, H., & Szolovits, P. (1993). What is a knowledge representation. *AI Magazine*, *14*, 17–33.

de Vos, W. (1990). Seven thoughts on teaching molecules. In P.L. Lijnse, P. Licht, W. de Vos, & A.J. Waarlo (Eds.). *Relating Macroscopic Phenomena to Microscopic Particles*. Utrecht, Centre for Science and Mathematics Education.

di Sessa, A.A. (1982). Unlearning Aristotelian physics: A study of knowledge-based learning. *Cognitive Science, 6*, 37–75.

Engel Clough, E., & Driver, R. (1986). A study of consistency in the use of students' conceptual frameworks across different task context. *Science Education, 70*, 471–496.

Falkenhainer, B. (1990). A unified approach to explanation and theory formation. In J. Shrager, & P. Langley (Eds.). *Computational Models of Scientific Discovery and Theory Formation* (pp. 157–196). San Mateo, CA: Morgan Kaufmann.

Falkenhainer, B., Forbus, K.D., & Gentner, D. (1989). The structure-mapping engine: An algorithm and examples. *Artificial Intelligence, 41*, 1–63.

Forbus, K.D., Ferguson, R., & Gentner, D. (1994). Incremental structure-mapping. In A. Ram, & K. Eislet (Eds.). *Proceedings of the Sixteenth Annual Conference of the Cognitive Science Society* (pp. 313–318). Hillsdale, NJ: Erlbaum.

Gentner, D. (1983). Structure-mapping: A theoretical framework for analogy. *Cognitive Science, 7*, 155–170.

Gentner, D., Brenn, S., Ferguson, R.W., Markman, A.B., Levidow, B.B., Wolff, P., & Forbus, K. (1997). Analogical reasoning and conceptual change: a case study of Johannes Kepler. *Learning and Instruction, 7*, 3–39.

Holyoak, K.J., & Thagard, P.R. (1989). Analogical mapping by constraint satisfaction. *Cognitive Science, 13*, 295–355.

Kaput, J.J. (1989). Linking representations in the symbol systems of algebra. In S. Wagner, & C. Kieran (Eds.). *Research Issues in the Learning and Teaching of Algebra*. Hillsdale, NJ.: Erlbaum.

Keane, M.T.G. (1990). Incremental analogizing: theory and model. In K.J. Gilhooly, M.T.G. Keane, R.H. Logie, & G. Erdos (Eds.). *Lines of Thinking* (pp. 221–235). Chichester: Wiley.

Kolodner, J.L. (1993). *Case-based Reasoning*. San Mateo, CA: Morgan Kaufmann.

Larkin, J.H., & Simon, H.A. (1987). Why a diagram is (sometimes) worth ten thousand words. *Cognitive Science, 11*, 65–99.

Löffler, G. (1996). Über die Grundlegung einer Verständigung über den Begriff des Atoms. Zur Erinnerung an Martin Wagenscheins 100. Geburtstag. *Chimica didactica, 22*, 296–322.

Mayer, R.E., & Anderson, R.B. (1991). Animations need narrations: an experimental test of a dual-coding hypothesis. *Journal of Educational Psychology, 83*, 484–490.

Newell, A. (1982). The knowledge level. *Artificial Intelligence, 34*, 87–127.

Novick, S., & Nussbaum, J. (1978). Junior high school pupils' understanding of the particulate nature of matter: An interview study. *Science Education, 62*, 273–281.

Novick, S., & Nussbaum, J. (1982). Pupil's understanding of the particulate nature of matter: a cross-age study. *Science Education, 66*, 187–196.

Paivio, A. (1986). *Mental Representations: A Dual Coding Approach*. Oxford: Oxford University Press.

Pfundt, H. (1981). Das Atom — letztes Teilungsstück oder erster Aufbaustein? Zu den Vorstellungen, die sich Schüler vom Aufbau der Stoffe machen. *Chimica didactica, 7*, 75–94.

Piaget, J. (1929). *The Child's Conception of the World*. London: Routledge and Keagan Paul.

Reimann, P., & Chi, M.T.H. (1989). Human expertise. In K.J. Gilhooly (Ed.). *Human and Machine Problem Solving* (pp. 161–192). London: Plenum Press.

Sèrè, M.G. (1985). The gaseous state. In R. Driver, E. Guesne, & A. Tiberghien (Eds.). *Children's Ideas in Science*. Milton Keynes: Open University Press.

Spiro, R., Feltovich, P., Coulson, R., & Anderson, D. (1989). Multiple analogies for complex concepts: antidotes for analogy-induced misconception in advanced knowledge acquisition. In S. Vosniadou, & A. Ortony (Eds.). *Similarity and Analogical Reasoning* (pp. 498–531). Cambrige: Cambridge University Press.

Stavy, R. (1988). Children's conception of gas. *International Journal of Science Education, 10*, 553–560.

Stavy, R. (1990). Children's conception of changes of states of matter: from liquid (or solid) to gas. *Journal of Research in Science Teaching, 27*, 247–266.

Tabachneck-Schijf, H.J.M., Leonardo, A.M., & Simon, H.A. (1997). CaMeRa: A computational model for multiple representations. *Cognitive Science, 21*, 305–350.

Tausch, M., & Wachtendonk, M. von (1996). *Stoff — Formel — Umwelt. Chemie 1*. Bamberg: Verlag C.C.Buchner.

Vosniadou, S. (1994). Capturing and modeling the process of conceptual change. *Learning and Instruction, 4*, 45–69.

White, B.Y., & Frederiksen, J.R. (1990). Causal model progressions as a foundation for intelligent learning environments. *Artificial Intelligence, 42*, 99–157.

Chapter 4

Ainsworth, S., Wood, D., & Bibby, P. (1996). Co-ordinating multiple representations in computer based learning environments. *Proceedings of EuroAIED*, pp. 336–341. (Lisbon, Portugal).

Bolton, J., & Ross, S. (in press). Developing students' physics problem solving skills. *Physics Education*.

Boyd, A., & Rubin, A. (1996). Interactive video: a bridge between motion and mathematics. *International Journal of Computers for Mathematical Learning, 1*(1), 57–93.

Brasell, H. (1987). The effect of real-time laboratory graphing on learning graphic representations of distance and velocity. *Journal of Research in Science Teaching, 24*(4), 385–395.

Champagne, A., Klopfer, L., & Anderson, J. (1980). Factors influencing the learning of classical mechanics. *American Journal of Physics, 48*(12), 1074–1079.

Chi, M., Feltovich, P., & Glaser, R. (1981). Categorization and representation of physics problems by experts and novices. *Cognitive Science, 5*, 121–152.

de Jong, T., & Ferguson-Hessler, M. (1986). Cognitive structure of good and poor problem solvers. *Journal of Educational Psychology, 78*, 279–288.

de Jong, T., & Ferguson-Hessler, M. (1991). Knowledge of problem situations in physics: A comparison of good and poor problem solvers. *Learning and Instruction, 1*, 298–302.

di Sessa, A., Hammer, D., Sherin, B., & Kolpakowski, T. (1991). Inventing graphing: metarepresentational expertise in children. *Journal of Mathematical Behaviour, 10*, 117–160.

Hatzipanagos, S. (1995). Fidelity and complexity in the design of computer based simulations in science. *Paper Presented at the the International Computer Assisted Learning Conference*. (Cambridge, UK).

Hayes-Roth, W. (1997, August). Graphing as social practice. *Paper Presented at the First European Science Education Research Association Conference*. (Rome, Italy).

Hennessy, S. (1993). Cognitive apprenticeship: implications for classroom learning. *Studies in Science Education, 22*, 1–41.

Janvier, C. (1978). The interpretation of complex Cartesian graphs: studies and teaching experiments. Unpublished PhD thesis, University of Nottingham.

Kaput, J.J. (1987). Towards a theory of symbol use in mathematics. In C. Janvier (Ed.). *Problems of Representation in the Teaching and Learning of Mathematics*. Hillsdale, NJ: Erlbaum.

Larkin, J. (1983). The role of problem representation in physics. In D. Gentner, & A. Stevens (Eds.). *Mental Models*. Hillsdale, NJ: Erlbaum.

Larkin, J., & Simon, H. (1987). Why a diagram is (sometimes) worth ten thousand words. *Cognitive Science, 11*, 65–99.

Leinhart, G., Zaslavsky, O., & Stein, M. (1990). Functions, graphs, and graphing: Tasks, learning and teaching. *Review of Educational Research, 60*(1), 1–64.

Linn, M.C., Layman, J., & Nachmias, R. (1987). Cognitive consequences of microcomputer-based laboratories: Graphing skills development. *Journal of Contemporary Educational Psychology, 12*, 244–253.

McClelland, J.A.G. (1985). Misconceptions in mechanics and how to avoid them. *Physics Education, 20*, 159–162.

McDermott, L.C., Rosenquist, E., & Van Zee, E. (1987). Student difficulties in connecting graphs and physics: examples from kinematics. *American Journal of Physics, 55*(6), 503–513.

Mokros, J.R., & Tinker, R.F (1987). The impact of microcomputer based labs on children's ability to interpret graphs. *Journal of Research in Science Teaching, 24*, 369–383.

Moschkovitz, J. (1997). Moving up and getting steeper: Negotiating shared descriptions of linear graphs. *Journal of the Learning Sciences, 5*(3), 239–278.

Preece, J. (1989). An investigation of graphing. Unpublished Ph.D. thesis, Open University.

Scanlon, E. (1990). Modelling physics problem solving. Open University internal document.

Scanlon, E. (1993). Solving the problem of physics problem solving. *International Journal of Mathematics Education for Science and Technology, 24*(3), 349–358.

Scanlon, E., & O'Shea, T. (1988) Cognitive economy in physics reasoning. In H. Mandl & A. Lesgold (Eds). *Learning Issues for Intelligent Tutoring Systems*. Berlin: Springer-Verlag.

Scanlon, E., O'Shea, T., Byard, M., Draper, S., Driver, R., Hennessy, S., Hartley, J.R., O'Malley, C., Mallen, C., Mohamed, G., & Twigger, D. (1993). Promoting conceptual change in children learning mechanics. *Proceedings of the Fifteenth Annual Cognitive Science Conference*. (Boulder, CO).

Stenning, K., & Oberlander, J. (1995). A cognitive theory of graphical and linguistic reasoning: logic and implementation. *Cognitive Science, 19*, 97–140.

Twigger, D., Byard, M., Draper, S., Driver, R., Hartley, R., Hennessy, S., Mohamed, R., O'Malley, C., O'Shea, T., & Scanlon, E. (1994). The conceptions of force and motion of students aged between 10 and 15 years: An interview study designed to guide instruction. *International Journal of Science Education, 16*(2), 215–229.

Whitelegg E.L. (1996, September). The supported learning in physics project, *Physics Education*.

Whitelegg, E., Scanlon, E., & Hatzipanagos, S. (1997). Multimedia Motion-motivating learners. *Journal of the Association for Learning Technology, 5*(1), 65–69.

Young, R., & O'Shea, T. (1981). Errors in children's subtraction. *Cognitive Science*, 5, 153–177.

Zajchowski, R., & Martin, J. (1993). Differences in the problem solving of stronger and weaker novices in physics: Knowledge, strategies or knowledge structure? *Journal of Research in Science Teaching, 30*(5), 459–470.

Chapter 5

Barwise, J., & Etchemendy, J. (1995). Heterogeneous logic. In J. Glasgow, N.H. Narayanan, & B. Chandrasekaran (Eds.). *Diagrammatic Reasoning; Cognitive and Computational Perspectives* (pp. 211–234). Cambridge, MA: AAAI and MIT Press.

Cheng, P.C.H. (forthcoming). Interactive law encoding diagrams for learning and instruction. In M.W. Dobson (Ed.), *Learning and Instruction (Special edition)*. Amsterdam: Elsevier.

Collins, A., Brown, J.S., & Newman, S.E. (1990). Cognitive apprenticeship: Teaching the crafts of reading, writing, and mathematics. In L.B. Resnick (Ed.), *Knowing, Learning, and Instruction: Essays in Honor of Robert Glaser* (pp. 453–494). Hillsdale, NJ: Lawrence Erlbaum Associates.

Dijkstra, S. (1997). Educational technology and media. In S. Dijkstra, N. Seel, F. Schott, & R.D. Tennyson (Eds.). *Instructional Design: International Perspectives,* Vol. II. Hillsdale, NJ: Erlbaum.

Dobson, M.W. (forthcoming). Information enforcement and learning: improving syllogistic reasoning skills. DPhil Thesis, Open University, UK.

Draper, S., Mohamed, R., Byard, M., Driver, R., Hartley, R., Mallen, C., Twigger, D., O'Malley, C., Hennessey, S., O'Shea, T.M.M., Scanlon, E., & Spensley. F. (1992). *Conceptual Change in Science* (CALRG Tech. Rep. no. 123). Milton Keynes: Open University, Institute of Educational Technology.

Elsom-Cook. M.T. (1990). *Guided Discovery Tutoring*. London: Chapman.

Euler, L. (1772). *Lettres à une princesse d'Allemagne, Vol 2. Sur divers sujets de physique et de philosophie*, Letters 102–108. [Letters to a princess of all the Russias, Vol 2. On subjects of science and philosophy]. Basel: Birkhauser .

Funt, B.V. (1995). Problem solving with diagrammatic representations. In J. Glasgow, N.H. Narayanan, & B. Chandrasekaran, (Eds.). *Diagrammatic Reasoning; Cognitive and Computational Perspectives* (pp. 33–69). Cambridge MA: AAAI and MIT Press.

Goel, V. (1995). *Sketches of Thought*. Cambridge, MA: MIT Press.

Green, T.R.G. (1989). Cognitive dimensions of notations. In A. Sutcliffe, & L. Macaulay (Eds.). *People and Computers, 5, HCI 1989. Proceedings of the Fifth Conference of the British Computer Society* (pp. 443–459). Cambridge: Cambridge University Press.

Green, T.R.G. (1996). Response to question at conference. *IEE Colloquium Digest, 51. Colloquium on Thinking with Diagrams.* (Savoy Place, London).

Johnson-Laird, P.N., & Byrne, R.M.J. (1991). *Deduction*. Hillsdale, NJ: Erlbaum.

Newstead, S.E. (1990). Conversion in syllogistic reasoning. In K.J. Gilhooly, M.T.G. Keane, R.G. Logie, & G. Erdos (Eds.), *Lines of Thinking Vol. 1* (pp. 73–84). London: Wiley.

Norman, D.A. (1993). Things that make us smart. Reading: MA: USA: Addison-Wesley.

Oakhill, J., Garnham, G., & Johnson-Laird, P.N. (1990). Belief bias effects in syllogistic reasoning. In K. Gilhooly, M.T.G Keane, R. Logie, & G. Erdos (Eds.). *Lines of Thinking, Vol. One* (pp. 125–138). London: Wiley.

O'Malley, C. (1990). Interface issues for guided discovery learning environments. In M.T. Elsom-Cook (Ed.). *Guided Discovery Tutoring* (pp. 24–42). London: Chapman.

Revlin, R., & Leirer, V. (1978). The effect of personal biases on syllogistic reasoning: Rational decisions from personalised representations. In R. Revlin & R.E. Mayer (Eds.), *Human Reasoning* (pp. 51–80). Washington, DC: Winston Wiley.

Stenning, K., & Oberlander, J. (1995). A cognitive theory of graphical and linguistic reasoning: Logic and implementation. *Cognitive Science, 19,* 97–140.

Stenning, K., & Tobin, R. (1993). Assigning information to modalities: Comparing graphical treatments of the syllogism. In E. Ejerhed, & S. Lindström (Eds.) *Action, Language and Cognition. Proceedings of International Conference on Dynamic Semantics*. Umea: Kluwer.
Wason, P.C., & Johnson-Laird, P.N. (1972). *Psychology and Reasoning: Structure and Content*. London: Batsford.

Chapter 6

Bredeweg, B., Schut, C., van den Heerik, K., & van Someren, M. (1992). Reducing ambiguity by learning assembly behaviour. In *Proceedings of the 6th International Workshop on Qualitative Reasoning*. (Herriot-Watt University, Edinburgh).
Chi, M.T.H., Bassok, M., Lewis, M., Reimann, P., & Glaser, R. (1989). Self-explanations: How students study and use examples to solve problems. *Cognitive Science*, *13*, 145–182.
Clark, P. & Matwin, S. (1993). Using qualitative models to guide inductive learning. In P. Utgoff (Ed.) *Proceedings of the Tenth International Conference on Machine Learning*. (pp. 49–56). San Mateo, CA: Kaufmann.
de Jong, T., & van Joolingen, W. (in press). Scientific discovery learning with computer simulations of conceptual domains. *Journal of Educational Research*.
Gerwin, D. (1974). Information processing, data inferences, and scientific generalization. *Behavioral Science*, *19*, 314–325.
Gerwin, D., & Newsted, P. (1977). A comparison of some inductive inference models. *Behavioral Science*, *22*, 1–11.
Klahr, D., & Dunbar, K. (1988). Dual space search during scientific reasoning. *Cognitive Science*, *12*, 1–48.
Klahr, D., Dunbar, K., & Fay, A.L. (1990). Designing good experiments to test bad hypotheses. In J. Shrager, & P. Langley (Eds.). *Computational Models of Scientific Discovery and Theory Formation*. San Mateo, CA: Kaufmann.
Klahr, D., Fay, A.L., & Dunbar, K. (1993). Heuristics for scientific experimentation: A developmental study. *Cognitive Psychology*, *25*, 111–146.
Langley, P. (1996). *Elements of Machine Learning*. San Francisco: Kaufmann.
Langley, P., & Zytkow, J.M. (1990). Data-driven approaches to empirical discovery. In J. Carbonell (Ed.). *Machine Learning: Paradigms and Models* (pp. 283–313). MIT/Elsevier.
Mayer, R. (1987). *Educational Psychology, a Cognitive Approach*. Boston, MA: Little, Brown.
Nordhausen, B., & Langley, P. (1990). An integrated approach to empirical discovery. In J. Shrager, & P. Langley (Eds.). *Computational Models of Scientific Discovery and Theory Formation*. San Mateo, CA: Kaufmann.
Ploetzner, R., & Spada, H. (1993). Multiple mental representations of information in physics problem solving. In G. Strube, & K.F. Wender (Eds.). *The Cognitive Psychology of Knowledge. The German Wissenspsychologie Project*. Amsterdam: Elsevier Publishers.
Ploetzner, R., & Spada, H. (1998). Constructing quantitative problem representations on the basis of qualitative reasoning. *Interactive Learning Environments*, *5*, 95–107.
Popper, K.R. (1959). *The Logic of Scientific Discovery*. London: Hutchinson.
Qin, Y., & Simon, H. (1995). Imagery and mental models. In J. Glasgow, N.H. Narayanan, & B. Chandrasekaran (Eds.). *Diagrammatic Reasoning — Cognitive and Computational Perspectives*. Cambridge, MA: AAAI Press/MIT Press.
Schauble, L., Glaser, R., Raghavan, K., & Reiner, M. (1991). Causal models and experimentation strategies in scientific reasoning. *The Journal of the Learning Sciences*, *1*, 201–239.

Simon, H.A., & Lea, G. (1974). Problem solving and rule induction: a unified view. In L.W. Gregg (Ed.). *Knowledge and Cognition*, Hillsdale, NJ: Erlbaum.
van Joolingen, W.R., & de Jong, T. (1997). An extended dual search space model of discovery learning. *Instructional Science*, *25*, 307–346.

Chapter 7

Ainsworth, S.E., Wood, D.J., & Bibby, P.A. (1996). Co-ordinating multiple representations in computer based learning environments. *Proceedings of the European Conference on Artificial Intelligence and Education*, (pp. 336–342). (Lisbon, Portugal).
Ainsworth, S.E., Wood, D.J., & Bibby, P.A. (1997). Evaluating principles for multi-representational learning environments. *7th EARLI Conference*. (Athens, Greece).
Ainsworth, S.E, Wood, D.J, & O'Malley, C. (1998). There is more than one way to solve a problem: Evaluating a learning environment that supports the development of children's multiplication skills. *Learning and Instruction*, *8*(2), 141–157.
Bibby, P.A., & Payne, S.J. (1993). Internalization and the use specificity of device knowledge. *Human–Computer Interaction*, *8*(1), 25–56.
Baroody, A.J. (1987). *Children's Mathematical Thinking: A Developmental Framework for Preschool, Primary and Special Education Teachers*. New York: Teachers College
Chi, M.T.H., Feltovich, P.J., & Glaser, R. (1981). Categorization and representation in physics problems by experts and novices. *Cognitive Science*, *5*, 121–152.
Confrey, J. (1992). *Function Probe(c)*. Santa Barbara, CA: Intellimation Library for the Macintosh.
Cox, R. (1996). Analytical reasoning with multiple external representations. Unpublished Ph.D. thesis, University of Edinburgh, UK.
Cox, R., & Brna, P. (1995). Supporting the use of external representations in problem solving: the need for flexible learning environments. *Journal of Artificial Intelligence in Education*, *2/3*, 239–302.
Cronbach, L.J., & Snow, R.E. (1977). *Aptitudes and Instructional Methods*. New York: Irvington.
Dufour-Janvier, B., Bednarz, N., & M. Belanger (1987). Pedagogical considerations concerning the problem of representation. In C. Janvier (Ed.), *Problems of Representation in the Teaching and Learning of Mathematics* (pp. 109–122). Hillsdale, NJ: Erlbaum.
Dugdale, S. (1982). Green Globs: A microcomputer application for graphing of equations. *Mathematics Teacher*, *75*, 208–214.
Ehrlich, K., & Johnson-Laird, P.N. (1982). Spatial descriptions and referential continuity. *Journal of Verbal Learning and Verbal Behaviour*, *21*, 296–306.
Gentner, D., & Toupon, C. (1986). Systematacity and surface similarity in the development of analogy. *Cognitive Science*, *10*, 277–300.
Gick, M.L, & Holyoak, K.J., (1980). Analogical problem solving. *Cognitive Psychology*, *12*, 306–355.
Gilmore, D.J., & Green, T.R.G. (1984). Comprehension and recall of miniature programs. *International Journal of Man–Machine Studies*, *21*, 31–48.
Hennessy, S., Twigger, D., Driver, R., O'Shea, T., O'Malley, C., Byard, M., Draper, S., Hartley, R., Mohamed, R., & Scanlon, E. (1995) Design of a computer-augmented curriculum for mechanics. *International Journal of Science Education*, *17*(1), 75–92.

Kaput, J.J. (1987). Towards a theory of symbol use in mathematics. In C. Janvier (Ed.). *Problems of Representation in the Teaching and Learning of Mathematics* (pp. 159–196). Hillsdale, NJ: Erlbaum.
Kaput, J.J. (1989). Linking representations in the symbol systems of algebra. In S. Wagner, & C. Kieran (Eds.). *Research Issues in the Learning and Teaching of Algebra.* Hillsdale, NJ: Erlbaum.
Lampert, M. (1986). Knowing, doing, and teaching multiplication. *Cognition and Instruction, 3,* 305–342.
Larkin, J.H., & Simon, H.A. (1987). Why a diagram is (sometimes) worth ten thousand words. *Cognitive Science, 11,* 65–99.
Oberlander, J., Cox, R., Monaghan, P., Stenning, K., & Tobin, R. (1996). Individual differences in proof structures following multimodal logic teaching. In G. W. Cottrell (Ed.). *Proceeding of the Eighteenth Annual Conference of the Cognitive Science Society,* (pp. 201–206). Hillsdale, NJ: Erlbaum.
Oliver, M., & O'Shea, T. (1996). Using the model-view-controller mechanism to combine representations of possible worlds for learners of modal logic. In *Proceedings of the European Conference of Artificial Intelligence in Education* (pp. 357–363). (Lisbon, Portugal).
Preece, J. (1983) Graphs are not straightforward. In T.R.G. Green, S.J. Payne, & G.C. van der Veer (Eds.). *The Psychology of Computer Use* (pp. 41–56). London: Academic Press.
Resnick, L., & Omanson, S. (1987). Learning to understand arithmetic. In R. Glaser (Ed.). *Advances in Instructional Psychology* (vol. 3, pp. 41–95). Hillsdale, NJ: Erlbaum.
Reys, R.E., Rybolt, J.F., Bestgen, B.J., & Wyatt, J.W. (1982). Processes used by good computational estimators. *Journal for Research in Mathematics Education, 13,* 183–201.
Schoenfeld, A. H., Smith, J. P., & Arcavi, A. (1993). Learning: the microgenetic analysis of one student's evolving understanding of a complex subject matter domain. In R. Glaser (Ed.). *Advances in Instructional Psychology* (vol. 4, pp. 55–175). Hillsdale, NJ: Erlbaum.
Schwartz, D.L. (1995). The emergence of abstract representations in dyad problem solving. *The Journal of the Learning Sciences, 4,* 321–354.
Snow, R.E., & Yalow, E. (1982). Education and intelligence. In R. J. Sternberg (Ed.). *A Handbook of Human Intelligence* (pp. 493–586). Cambridge: Cambridge University Press.
Sowder, J.T., & Wheeler, M.M. (1989). The development of concepts and strategies used in computational estimation. *Journal for Research in Mathematics Education, 20,* 130–146.
Tabachneck, H.J.M., Koedinger, K.R., & Nathan, M.J. (1994). Towards a theoretical account of strategy use and sense making in mathematical problem solving. In A. Ram, & K. Eiselt (Eds.). *16th Annual Conference of the Cognitive Science Society* (pp. 836–841). Hillsdale, NJ: Erlbaum.
Tabachneck, H.J.M., Leonardo, A.M., & Simon, H.A. (1994). How does an expert use a graph? A model of visual and verbal inferencing in economics. In A. Ram & K. Eiselt (Eds.). *16th Annual Conference of the Cognitive Science Society* (pp. 842–847). Hillsdale, NJ: Erlbaum.
Thompson, P.W. (1992). Notations, conventions and constraints: contributions to effective uses of concrete materials in elementary mathematics. *Journal for Research in Mathematics Education, 23,* 123–147.
Underwood, J., & Underwood, G. (1987). Data organisation and retrieval by children. *British Journal of Educational Psychology, 57,* 313–329.
Winn, B. (1987). Charts, graphs and diagrams in educational materials. In D.M. Willows, & H.A. Houghton (Eds.). *The Psychology of Illustration* (pp. 152–198). New York: Springer.
Yerushalmy, M. (1989). The use of graphs as visual interactive feedback while carrying out algebraic transformations. In *Proceedings of the 13th International Conference for the Psychology of Mathematics Education* (vol. 3, pp. 252–260). (Paris, France).

Yerushalmy, M. (1991). Student perceptions of aspects of algebraic function using multiple representation software. *Journal of Computer Assisted Learning, 7*, 42–57.

Chapter 8

Baum, D.R., & Jonides, J. (1979). Cognitive maps: Analysis of comparative judgments of distance. *Memory and Cognition, 7*, 462–468.

Cumming, T., & Worley, C. (1993). *Organization Development and Change*. St Paul, MN, West Publ.

Farah, M.J. (1990). *Visual Agnosia: Disorders of Object Recognition and What they Tell us about Normal Vision*. Cambridge, MA: MIT Press.

Gaba, D.M. (1992). Dynamic decision-making in anesthesiology: Cognitive models and training approaches. In D.A. Evans, & V.L. Patel (Eds.). *Advanced Models of Cognition for Medical Training and Practice* (pp. 123–147). Berlin: Springer-Verlag.

Hayes, J.R., & Simon, H.A. (1974–1976).Understanding. In H.A. Simon (Ed.). *Models of Thought*, chapter 7 (pp. 445–512). New Haven, CT: Yale University Press.

Kaplan, C.A., & Simon, H.A. (1990). In search of insight. *Cognitive Psychology, 22*, 374–419.

Larkin, J.H., & Simon, H.A. (1987). Why a diagram is (sometimes) worth ten thousand words. *Cognitive Science, 11*, 65–99.

Sachs, O.W. (1995). *An Anthropologist on Mars: Seven Paradoxical Tales*. New York: Knopf.

Schraagen, J.M.C. (1994). The generality and specificity of expertise. PhD Thesis, University of Amsterdam. Amsterdam.

Tabachneck, H.J.M. (1992). Computational differences in mental representations: Effects of mode of data presentation on reasoning and understanding. Doctoral Dissertation. Carnegie Mellon University, Pittsburgh, PA.

Tabachneck, H.J.M., & Simon, H.A. (1995). Alternative representations of instructional material. In D. Peterson (Ed.). *Forms of Representation*. London: Intellect Books.

Tabachneck, H.J.M., Leonardo, A.M., & Simon, H.A. (1997). How does an expert use a graph? CaMeRa: A computational model of multiple representations. *Cognitive Science, 21*, 305–350.

Chapter 9

Alpay L. (1996). *Modelling of Reasoning Strategies, and Representation through Conceptual Graphs: Application to Accidentology*. INRIA Research Report no.2810. Sophia-Antipolis, France: INRIA.

Alpay, L., Amergé, C., Corby, O., Dieng, R., Giboin, A., Labidi, S., Lapalut, S., Després, S., Ferrandez, F., Fleury, D., Girard, Y., Jourdan, J-L., Lechner, D., Michel, J-E., & Van Elslande, P. (1996). *Acquisition et modélisation des connaissances dans le cadre d'une coopération entre plusieurs experts : Application à un système d'aide à l'analyse de l'accident de la route*. Rapport final du contrat DRAST n. 93.0033.

Bromme, R. (1997). Beyond one's own perspective: The psychology of cognitive interdisciplinarity. University of Munster, Report Nr. 34. In P. Weingart & N. Stehr (Eds.). *Practicing Interdisciplinarity*. Toronto: Toronto University Press.

Clark, H.H. (1992). *Arenas of Language Use*. Chicago, IL: The University of Chicago Press.

Clark, H.H., & Brennan, S.E. (1991). Grounding in communication. In L.B. Resnick, J.M. Levine, & S.D. Teasley (Eds.). *Perspectives on Socially Shared Cognition* (pp. 97–129). Washington, DC: APA.

Dieng, R., Giboin, A., Amergé, C., Corby, O., Després, S., Alpay, L., Labidi, S., & Lapalut, S. (1996). Building of a corporate memory for traffic accident analysis. *Proceedings of the 10th Knowledge Acquisition for Knowledge-Based Systems Workshop (KAW'96)*. (Banff, Canada). Also in *http://ksi.cpsc.ucalgary.ca/KAW/KAW96/dieng/dieng-kaw96.html*

Ferrandez, F., Brenac, T., Girard, Y., Jourdan, J-L., Lechner, D., Michel, J-E., & Nachtergaele, C. (1995). *L'Étude détaillée d'accidents orientée vers la sécurité primaire: Méthodologie de recueil et de pré-analyse*. France: INRETS, Presses de l'E.N.P.C.

Fleury, D. (1990). Accident analysis methods and research prospects. *IATSS Research, 14*(1).

Fleury, D., Fline, C., & Peytavin, J.F. (1991). Le diagnostic local de sécurité, outils et méthodes. Guide méthodologique. Éditions du Service d'Études. *Techniques des Routes et Autoroutes (SETRA), Collection Etudes de sécurité.*

Gaines, B.R., & Shaw, M.L.G. (1989). Comparing the conceptual systems of experts. *Proceedings of the 9th IJCAI* (pp. 633–638). (Detroit, USA).

George, C. (1983). *Apprendre par l'action (Learning by Doing)*. Paris: PUF.

Giboin, A. (1995). ML comprehensibility and KBS explanation: stating the problem collaboratively. *Proceedings of the IJCAI'95 Workshop on Machine Learning and Comprehensibility* (pp. 1–11). (IJCAI, Montréal, Québec, Canada).

Girard, Y., & Michel, J.-E. (1991a). *L'Accidentologie des petits véhicules légers rapides par l'analyse approfondie des accidents. 1ère partie: Analyse de fichier détaillé.* Rapport sur convention d'études DSCR/INRETS n. 90/115RB.

Girard, Y., & Michel, J.-E. (1991b). *L'Accidentologie des petits véhicules légers rapides par l'analyse approfondie des accidents. 2ème partie: Analyse clinique de données approfondies.* Rapport sur convention d'études DSCR/INRETS n. 90/115RB.

Guindon, R., Krasner, H., & Curtis, B. (1987). Breakdowns and processes during the early phases of software design by professionals. In G. Olson, S. Sheppard and E. Soloway, (Eds.). *Empirical Studies of Programmers: Second Workshop* (pp. 65–82). Norwood, NJ: Ablex.

Lechner, D. *et al.* (1986). *La Reconstitution cinématique des accidents.* Rapport Inrets no. 21.

Lechner, D., & Ferrandez, F. (1990). *Analysis and Reconstruction of Accident Sequences.* Technical Paper n. 905121. XXIII FISITA Congress, 1, pp. 931–939.

Norman, D.A. (1982). *Learning and Memory*. San Francisco, CA: Freeman.

Norman, D.A. (1993). *Things that Make us Smart*. Reading, MA: Addison-Wesley.

Ostwald, J. (1995). Supporting collaborative design with representations for mutual understanding. *CHI95 Doctoral Consortium, http://www.cs.colorado.edu/~ostwald/interests/two-pager.html*

Ostwald, J. (1996). Knowledge construction in software development: The evolving artifact approach. Doctoral Dissertation, University of Colorado, Boulder, *http://www.cs.colorado.edu/~ostwald/thesis/*

Rumelhart, D., & Norman, D. (1981). Analogical processes in learning. In J.R. Anderson (Ed.). *Cognitive Skills and their Acquisition*. Hillsdale, NJ: Erlbaum.

Shaw, M.L.G., and Gaines, B.R. (1989). A methodology for recognizing conflict, correspondence, consensus and contrast in a knowledge acquisition system. *Knowledge Acquisition, 1*(4), 341–363.

Sumner, T. (1995). The high-tech toolbelt: A study of designers in the workplace. *Proceedings of CHI'95 (Computers and Human Interactions)*. Also in *http://www.acm.org/sigs/sigchi/chi95/proceedings/papers/trs_bdy.htm*

Tabachneck, H.J.M., Leonardo, A.M., & Simon, H.A. (1997). CaMeRa: A computational model of multiple representations. *Cognitive Science*, *21*, 305–350.
Van Elslande, P. (1992). Les erreurs d'interprétation en conduite automobile : Mauvaise catégorisation ou activation erronée de schémas ? *Intellectica*, *15*, 125–149.

Chapter 10

Atkinson, P. (1995). *Medical Talk and Medical Work*. Sage: London.
Baumjohann, M., & Middendorf, S. (1997). Interdisziplinäre Kommunikation zwischen Ärzten und Pflegekräften einer kinderonkologischen Station und Ambulanz. Eine Untersuchung zum Konzept der Perspektivenübernahme. Unpublished thesis (Diplomarbeit). University of Muenster.
Blickensderfer, L. (1996). Nurses and physicians. Creating a collaborative environment. *Journal of Intravenous Nursing*, *19*, 127–131.
Boshuizen, H.P.A., Schmidt, H.G., Custers, E.J.F.M., & van de Wiel, M.W.J. (1995). Knowledge development and restructuring in the domain of medicine; The role of theory and practice. Special issue on fusing theory and experience: The structure of professional knowledge. *Learning and Instruction*, *5*, 269–289.
Bromme, R. (1992). *Der Lehrer als Experte. Zur Psychologie des professionellen Wissens*. Bern: Huber.
Bromme, R. (in press). Beyond one's own perspective: The psychology of cognitive interdisciplinarity. In P. Weingart, & N. Stehr (Eds.). *Practicing Interdisciplinarity*. Toronto: Toronto University Press.
Bromme, R., & Tillema, H. (1995). Fusing experience and theory: The structure of professional knowledge. Special issue on fusing theory and experience: the structure of professional knowledge. *Learning and Instruction*, *5*, 261–269.
Chi, M., Glaser, R., & Farr, M. (Eds.). (1988). *The Nature of Expertise*. Hillsdale, NJ: Erlbaum.
Clark, H.H. (1992). *Arenas of Language Use*. Chicago, IL: University of Chicago Press.
Clark, H.H. (1996). *Using Language*. Cambridge: Cambridge University Press.
Ericsson, K.A., & Smith, J. (Eds.). (1991). *Toward a General Theory of Expertise. Prospects and Limits*. Cambridge, MA: Cambridge University Press.
Evans, D.A., & Patel, V.L.E. (Eds.). (1992). *Advanced Models of Cognition for Medical Training and Practice*. Berlin: Springer-Verlag.
Fleck, L. (1979). *Genesis and Development of a Scientific Fact*. Chicago: University of Chicago Press (Original Published 1935).
Hoffman, R.R. (Ed.). (1992). *The Psychology of Expertise. Cognitive Research and Empirical Artificial Intelligence*. New York: Springer.
Johnson-Laird, P. (1981). Mutual ignorance: comments on Clark and Carlson's paper. In N.V. Smith (Ed.). *Mutual Knowledge* (pp. 40–45). New York: Academic Press.
Keysar, B. (1994). The illusory transparancy of intention: linguistic perspective taking in text. *Cognitive Psychology*, *26*, 165–208.
Kitchener, K.S. (1983). Cognition, metacognition, and epistemic cognition: A three level model of cognitive processing. *Human Development*, *26*, 222–232.
Krauss, R.M., & Fussell, S.R. (1990). Mutual knowledge and communicative effectiveness. In J. Galegher, R. Kraut, & C. Egido (Eds.). *Intellectual Teamwork: Social and Technological Foundations of Cooperative Work* (pp. 111–146). Hillsdale, NJ: Erlbaum.

Krauss, R.M. & Fussell, S.R. (1991). Perspektive-taking in communication: Representations of others' knowledge in reference. *Social Cognition, 9*, 2–24.

Kuhn, D. (1991). *The Skills of Argument*. New York: Cambridge University Press.

Larkin, J.H. (1983). The role of problem representation in physics. In D. Gentner, & A.L. Stevens (Eds.). *Mental Models* (pp. 75–98). Hillsdale, NJ: Erlbaum.

Larkin, J.H., McDermott, J., Simon, D.P. & Simon, H.A. (1980). Models of competence in solving physics problems. *Cognitive Science, 12*, 317–345.

Mayring, P. (1995). *Qualitative Inhaltsanalyse. Grundlagen und Techniken*. Weinheim: Deutscher Studien Verlag.

Mead, G.H. (1934). Mind, self and society. In C.H. Morris (Ed.). *Mind, Self and Society*. Chicago, IL: University of Chicago Press.

Schön, D. (1983). *The Reflective Practitioner*. New York: Basic Books.

Stein, L.I., Watts, D.T., & Howell, T. (1990). The doctor–nurse game revisited. *New England Journal of Medicine, 322*, 546–549.

van Someren, M.W., & Reimann, P. (1995). Multi-objective learning with multiple representations. In P. Reimann, & H. Spada (Eds.). *Learning in Humans and Machines: Towards an Interdisciplinary Learning Science* (pp. 130–153). Oxford: Elsevier.

Chapter 11

Atwood, G. (1971). An experimental study of visual imagination and memory. *Cognitive Psychology, 2*, 290–299.

Baddeley, A.D. (1986). *Working Memory*. Oxford: Clarendon Press.

Baylor, G.W. (1971). A treatise on the mind's eye — an empirical investigation of visual mental imagery. Unpublished doctoral dissertation, Carnegie Mellon University, Pittsburgh, PA.

Brooks, L.R. (1967). The suppression of visualization by reading. *Quarterly Journal of Experimental Psychology, 19*, 289–299.

Brooks, L.R. (1968). Spatial and verbal components in the act of recall. *Canadian Journal of Psychology, 22*, 349–368.

Ericsson, K.A., & Simon, H.A. (1993). *Protocol Analysis: Verbal Reports as Data (revised edition)*. Cambridge, MA: MIT Press.

Finke, R.A., & Shepard, R.N. (1986). Visual functions of mental imagery. In K.T. Boff, L. Kaufman, & J.P. Thomas (Eds.). *Handbook of Perception and Human Performance: Vol. 2*, (Chapter 37). New York: Wiley-Interscience.

Glasgow, J.I. (1993). The imagery debate revisited: A computational perspective. *Computational Intelligence, 9*(4), 309–333.

Just, M.A., & Carpenter, P.A. (1985). Cognitive coordinate systems: Accounts of mental rotation and individual differences in spatial abilities. *Psychological Review, 92*(2), 137–172.

Kaput, J.J. (1989). Linking representations in the symbol systems of algebra. In S. Wagner, & C. Kieran (Eds.). *Research Issues in the Learning and Teaching of Algebra*. Reston, VA: NCTM.

Kosslyn, S.M. (1980). *Image and Mind*. Cambridge, MA: Harvard University Press.

Kosslyn, S.M. (1994). *Image and Brain: The Resolution of the Imagery Debate*. Cambridge, MA: The MIT Press.

Kosslyn, S.M., & Koenig, O. (1992). *Wet Mind: The New Cognitive Neuroscience*. New York: The Free Press.

Larkin, J.H., & Simon, H.A. (1987). Why a diagram is (sometimes) worth ten thousand words. *Cognitive Science, 11*, 65–99.

Marshall, A. (1947). *Principles of Economics: An Introductory Volume (eighth edition)*. New York: Macmillan.

Paivio, A. (1971). *Imagery and Verbal Processes*. New York: Holt.

Richman, H.B., Staszewski, J.J., & Simon, H.A. (1995). Simulation of expert memory using EPAM IV. *Psychological Review, 102,* 305–330.

Riding, R.J., & Calvey, I. (1981). The assessment of verbal-imagery learning styles and their effect on the recall of concrete and abstract prose passages by 11-year-old children. *British Journal of Psychology, 72*, 59–64.

Shute, V.J., Glaser, R., & Raghavan, K. (1989) Inference and discovery in an exploratory laboratory. In P.L. Ackerman, R.J. Sternberg, & R. Glaser (Eds.). *Learning and Individual Differences*. New York: Freeman.

Tabachneck, H.J.M. (1992). Computational differences in mental representations: Effects of mode of data presentation on reasoning and understanding. Doctoral dissertation, Department of Psychology, Carnegie Mellon University, Pittsburgh, PA.

Tabachneck, H.J.M., & Simon, H. A. (1992). Effect of mode of data presentation on reasoning about economic markets. In *Working Notes of the 1992 AAAI Spring Symposium on Diagrammatic Reasoning*. Stanford University, CA. Republished in (1994) N.H. Narayanan (Ed.). *Reasoning with Diagrammatic Representations*, AAAI Press, AAAI Technical Report Series #SS-92-02.

Tabachneck, H.J.M., & Simon, H.A. (1996). Alternative representations of instructional material. In D. Peterson (Ed.). *Forms of Representation: An Interdisciplinary Theme for Cognitive Science.*

Tabachneck-Schijf, H.J.M., Leonardo, A.M., & Simon, H.A. (1997). CaMeRa: A computational model of multiple representations. *Cognitive Science*, *21*, 305–350.

Zhang, G., & Simon, H.A. (1985). STM capacity for Chinese words and idioms: chunking and acoustical loop hypotheses. *Memory and Cognition*, *13*, 193–201.

Zhang, J., & Norman, D.A. (1994). Representations in distributed cognitive tasks. *Cognitive Science*, *18*, 87–122.

Chapter 12

Anderson, J.R. (1987). Skill acquisition: Compilation of weak-method problem solutions. *Psychological Review*, *94*, 192–210.

Anderson, J.R. (1993). *Rules of the Mind*. Hillsdale NJ: Erlbaum.

Baddeley, A. (1997). *Human Memory; Theory and Practice* (revised edition). Hove: Psychology Press.

Boshuizen, H.P.A. (1989). De ontwikkeling van medische expertise; Een cognitief-psychologische benadering (The development of medical expertise: a cognitive psychological approach). Academisch proefschrift Rijksuniversiteit Limburg. Haarlem: Theseus.

Boshuizen, H.P.A., & Schmidt, H.G. (1992). On the role of biomedical knowledge in clinical reasoning by experts, intermediates and novices. *Cognitive Science*, *16*, 153–184.

Boshuizen, H.P.A., Schmidt, H.G., Custers, E.J.F.M., & van de Wiel, M.W.J. (1995). Knowledge development and restructuring in the domain of medicine; the role of theory and practice. Special issue on fusing theory and experience: The structure of professional knowledge. *Learning and Instruction*, *5*, 269–289.

Dawson-Saunders, B., Feltovich, P.J., Coulson, R.L., & Steward D. (1990). A survey of medical school teachers to identify basic biomedical concepts medical students should understand. *Academic Medicine, 7*, 448–454.

Groothuis, S., Boshuizen, H.P.A., & Talmon, J.L. (1998). Is endocrinilogy as easy as they say? An analysis of the conceptual difficulties of the domain. *Teaching and Learning in Medicine, 10*(4).

Jones, J.G., Calson, G.J., & Calson, C. (1986). The acquisition of cognitive knowledge through clinical experiences. *Medical Education, 20*, 10–13.

Kintsch, W. (1980). The role of knowledge in discourse comprehension: A construction-integration model. *Psychological Review, 95*(2), 163–182.

Koedinger, K.R., & Anderson, J.R. (1990). Abstract planning and perceptual chunks: Elements of expertise in geometry. *Cognitive Science, 14*, 511–550.

Kosslyn, S.M. (1980). *Image and Mind.* Cambridge MA: Harvard University Press.

Laird, J.E., Rosenbloom, P.S., & Newell, A. (1986). Chunking in soar; the anatomy of a general learning mechanism. *Machine Learning, 1*, 11–45.

Mandl, H., Gruber, H., & Renkl, A. (1993). Problems of knowledge utilization in the development of expertise. In W.J. Nijhof, & J.N. Streumer (Eds), *Flexibility and Cognitive Structure in Vocational Education.* (pp. 291–305). Utrecht: Lemma.

Mani, K., & Johnson-Laird, P.N. (1982). The mental representation of spatial descriptions. *Memory and Cognition, 10*(2), 181–187.

Miller, G.A. (1956). The magical number seven, plus or minus two: Some limits on our capacity for processing information. *Psychological Review, 63*, 81–97.

Norman, D.A., & Shallice, T. (1986). Attention to action: Willed and automatic control of behavior. In R.J. Davidson, G.E. Schwarts, & D. Shapiro (Eds.). *Consciousness and Self-regulation. Advances in Research and Theory* (Vol. 4, pp. 1–18). New York: Plenum Press.

Patrick, J. (1992). *Training: Research and Practice.* London: Academic Press.

Reber, A.S. (1989). Implicit learning and tacit knowledge. *Journal of Experimental Psychology: General, 118*, 219–235.

Schmidt, H.G., & Boshuizen, H.P.A. (1993a). On the origin of intermediate effects in clinical case recall. *Memory and Cognition, 21*, 338–351.

Schmidt, H.G., & Boshuizen, H.P.A. (1993b). On acquiring expertise in medicine. *Educational Psychology Review, 5*, 1–17.

Schmidt, H.G., Norman, G.R., & Boshuizen, H.P.A. (1990). A cognitive perspective on medical expertise: Theory and implications. *Academic Medicine, 65*, 611–621.

Shepard, R.N., & Metzler, J. (1971). Mental rotation of three-dimentional objects. *Science, 171*, 701–703.

Sweller, J., & Chandler, P. (1991). Evidence for cognitive load theory. *Cognition and Instruction, 8*(4), 351–362.

van de Wiel, M.W.J. (1997). *Knowledge Encapsulation: Studies on the Development of Medical Expertise.* Doctoral dissertation Maastricht University. Maastricht.

Vander, A.J., Sherman, J.H., & Luciano, D.S. (1990). *Human Physiology; The Mechanisms of Body Function* (5th edition). New York: McGraw-Hill.

Chapter 13

Alexander, P.A., & Judy, J.E. (1988). The interaction of domain-specific and strategic knowledge in academic performance. *Review of Educational Research, 58*, 375–404.

Anderson, J.R. (1983). *The Architecture of Cognition*. Cambridge, MA: Harvard University Press.

Anderson, T., Tolmie, A., Howe, C., Mayes, T., & Mackenzie, M. (1992). Mental models of motion. In Y. Rogers, A. Rutherford, & P.A. Bibby (Eds.). *Models in the Mind* (pp. 57–71). London: Academic Press.

Boshuizen, H.P.A., & Schmidt, H.G. (1992). On the role of biomedical knowledge in clinical reasoning by experts, intermediates and novices. *Cognitive Science*, *16*, 153–184.

Chandrasekaran, B., & Mittal, S. (1984). Deep versus compiled knowledge approaches to diagnostic problem-solving. In M.J. Coombs (Ed.). *Developments in Expert Systems* (pp. 23–34). London: Academic Press.

Chi, M.T.H., & Bassok, M. (1989). Learning from examples via self-explanations. In L.B. Resnick (Ed.). *Knowing, Learning, and Instruction: Essays in Honor of Robert Glaser* (pp. 251–282). Hillsdale, NJ: Erlbaum.

Chi, M.T.H., Feltovich, P.J., & Glaser, R. (1981). Categorization and representation of physics problems by experts and novices. *Cognitive Science*, *5*, 121–152.

Chiesi, H.L., Spilich, G.J., & Voss, J.F. (1979). Acquisition of domain-related information in relation to high and low domain knowledge. *Journal of Verbal Learning and Verbal Behavior*, *18*, 257–273.

Chinn, C.A., & Brewer, W.F. (1993). The role of anomalous data in knowledge acquisition: A theoretical framework and implications for science instruction. *Review of Educational Research*, *63*, 1–51.

Craik, F.I.M., & Lockhart, R.S. (1972). Levels of processing: a framework for memory research. *Journal of Verbal Learning and Verbal Behaviour*, *11*, 671–684.

de Groot, A.D. (1946). *Het Denken van den Schaker [Thought in Chessplayers]*. Published doctoral dissertation, Amsterdam: Noordhollandse Uitgeversmaatschappij.

de Groot, A.D., & Gobet, F. (1996). *Perception and Memory in Chess, Studies in the Heuristics of the Professional Eye*. Assen, The Netherlands: van Gorcum.

de Jong, T. (1986). Kennis en het oplossen van vakinhoudelijke problemen [Knowledge based problem solving]. Unpublished doctoral dissertation. Eindhoven University of Technology, The Netherlands.

de Jong, T., & Ferguson-Hessler, M.G.M. (1991). Knowledge of problem situations in physics: a comparison of good and poor novice problem solvers. *Learning and Instruction*, *1*, 289–302.

de Jong, T., & Ferguson-Hessler, M.G.M. (1996). Types and qualities of knowledge. *Educational Psychologist*, *31*, 105–113.

de Kleer, J., & Brown, J.S. (1981). Mental models of physical mechanisms and their acquisition. In J.R. Anderson (Ed.). *Cognitive Skills and their Acquisition* (pp. 285–309). Hillsdale, NJ: Erlbaum.

de Kleer, J., & Brown, J.S. (1983). Assumptions and ambiguities in mechanics. In D. Gentner, & A.L. Stevens (Eds.). *Mental Models* (pp. 155–190). Hillsdale, NJ: Erlbaum.

di Sessa, A.A. (1983). Phenomenology and the evolution of intuition. In D. Gentner, & A.L. Stevens (Eds.). *Mental Models* (pp. 15–33). Hillsdale, NJ: Erlbaum.

di Sessa, A.A. (1993). Toward an epistemology of physics. *Cognition and Instruction*, *10*, 105–225.

Duncker, K. (1945). On problem-solving. *Psychological Monographs*, *58*(270), 1–113.

Egan, D.E., & Schwarz, B.J. (1979). Chunking in recall of symbolic drawings. *Memory and Cognition*, *7*, 149–158.

Elio, R., & Scharf, P.B. (1990). Modelling novice-to-expert shifts in problem solving strategy and knowledge organization. *Cognitive Science*, *14*, 579–639.

Ericsson, K.A., & Simon, H.A. (1993) *Protocol Analysis* (revised edition), Cambridge, MA: MIT Press.

Ferguson-Hessler, M.G.M. (1989). Over kennis en kunde in de fysica [On knowledge and expertise in physics]. Unpublished doctoral dissertation. Eindhoven University of Technology, The Netherlands.

Gentner, D., & Stevens, A.L. (Eds.). (1983). *Mental Models*. Hillsdale, NJ: Erlbaum.

Goei, S.L. (1994). Mental models and problem solving in the domain of computer numerically controlled programming. Unpublished doctoral dissertation. Twente University, Enschede, The Netherlands.

Hammond, K.J. (1989). *Case-based Planning: Viewing Planning as a Memory Task*. Boston, MA: Academic Press.

Hegarty, M., Mayer, R.E., & Monk, C.E. (1995). Comprehension of arithmetic word problems: A comparison of successful and unsuccessful problem solvers. *The Journal of Educational Psychology*, *87*, 18–31.

Jacobson, M.J., & Spiro R.J. (1995). Hypertext learning environments, cognitive flexibility, and the transfer of complex knowledge: An empirical investigation. *Journal of Educational Computing Research*, *12*, 301–333.

Johnson-Laird, P.N. (1983). *Mental Models: Towards a Cognitive Science of Language, Inference and Conciousness*. Cambridge: Cambridge University Press.

Kuhn, D., Schauble, L., & Garcia-Mila, M. (1992). Cross-domain development of scientific reasoning. *Cognition and Instruction*, *9*, 285–327.

Larkin, J.H. (1983). The role of problem representations in physics. In D. Gentner, & A.L. Stevens (Eds.). *Mental Models* (pp. 75–98). Hillsdale, NJ: Erlbaum.

Marshall, S.P. (1995). *Schemas in Problem Solving*. Cambridge: Cambridge University Press.

McDermott, J., & Larkin, J.H. (1978). *Re-representing Textbook Physics Problems*. Proceedings of the 2nd National Conference of the Canadian Society for Computational Studies of Intelligence.

Newell, A., & Simon, H.A. (1972). *Problem Solving*. Englewood Cliffs, NJ: Prentice-Hall.

Paivio, A. (1986). *Mental Representations: A Dual Coding Approach*. Oxford: Oxford University Press.

Polya, G. (1945). *How to Solve it*. Princeton, NJ: Princeton University Press.

Plötzner, R. (1993). The integrative use of qualitative and quantitative knowledge in physics problem solving. Unpublished doctoral dissertation. University of Freiburg, Department of Psychology.

Plötzner, R., & Spada, H. (1993). Multiple mental representations of information in physics problem solving. In G. Strube, & K.F. Wender (Eds.). *The Cognitive Psychology of Knowledge* (pp. 285–312). Amsterdam: Elsevier Science Publishers.

Rumelhart, D.E., & Ortony, A. (1977). The representation of knowledge in memory. In R.C. Anderson, R.J. Spiro, & W.E. Montague (Eds.). *Schooling and the Acquisition of Knowledge* (pp. 99–137). Hillsdale, NJ: Erlbaum.

Savelsbergh, E.R., de Jong, T., & Ferguson-Hessler, M.G.M. (1996). *Forms of Problem Representation in Physics* (IST-MEMO-96-01). Enschede, The Netherlands: University of Twente, Faculty of Educational Science and Technology, Department of Instructional Technology.

Savelsbergh, E.R., de Jong, T., & Ferguson-Hessler, M.G.M. (submitted). *Forms of Problem Representation in Physics: An Exploratory Study*.

Schank, R.A., & Abelson, R. (1977). *Scripts, Plans, Goals and Understanding*. Hillsdale, NJ: Erlbaum.

Sloman, (1995). Musings on the roles of logical and non-logical representations in intelligence. In J. Glasgow, N.H. Narayanan, & B. Chandrasekaran (Eds.). *Diagrammatic Reasoning:*

Cognitive and Computational Perspectives (pp. 7–32). Menlo Park, CA: AAAI Press/The MIT Press.
Stenning, K. (1992). Distinguishing conceptual and empirical issues about mental models. In Y. Rogers, A. Rutherford, & P.A. Bibby (Eds.). *Models in the Mind* (pp. 29–48). London: Academic Press.
Sweller, J., Mawer, R.F., & Ward, M.R. (1983). Development of expertise in mathematical problem solving. *Journal of Experimental Psychology: General*, *112*, 639–661.
Tulving, E. (1983). *Elements of Episodic Memory*. New York: Oxford University Press.
van Hiele, P.M. (1986). *Structure and Insight; A Theory of Mathematics Education*. Orlando, FL: Academic Press.
Vermunt, J.D. (1991). Leerstrategieen van studenten in een zelfinstructie-leeromgeving [Learning strategies of students in a self-instructional learning environment]. *Pedagogische Studiën*, *68*, 315–325.

Chapter 14

Amarel, S. (1968). On representations of problems of reasoning about actions. In D. Michie (Ed.). *Machine Intelligence 3*. Edinburgh: Edinburgh University Press.
Anderson, J.R. (1990). *The Adaptive Character of Thought*. Hillsdale, NJ: Erlbaum.
Anderson, J.R. (1991). The place of cognitive architectures in a rational analysis. In K. VanLehn (Ed.). *Architectures for Cognition*. Hillsdale, NJ: Erlbaum.
Anderson, J.R. (1993). *Rules of the Mind*. Hillsdale, NJ: Erlbaum.
Anzai, Y., & Simon, H.A. (1979). The theory of learning by doing. *Psychological Review*, *86*, 124–140.
Boshuizen, H.P.A. (1992). The role of biomedical knowledge in clinical reasoning by experts, intermediates and novices. *Cognitive Science*, *16*, 153–184.
Dasarathy, B. (1991). *Nearest Neighbor (NN) Norms: NN Pattern Classification Technique*. Los Alamitos, CA: IEEE Computer Society Press.
de Jong, G., & Mooney, R. (1986). Explanation-based learning: An alternative view. *Machine Learning*, *1*, 145–176.
Francis, A.G., & Ram, A. (1985). A comparative utility analysis of case-based reasoning and control-rule learning systems. In N. Lavrac, & S. Wrobel (Eds.). *Machine Learning: ECML-95*. Berlin: Springer Verlag.
Keller, R.M. (1988). Defining operationality for explanation-based learning. *Artificial Intelligence*, *35*, 227–241.
Kosslyn, S.M. (1975). Information representation in visual images. *Cognitive Psychology*, *7*, 341–370.
Laird, J.E., Rosenbloom, P.S., & Newell, A. (1986). Chunking in SOAR: the anatomy of a general learning mechanism. *Machine Learning*, *1*, 11–46.
Langley, P. (1996). *Elements of Machine Learning*. San Mateo, CA: Kaufmann.
Larkin, J.H., & Simon, H.A. (1987). Why a diagram is (sometimes) worth a thousand words. *Cognitive Science*, *11*, 65–99.
Marr, D. (1982). *Vision: A Computational Investigation into the Human Representation and Processing of Visual Information*. New York: W.H. Freeman.
Minton, S.N. (1990). Quantitative results concerning the utility of explanation-based learning. *Artificial Intelligence*, *42*, 363–391.
Mitchell, T.M., Keller, R., & Kedar-Cabelli, S. (1986). Explanation-based generalization: a unifying view. *Machine Learning*, *1*, 47–80.

Portinale, L., & Torasso, P. (1995). ADAPtER, an integrated diagnsotic system combining case-based and abductive reasoning. In: *Proceeding of the First International Conference on Case-Based Reasoning — ICCBR 94, Lecture Notes in Artificial Intelligence 1010*, 277–288.

Portinale, L., & Torasso, P. (1996). On the usefulness of re-using diagnostic solutions. In W. Wahlster (Ed.). *Proceedings 12th European Conference on Artificial Intelligence ECAI-96* (pp. 137–141). Chichester: Wiley.

Portinale, L., & Torasso, P. (1998). Performance issues in ADAPtER — a combined CBR–MBR diagnostic architecture. In *Proceedings AAAI Spring Symposium on Multi-Modal Reasoning*, Stanford, CT.

Pylyshyn, Z. (1989). Computing in cognitive science. In M. Posner (Ed). *Foundations of Cognitive Science*. Cambridge, MA: MIT Press.

Simon, H.A. (1975). The functional equivalence of problem solving skills. *Cognitive Psychology*, *7*, 268–288.

Simon, H.A. (1981). *The Sciences of the Artificial*. Cambridge, MA: MIT Press.

Smyth B., & Keane, M. (1995). Remembering to forget. In *Proceedings of the Fourteenth International Joint Conference on Artificial Intelligence* (pp. 377–383). San Mateo, CA: Kaufmann.

Straatman, R., & Beys, P. (1995). A performance model for knowledge based systems. In M. Ayel, & M.C. Rousse (Eds.). *Proceedings European Workshop on Validation and Verification of Knowledge-Based Systems EUROVAV-97* (pp. 253–263). Université de Savoie, Chambery.

Tambe, M., Newell, A., & Rosenbloom, P. (1990). The problem of expensive chunks and its solution by restricting expressiveness. *Machine Learning*, *5*, 299–348.

Torasso, P., Console, L., Portinale, L., & Theseider Dupre, D. (1993). Combining heuristic and causal reasoning in diagnostic problem solving. In J.M. David, J.P. Krivine, & R. Simmons (Eds). *Second Generation Expert Systems*, Berlin: Springer-Verlag.

Tulving, E. (1972). Episodic and semantic memory. In E. Tulving, & W. Donaldson (Eds.). *Organization of Memory*. New York: Academic Press.

VanLehn, K. (1989). Problem solving and cognitive skill acquisition. In M. Posner (Ed.). *Foundations of Cognitive Science*. Cambridge, MA: MIT Press.

van Someren, M.W., Barnard, Y., & Sandberg, J. (1993). *The Think Aloud Method*. London: Academic Press.

Chapter 15

Anderson, J.R. (1993). *Rules of the Mind*. Hillsdale, NJ: Erlbaum.

Boshuizen, H.P.A., & Schmidt, H.G. (1992). On the role of biomedical knowledge in clinical reasoning by experts, intermediates, and novices. *Cognitive Science*, *16*, 153–184.

Carraher, T.N., & Schliemann, A.D. (1988). Research into practice: Using money to teach about the decimal system. *Arithmetic Teacher*, *36*(4), 42–43.

Chandler, D. (1995). *Processes of Mediation*. On-line course notes from a media course taught at University of Wales, Aberystwyth. *http://www.aber.ac.uk/~dgc/process.html*

Dunbar, K., & Klahr, D. (1989). Developmental differences in scientific discovery processes. In D. Klahr, & K. Kotovsky (Eds.). *Complex Information Processing: The Impact of Herbert Simon* (pp. 109–143). Hillsdale, NJ: Erlbaum.

Grubb, W.N. (1996). *Learning to Work: The Case for Reintegrating Job Training and Education*. New York: Sage.

Gott, S.P., Lesgold, A.M., & Kane, R.S. (1997). Promoting the transfer of technical competency. In S. Dijkstra, et al. (Eds.). *Instructional Design: International Perspectives* (vol. 2). Hillsdale, NJ: Erlbaum.

Katz, S. (1997). Peer and student–mentor interaction in a computer-based training environment for electronic fault diagnosis. *LRDC Technical Report*, University of Pittsburgh.

Lesgold, A.M. (1984). Acquiring expertise. In J.R. Anderson, & S.M. Kosslyn (Eds.). *Tutorials in Learning and Memory: Essays in Honor of Gordon Bower*. San Francisco, CA: W.H. Freeman.

Lesgold, A.M., Lajoie, S.P., Bunzo, M., & Eggan, G. (1992). SHERLOCK: A coached practice environment for an electronics troubleshooting job. In J. Larkin, & R. Chabay (Eds.). *Computer Assisted Instruction and Intelligent Tutoring Systems: Shared Issues and Complementary Approaches* (pp. 201–238). Hillsdale, NJ: Erlbaum.

Lévi-Strauss, C. (1974). *The Savage Mind*. London: Weidenfeld & Nicolson.

Newell, A. (1990). *Unified Theories of Cognition*. Cambridge, MA: Harvard University Press.

Patel, V.L., & Kaufman, D.R. (1995) Clinical reasoning and biomedical knowledge. In J. Higgs, & M. Jones (Eds.). *Clinical Reasoning in the Health Professions* (pp. 117–128). Oxford: Butterworth-Heinemann.

Rasmussen, J., Pejtersen, A.M., & Goodstein, L.P. (1994). *Cognitive Systems Engineering*. New York: Wiley.

Roschelle, J. (1992). Learning by collaborating: Convergent conceptual change. *Journal of the Learning Sciences*, *2*(3), 235–276.

Rumelhart, D.E., & Norman, D.A. (1981). An activation-trigger-schema model for the simulation of skilled typing. *Proceedings of the Berkeley Conference of the Cognitive Science Society*. (Berkeley, CA).

Spiro, R.J., Feltovich, P.J., Coulson, R.L., & Anderson, D.K. (1989). Multiple analogies for complex concepts: Antidotes for analogy-induced misconception in advanced knowledge acquisition. In S. Vosniadou, & A. Ortony (Eds.). *Similarity and Analogical Reasoning* (pp. 498–531). Cambridge: Cambridge University Press.

Suthers, D., Erdošne Toth, E., & Weiner, A. (1997, December). An integrated approach to implementing collaborative inquiry in the classroom. *Proceedings of the Conference on Computer Supported Collaborative Learning (CSCL'97)*. (Toronto, Canada).

Sweller, J. (1994). Cognitive load theory, learning difficulty and instructional design. *Learning and Instruction*, *4*, 295–312.

Chapter 16

Oberlander J., Cox R., Tobin R., Stenning K., & Monaghan, P. (1996). Individual differences in proof structures following multimodal logic teaching. In G.W. Cottrell (Ed.). *Proceedings of the 18th Annual Conference of the Cognitive Science Society*. (UCSD, La Jolla, CA).

Stenning, K. (in press). The cognitive consequences of modality assignment for educational communication: the picture in logic. *Learning and Instruction*.

Stenning, K., Cox, R., & Oberlander, J. (1995). Contrasting the cognitive effects of graphical and sentential logic teaching: reasoning, representation and individual differences. *Language and Cognitive Processes*, *10*, 333–354.

van der Pal, J. (1995). *The Balance of Situated Action and Formal Instruction for Learning Conditional Reasoning*. Thesis, University of Twente, Enschede.

Index